Scientific Computation

Hiqmet Kamberaj

Computer Simulations in Molecular Biology

Springer

Hiqmet Kamberaj
Computer Engineering
International Balkan University
Skopje, North Macedonia

ISSN 1434-8322 ISSN 2198-2589 (electronic)
Scientific Computation
ISBN 978-3-031-34841-9 ISBN 978-3-031-34839-6 (eBook)
https://doi.org/10.1007/978-3-031-34839-6

This Springer imprint is published by the registered company Springer Nature Switzerland AG
The registered company address is: Gewerbestrasse 11, 6330 Cham, Switzerland

To the memory of my Mother, Father, and Brother

Preface

Classical and quantum mechanics molecular dynamics simulations are often used to study the dynamics of molecular biological systems, such as proteins, DNA/RNA, membranes, and their complexes, which include transitions between various conformational states due to atomic/electronic motions. Besides, they are used to predict the properties of the materials. A common problem of standard molecular dynamics simulation is the time and size scale limitations of the technique in studying the slow conformational dynamics of those systems. Therefore, the considerable time and size scale of physical and chemical phenomena will require new statistical and computational approaches to be studied efficiently.

Thus, computer simulations are essential in understanding physical and chemical properties in biophysics and material science. This textbook presents computer simulation models and methods used in quantum mechanics and molecular mechanics molecular dynamics simulations of biological systems. Besides, I introduce an interactive molecular dynamics simulation virtual laboratory software for computer simulation and modelling of biological systems using classical and quantum mechanics methods, which can be used in the classrooms for interactive teaching and development of research methodologies. I will also discuss different approaches to enhancing conformational sampling of molecular dynamics simulations.

Furthermore, I discuss the artificial neural network methods employed to build up deep learning architectures for molecular modelling, applied to classical and quantum mechanical molecular dynamics simulations. I focus on predicting molecular properties and their role in enabling rational design technologies.

Moreover, in this textbook, the reader will find some advanced concepts of computer simulation models used in statistical physics and a particular understanding of biological and physical systems. It discusses the molecular dynamics approach in detail to understand the use of this method in computational chemistry and biophysics.

The textbook is organised as the following. In Chap. 1, I introduce quantum mechanics' molecular dynamics simulation methods and computational modelling of electronic calculations. Chapter 2 describes the basis set functions for molecules and periodic systems, such as solids and metals.

In Chap. 3, I describe the advances in developing semi-empirical quantum mechanics molecular dynamics simulations. Then, in Chap. 4, I present the machine learning quantum mechanics molecular dynamics approach. In the next chapter, Chap. 5, I introduce the classical molecular dynamics methods used to study molecular systems. In particular, I will present the equations of motion in the most relevant statistical ensembles used in molecular dynamics simulations. Then, Chap. 6 describes the molecular force field development strategies at the atomistic level and the force field problems. Chapter 7 introduces the coarse-grained force fields for the proteins and DNAs.

In Chap. 8, I briefly describe the practical aspects of molecular dynamics simulations, such as designing the constraints, periodic boundary conditions, spherical cut-offs, treatment of the long-range interactions, and identifying the equilibrium states of the simulations.

In Chap. 9, I describe a Python Graphical User Interface (GUI) for the CHARMM package for performing molecular dynamics simulations and analysing data from the simulations.

Chapter 10 describes the methods used to equilibrate the system in explicit solvent, and Chap. 11 shows how to calculate the normal modes of a protein. Chapter 12 discusses molecular biological systems' standard molecular dynamics simulations and calculations of the solvent and solute properties from a trajectory. Finally, Chap. 13 presents the principal component analysis of a biological system using a quasi-harmonic approach.

The textbook is aimed at graduate students and research scientists in theoretical and computational biophysics, physics, chemistry, and material science. In addition, postgraduate students of other branches, such as applied mathematics, computer sciences, and bioinformatics, can use the book.

Skopje, North Macedonia Hiqmet Kamberaj
April 2023

Acknowledgements I thank my family for their continuous support: Nera (my wife), Jon (my son), and Lina (my daughter).

Contents

Chapter 1
Quantum Mechanics Computer Simulations

The chapter aims to introduce different quantum mechanics computer simulation methods and numerical molecular electronic calculations. In particular, the focus is on quantum mechanics models for molecular dynamics.

In this chapter, we introduce the quantum mechanics molecular models and quantum mechanics molecular dynamics simulations of molecules and biomolecular systems. Besides, the numerical aspects of the non-relativistic electronic structure of molecular systems will be discussed.

1.1 Introduction

Quantum mechanics (QM) calculations have an important contribution in studying materials properties. Thus, their popularity in predicting properties of new materials is increasing progressively. These properties include the energy, geometrical and band structure of materials. Besides, with the increase in computer power and speed, prediction of other properties are made possible, such as piezoelectric response, photon absorption, nuclear magnetic resonance shielding, and infrared Raman spectrum. Moreover, the improvement in the QM methods have made possible that QM calculations to reach the accuracy of the experimental measurements.

There exist several QM methods and modelling strategies of the electrons and atoms in the structure of molecules and materials. These include the methods that are based on the wave functions (Hehre et al. 1986) and density functional theory (Hohenberg and Kohn 1964; Kohn and Sham 1965; Schwarz and Blaha 2017). Furthermore, improvement on the speed and prediction efficiency of calculates are seen by QM cluster methods, which consider the specific arrangement of atoms

---

Supplementary Information The online version contains supplementary material available at https://doi.org/10.1007/978-3-031-34839-6_1.

H. Kamberaj, *Computer Simulations in Molecular Biology*, Scientific Computation,
https://doi.org/10.1007/978-3-031-34839-6_1

in the structure (Schwarz and Blaha 2017). This chapter aims to discuss in details QM molecular models and QM molecular dynamics simulations of molecules and biomolecular systems.

The time-dependent Schrödinger equation of a system of N quantum particles at the configuration $\mathbf{r} = (\mathbf{r}_1, \ldots, \mathbf{r}_N)$ is (Atkins and Friedman 2005):

$$i\hbar \frac{\partial \Psi(\mathbf{r}, t)}{\partial t} = \hat{H}\Psi(\mathbf{r}, t) \tag{1.1}$$

where $\hat{H}$ is the Hamiltonian operator for the system of quantum particles (with masses m_k, for $k = 1, \ldots, N$):

$$\hat{H} = \underbrace{-\sum_{k=1}^{N} \frac{\hbar^2}{2m_k}\nabla_k^2}_{\text{kinetic energy}} + \underbrace{U(\mathbf{r}_1, \ldots, \mathbf{r}_N, t)}_{\text{potential energy}} \tag{1.2}$$

Here, $\hbar = h/2\pi$ is the reduced Planck's constant, $i = \sqrt{-1}$, and Ψ is the wave-function describing the particles motion, used to derive different properties of the system. In Eq. (1.2), ∇_k^2 is the operator

$$\nabla_k^2 = \frac{\partial^2}{\partial \mathbf{r}_k^2} \equiv \frac{\partial^2}{\partial x_k^2} + \frac{\partial^2}{\partial y_k^2} + \frac{\partial^2}{\partial z_k^2} \tag{1.3}$$

If $U(\mathbf{r}_1, \ldots, \mathbf{r}_N, t)$ is independent of time, $U(\mathbf{r}_1, \ldots, \mathbf{r}_N)$, then the wave-function can be written as

$$\Psi(\mathbf{r}, t) = \psi(\mathbf{r})T(t) \tag{1.4}$$

In this case, the following time-independent Schrödinger equation is obtained (Atkins and Friedman 2005):

$$\hat{H}\psi(\mathbf{r}) = E\psi(\mathbf{r}) \tag{1.5}$$

where $\hat{H}$ is the time-independent Hamiltonian operator and E is the energy of the particles:

$$E = \frac{\int \psi^*(\mathbf{r})\hat{H}\psi(\mathbf{r})\, d\mathbf{r}}{\int \psi^*(\mathbf{r})\psi(\mathbf{r})\, d\mathbf{r}} \tag{1.6}$$

where ψ^* denotes the complex conjugate of ψ. (Note that if ψ is a real function, which is often required, then $\psi^* = \psi$.)

1.2 The Schrödinger Equations of (Bio)Molecular Systems

In particular, consider an N-electron (bio)molecule of M nuclei. The Hamiltonian operator of that system is given as (Iftime et al. 2005)

$$\begin{aligned}\hat{H} = &-\frac{\hbar^2}{2m_e}\sum_{k=1}^{N}\nabla_k^2 + \sum_{k=1}^{N}\sum_{j>k}^{N} k_e \frac{e^2}{\mid \mathbf{r}_j - \mathbf{r}_k \mid} \\ &-\sum_{\mu=1}^{M}\frac{\hbar^2}{2M_\mu}\nabla_\mu^2 + \sum_{\mu=1}^{M}\sum_{\nu>\mu}^{M} k_e \frac{Z_\mu Z_\nu e^2}{\mid \mathbf{R}_\nu - \mathbf{R}_\mu \mid} \\ &-\sum_{k=1}^{N}\sum_{\nu=1}^{M} k_e \frac{Z_\nu e^2}{\mid \mathbf{R}_\nu - \mathbf{r}_k \mid}\end{aligned} \tag{1.7}$$

where the electronic configuration is represented by coordinates

$$\mathbf{r} \equiv (\mathbf{r}_1, \mathbf{r}_2, \ldots, \mathbf{r}_N) \tag{1.8}$$

and the nuclei configuration is represented by coordinates

$$\mathbf{R} \equiv (\mathbf{R}_1, \mathbf{R}_2, \ldots, \mathbf{R}_M) \tag{1.9}$$

In Eq. (1.7), the first term is the kinetic energy of electrons, the second term is the interaction potential energy between electrons, the third term is the kinetic energy of nuclei, the fourth term is the interaction repulsion potential energy between nuclei, and the last term is the interaction energy between electrons and nuclei. Here, e is the charge of electron, $k_e = 1/4\pi\epsilon_0$ (with ϵ_0 being the vacuum permittivity), M_μ (for $\mu = 1, 2, \ldots, M$) are the nuclei masses, m_e mass of electron, and Z_μ (for $\mu = 1, 2, \ldots, M$) are the atomic numbers (equal to the number of protons in a nucleus).

The wave-function, characterising the quantum (bio)molecular system, can exactly be expressed in terms of time-independent electronic wave-functions $\psi_n(\mathbf{r}; \mathbf{R})$ (which are functions of electronic configuration $\mathbf{r}$ and parametric on $\mathbf{R}$) and time-dependent nuclear wave-functions $T_n(\mathbf{R}; t)$ (which are functions of $\mathbf{R}$ and parametric on t) as (Paquet and Viktor 2018):

$$\Psi(\mathbf{r}, \mathbf{R}; t) = \sum_{n=0}^{\infty} \psi_n(\mathbf{r}; \mathbf{R}) T_n(\mathbf{R}; t) \tag{1.10}$$

The Schrödinger's time-dependent equation (Eq. (1.1)) is

$$i\hbar \frac{\partial \Psi(\mathbf{r}, \mathbf{R}; t)}{\partial t} = \hat{H}\Psi(\mathbf{r}, \mathbf{R}; t) \tag{1.11}$$

The time-independent Schrödinger equation becomes

$$\hat{\mathbb{H}}\psi_n(\mathbf{r};\mathbf{R}) = E_n(\mathbf{R})\psi_n(\mathbf{r};\mathbf{R}) \tag{1.12}$$

where $\hat{\mathbb{H}}$ defines the Hamiltonian operator of electronic configuration $\mathbf{r}$:

$$\begin{aligned}\hat{\mathbb{H}} = &-\frac{\hbar^2}{2m_e}\sum_{i=1}^{N}\nabla_k^2 + \sum_{k=1}^{N}\sum_{j>k}^{N} k_e\frac{e^2}{\mid \mathbf{r}_j - \mathbf{r}_k \mid} \\ &-\sum_{k=1}^{N}\sum_{\nu=1}^{M} k_e\frac{Z_\nu e^2}{\mid \mathbf{R}_\nu - \mathbf{r}_k \mid}\end{aligned} \tag{1.13}$$

In Eq. (1.12), n is the set of quantum numbers labelling the electronic states (that is, the eigenvectors or wave-functions ψ_n and eigenvalues or energies E_n of electrons). Substituting Eq. (1.10) into Eq. (1.11), we obtain the system of equations representing the time evolution of the nuclear wave-functions:

$$\begin{aligned}i\hbar\frac{\partial T_n(\mathbf{R};t)}{\partial t} = &\left[-\sum_{\mu=1}^{M}\frac{\hbar^2}{2M_\mu}\nabla_\mu^2 + E_n(\mathbf{R}) + U_{\mathrm{NN}}(\mathbf{R})\right]T_n(\mathbf{R};t) \\ &+\sum_{m=0}^{\infty}\kappa_{nm}T_m(\mathbf{R};t)\end{aligned} \tag{1.14}$$

where $U_{\mathrm{NN}}(\mathbf{R})$ is the nuclear-nuclear Coulomb's repulsion potential energy (see Eq. (1.7)) and the matrix elements κ_{nm} give the strength of coupling between the nuclear wave-functions and electronic wave-functions, determined as (Paquet and Viktor 2018):

$$\begin{aligned}\kappa_{nm} = &\int \psi_n^*(\mathbf{r};\mathbf{R})\left[-\sum_{\mu=1}^{M}\frac{\hbar^2}{2M_\mu}\nabla_\mu^2\right]\psi_m(\mathbf{r};\mathbf{R})\,d\mathbf{r} \\ &-\sum_{\mu=1}^{M}\frac{\hbar^2}{M_\mu}\left[\int \psi_n^*(\mathbf{r};\mathbf{R})\nabla_\mu\psi_m(\mathbf{r};\mathbf{R})\,d\mathbf{r}\right]\nabla_\mu\end{aligned} \tag{1.15}$$

where ψ_n^* denotes the complex conjugate of ψ_n. The Schrödinger equations, given above, can be solved exactly for only a few problems, such as a particle in the box, the harmonic oscillator, the particle in a ring, the particle in a sphere, and the hydrogen atom (Atkins and Friedman 2005). Therefore, the following approximations are necessary to decrease the computational complexity.

1.3 Adiabatic and Born–Oppenheimer (BO) Approximations

In the adiabatic approximation, it is assumed that the electronic wave-functions adjust almost instantaneously to any changes in the positions of the nuclei, which is known as the timescale separation; that is, fast electrons dynamics and slow nuclei motion. That is justifiable since $M_\mu \gg m_e$ (for $\mu = 1, 2, \ldots, M$) and assuming that electrons are non-relativistic. (Note that the relativistic effects on the electron motions may be present for electrons close to the nuclei.) Under the adiabatic approximation, we have

$$\nabla_\mu \psi_m(\mathbf{r}; \mathbf{R}) = 0 \tag{1.16}$$

That is, during the fast electronic motion the nuclei are assumed fixed at the configuration $\mathbf{R}$. Thus, Eq. (1.15) reduces to

$$\kappa_{nm} = \begin{cases} -\sum\limits_{\mu=1}^{M} \dfrac{\hbar^2}{2M_\mu} \left[\int \psi_n^*(\mathbf{r}; \mathbf{R}) \nabla_\mu^2 \psi_n(\mathbf{r}; \mathbf{R})\, d\mathbf{r}\right] & \text{if } n = m, \\ 0 & \text{if } n \neq m \end{cases} \tag{1.17}$$

Next approximation is Born and Oppenheimer (1927), which relies on the fact that nuclei and electrons have different timescales of motion. That is, the nuclei remain fixed on the electrons timescale of motion, which also is due to the differences on the masses of electrons and nuclei. Therefore, the motion of a (bio)molecule is divided into two levels, namely the electrons and nuclei (Leach 2001; Atkins and Friedman 2005). Using BO approximation, Eq. (1.10) is approximated as

$$\Psi(\mathbf{r}, \mathbf{R}; t) \approx \psi_n(\mathbf{r}; \mathbf{R}) T_n(\mathbf{R}; t) \tag{1.18}$$

Thus, there is a single state of the (bio)molecular system (Paquet and Viktor 2018). The electrons are the independent degrees of freedom of $\hat{H}$, while the nuclei are assumed fixed. Therefore, the wave-function depends only on the positions of the nuclei but not on their momenta. Besides, the eigenstate energy E_n is much greater than the coupling matrix elements κ_{nn}, and thus Eq. (1.14) reduces to

$$i\hbar \frac{\partial T_n(\mathbf{R}; t)}{\partial t} = \left[-\sum_{\mu=1}^{M} \frac{\hbar^2}{2M_\mu} \nabla_\mu^2 + E_n(\mathbf{R}) + U_{\mathrm{NN}}(\mathbf{R}) \right] T_n(\mathbf{R}; t) \tag{1.19}$$

Here, $E_n(\mathbf{R})$ gives the electronic energy level, or the so-called *potential energy surface* (PES).

Furthermore, the electrons are non-excited, and hence they only occupy the so-called *ground state energy* level; that is, $n = 0$, which implies that (Leach 2001; Iftime et al. 2005; Paquet and Viktor 2018):

$$\hat{\mathbb{H}}\psi_0(\mathbf{r};\mathbf{R}) = E_0(\mathbf{R})\psi_0(\mathbf{r};\mathbf{R}) \tag{1.20}$$

where E_0 represents the ground state energy of electrons at the nuclei configuration $\mathbf{R}$. This is the potential energy of Hamiltonian used to investigate the dynamics of nuclei in quantum mechanics simulations (Leach 2001). Then, Eq. (1.19) takes the form:

$$i\hbar\frac{\partial T_0(\mathbf{R};t)}{\partial t} = \left[-\sum_{\mu=1}^{M}\frac{\hbar^2}{2M_\mu}\nabla_\mu^2 + E_0(\mathbf{R}) + U_{\text{NN}}(\mathbf{R})\right]T_0(\mathbf{R};t) \tag{1.21}$$

1.4 Self-consistent Field Method

Note that the electronic wave-function $\psi_0(\mathbf{r};\mathbf{R})$ is the solution of Eq. (1.20) with Hamiltonian given by Eq. (1.13). Besides, $\hat{\mathbb{H}}$ includes the electron-electron potential energy, which depends on the electron-electron separation $|\mathbf{r}_j - \mathbf{r}_i|$, and it is crucial on the electronic calculations. The Hamiltonian $\hat{\mathbb{H}}$, Eq. (1.13), can also be written as:

$$\begin{aligned}\hat{\mathbb{H}} &= \sum_{i=1}^{N}\left[-\frac{\hbar^2}{2m_e}\nabla_i^2 + \sum_{j>i}^{N}k_e\frac{e^2}{|\mathbf{r}_j - \mathbf{r}_i|} - \sum_{\nu=1}^{M}k_e\frac{Z_\nu e^2}{|\mathbf{R}_\nu - \mathbf{r}_i|}\right] \\ &\equiv \sum_{i=1}^{N}\hat{h}_i(\mathbf{r};\mathbf{R})\end{aligned} \tag{1.22}$$

where $\hat{h}_i$ is the one-electron Hamiltonian operator. Since the electrons are considered independent and indistinguishable, $\hat{h}_i(\mathbf{r};\mathbf{R})$ is denoted as $\hat{h}_i(1)$ to indicate single-electron Hamiltonian operator occupying orbital i. Thus, the N-electron equations, namely Eqs. (1.20) and (1.22), can be split into n one-electron equations, where n is the number of wave-functions (or orbitals) each with occupancy s_i (which gives the number of electrons occupying the orbital i). We can write $\psi_0(\mathbf{r};\mathbf{R})$ as a product of n one-electron wave-functions $\psi_i^0(\mathbf{r};\mathbf{R})$ (the so-called *spatial orbitals*, which depend on the position of an electron $\mathbf{r}$ occupying orbital i and parametrically on the configuration of nuclei $\mathbf{R}$) (Atkins and Friedman 2005). The wave-function of one-electron spatial orbital i, $\psi_i^0(\mathbf{r};\mathbf{R})$, is the solution of the following equation:

$$\hat{h}_i(1)\psi_i^0(\mathbf{r};\mathbf{R}) = E_i^0\psi_i^0(\mathbf{r};\mathbf{R}) \tag{1.23}$$

where E_i^0 denotes the energy of single-electron orbital i. Note that in Eq. (1.23) $\mathbf{r}$ indicates coordinate vector of any of N electrons of system, which acts as the origin of the coordinate system, $\mathbf{r} \in \{\mathbf{r}_1, \mathbf{r}_2, \ldots, \mathbf{r}_N\}$. This model characterises an electron-independent model in which electrons are indistinguishable. If we assume that we have n one-electron orbitals, then

$$\psi_0(\mathbf{r};\mathbf{R}) = \underbrace{\psi_a^0(\mathbf{r}_1;\mathbf{R})\psi_b^0(\mathbf{r}_2;\mathbf{R})\cdots\psi_z^0(\mathbf{r}_N;\mathbf{R})}_{n\text{-orbitals}} \tag{1.24}$$

Then, the total energy of the electrons in the ground state (the minimum energy) is given as

$$E_0 = \sum_{i=1}^{n} s_i E_i^0 \tag{1.25}$$

The requirement that the wave-functions of electrons obey the Pauli principle and Fermi-Dirac statistics introduces the so-called *spin-orbitals* by taking into account the spin of electron. In particular, the spin-orbital is product of the spin function $\chi(\uparrow\downarrow)$ and spatial orbital wave-function $\psi_a^0(\mathbf{r};\mathbf{R})$: $\psi_a(\mathbf{x};\mathbf{R}) = \chi(\tau)\cdot\psi_a^0(\mathbf{r};\mathbf{R})$ (Leach 2001; Atkins and Friedman 2005) (where τ is the spin space variable, either $\uparrow$ or $\downarrow$ spin, and $\mathbf{x}$ is the joint spin and spatial spaces of the electron occupying orbital i). Thus, the overall electronic wave-function is written as

$$\psi_0(\mathbf{x};\mathbf{R}) = \frac{1}{\sqrt{N!}}\det\mid\psi_a(\mathbf{x}_1;\mathbf{R})\psi_b(\mathbf{x}_2;\mathbf{R})\cdots\psi_z(\mathbf{x}_N;\mathbf{R})\mid \tag{1.26}$$

where the Slater determinant is used (Atkins and Friedman 2005). Besides, the spin orbitals are orthonormal, which is written in terms of overlap integrals between two spin orbitals as follows (Leach 2001):

$$\int \psi_i^*(\mathbf{x};\mathbf{R})\psi_j(\mathbf{x};\mathbf{R})\,d\mathbf{x} = \delta_{ij} \tag{1.27}$$

where the integration is over the joint spin and spatial space and δ_{ij} is the Kronecker's delta: $\delta_{ij} = 1$, if $i = j$, and $\delta_{ij} = 0$, if $i \neq j$.

In general, two approaches are used to treat either closed-shell or open-shell molecular systems, namely the spin-restricted and spin-unrestricted systems (Leach 2001). In the spin-restricted theory, a combination of the singly and doubly electron occupied molecular orbitals are used, which corresponds to the closed-shell approach. Both electrons (α and β) occupying the same spin orbital have the same spatial functions (but different spins). Whereas, in the spin-unrestricted theory, two different sets of molecular orbitals are used; that is, one for electrons with spin α (namely $\psi_i^\alpha(\mathbf{r};\mathbf{R})$) and the other (namely $\psi_i^\beta(\mathbf{r};\mathbf{R})$) for electrons with spin β, and all the equations that follow below should be written and solved for each spatial orbital separately. In all calculations shown in the following, we are going to use the spatial orbital notion $\psi_i(\mathbf{r};\mathbf{R})$ for the orbital i corresponding to the spatial function of either electron with spin α or β; therefore, the occupation number $s_i = 1$ and $n = N$ for spin-unrestricted systems (and hence there are two sets of spatial orbitals), whereas for spin-restricted systems $s_i = 2$ and $n = N/2$ (and hence the same spatial orbital is used for both spins) (Iftime et al. 2005).

The constraint, Eq. (1.27), is given as

$$g = \sum_{i=1}^{n} \sum_{j=1}^{n} \left[\int \psi_i^*(\mathbf{r}; \mathbf{R}) \psi_j(\mathbf{r}; \mathbf{R})\, d\mathbf{r} - \delta_{ij} \right] = 0 \tag{1.28}$$

The constraint given by Eq. (1.28) is tackled using Lagrange multipliers method (Leach 2001). According to this method, the derivatives of the constraint g (see Eq. (1.28)) multiplied by Lagrange multiplier is added to the derivative of the function to be minimised, and the sum is equal to zero. In general, for the Hamiltonian formalism, that is equivalent to adding a penalty potential function as follows (Iftime et al. 2005; Paquet and Viktor 2018):

$$U_{\text{penalty}} = \frac{1}{2} \sum_{i=1}^{n} \sum_{j=1}^{n} \lambda_{ij} \left(\langle \psi_i | \psi_j \rangle - \delta_{ij} \right)^2 \tag{1.29}$$

where $\langle \psi_i | \psi_j \rangle \equiv \int \psi_i^*(\mathbf{r}; \mathbf{R}) \psi_j(\mathbf{r}; \mathbf{R})\, d\mathbf{r}$.

Next, the single-electron spatial-orbital $\psi_i(\mathbf{r}; \mathbf{R})$ of the orbital i is written as a linear combination of basis set functions (Leach 2001; Atkins and Friedman 2005):

$$\psi_i(\mathbf{r}; \mathbf{R}) = \sum_{\mu=1}^{K} c_{\mu i} \phi_\mu(\mathbf{r}; \mathbf{R}) \tag{1.30}$$

Here, ϕ_μ is the basis set of a finite K functions and $c_{\mu i}$ are the linearity coefficients. The optimum values of the coefficients are that for which the energy is a minimum (Leach 2001) determined as follows:

$$\frac{\partial E_0}{\partial c_{\mu i}} + \frac{\partial}{\partial c_{\mu i}} \sum_{k=1}^{n} \sum_{j=1}^{n} \lambda_{kj} g_{kj} = 0, \quad \mu = 1, 2, \ldots, K; \; i = 1, 2, \ldots, n \tag{1.31}$$

where

$$g_{kj} = \langle \psi_k | \psi_j \rangle - \delta_{kj} \tag{1.32}$$

That is known as *minimisation of electronic energy*, which is computationally very demanding for biological molecular systems. In principle, an infinite basis set should be used for a precise calculation of molecular orbitals in Eq. (1.30), $K = \infty$; however, in practice only a finite basis set is used (Lehtola 2019b; Reyes et al. 2019).

1.5 Ehrenfest Molecular Dynamics

Equation (1.21) gives the Schrödinger equation for determining the nuclei wave-functions. However, it is possible to assume that the slow nucleus dynamical motion can be described semi-classically by second law of Newton. For that, the so-called Ehrenfest theorem is used (Hall 2013; Ding et al. 2015; Jensen 2017). According to Ehrenfest theorem, the average nuclear motion is governed classically by the second law of Newton with a potential equal to U_{NN} plus the quantum expectation of the (bio)molecule electrons Hamiltonian with the electronic wave-functions as:

$$\langle \hat{\mathbb{H}} \rangle = \langle \psi | \hat{\mathbb{H}} | \psi \rangle = \int \psi^*(\mathbf{r}; \mathbf{R}) \hat{\mathbb{H}}(\mathbf{r}; \mathbf{R}) \psi(\mathbf{r}; \mathbf{R}) \, d\mathbf{r} \tag{1.33}$$

where $\psi(\mathbf{r}; \mathbf{R})$ is the electronic wave-function at the nuclei configuration $\mathbf{R}$.

Therefore, the dynamics of the nuclei are described by the following Newton's second law:

$$M_\mu \ddot{\mathbf{R}}_\mu = -\nabla_\mu \left[\langle \hat{\mathbb{H}} \rangle + U_{\mathrm{NN}}(\mathbf{R}) \right], \quad \text{for } \mu = 1, 2, \ldots, M \tag{1.34}$$

and the electronic motion is described by the time-dependent Schrödinger equation:

$$i\hbar \frac{\partial \psi(\mathbf{r}; \mathbf{R})}{\partial t} = \hat{\mathbb{H}} \psi(\mathbf{r}; \mathbf{R}) \tag{1.35}$$

Equation (1.35) guaranties that the orthogonality of electronic wave-functions is maintained since $\hat{\mathbb{H}}$ is a Hermitian operator (Jensen 2017).

The disadvantage is that analytical solutions are feasible for small molecules only; therefore, standard approximations are used to decrease the computational complexity, as shown in the following discussion.

1.6 Born–Oppenheimer Molecular Dynamics

Using the adiabatic and Born–Oppenheimer approximations, the expectation of the electronic Hamiltonian (see also Eq. (1.33)) can be obtained as

$$\langle \psi_0 | \hat{\mathbb{H}} | \psi_0 \rangle = \int \psi_0^*(\mathbf{r}; \mathbf{R}) \hat{\mathbb{H}}(\mathbf{r}; \mathbf{R}) \psi_0(\mathbf{r}; \mathbf{R}) \, d\mathbf{r} \tag{1.36}$$

The Born–Oppenheimer molecular dynamics is characterised by the following (classical) Hamiltonian function:

$$\mathcal{H}_{\mathrm{BO}} = \sum_{\mu=1}^{M} \frac{\mathbf{P}_{\mu}^{2}}{2M_{\mu}} + \langle\psi_0|\hat{\mathbb{H}}|\psi_0\rangle + U_{\mathrm{NN}}(\mathbf{R}) + U_{\mathrm{penalty}} \tag{1.37}$$

where the first term corresponds to the kinetic energy of nuclei, the second term is the minimum electronic energy (see Eq. (1.25)), obtained using Ehrenfest theorem, the third term is the nucleus-nucleus repulsion Coulombic interaction energy, and the fourth term is a penalty potential energy to ensure the orthonormality of spatial orbitals at all times. Here, $\mathbf{P}_{\mu}$ (for $\mu = 1, 2, \ldots, M$) are the nuclei momenta.

Then, the dynamics of the system are described by the Hamiltonian equations of motion:

$$\begin{aligned} \dot{\mathbf{R}}_{\mu} &= \frac{\partial \mathcal{H}_{\mathrm{BO}}}{\partial \mathbf{P}_{\mu}} \\ \dot{\mathbf{P}}_{\mu} &= -\frac{\partial \mathcal{H}_{\mathrm{BO}}}{\partial \mathbf{R}_{\mu}}, \quad \text{for } \mu = 1, 2, \ldots, M \\ \dot{\psi}_i^{*} &= \frac{\partial \mathcal{H}_{\mathrm{BO}}}{\partial P_{\psi_i}} \\ \dot{P}_{\psi_i} &= -\frac{\partial \mathcal{H}_{\mathrm{BO}}}{\partial \psi_i^{*}}, \quad \text{for } i = 1, 2, \ldots, n \end{aligned} \tag{1.38}$$

The dynamics of the nuclei are as follows:

$$\begin{aligned} \dot{\mathbf{R}}_{\mu} &= \frac{\mathbf{P}_{\mu}}{M_{\mu}} \\ \dot{\mathbf{P}}_{\mu} &= -\Big[\langle\psi_0|\nabla_{\mu}\hat{\mathbb{H}}|\psi_0\rangle + \nabla_{\mu}U_{\mathrm{NN}}(\mathbf{R})\Big], \quad \text{for } \mu = 1, 2, \ldots, M \end{aligned} \tag{1.39}$$

where $\mathbf{R}$ is the nuclei positions vector and the gradient $\langle\psi_0|\nabla_{\mu}\hat{\mathbb{H}}|\psi_0\rangle$ is calculated in the limit of the electronic ground state (that is, minimum electronic energy, $\nabla_{\mu}E_0(\mathbf{R}) = \langle\psi_0|\nabla_{\mu}\hat{\mathbb{H}}|\psi_0\rangle$). The equations of motion for the single-electron spatial orbitals i are

$$\hat{h}_i(1)\psi_i(\mathbf{r};\mathbf{R}) = \sum_{j=1}^{n} \lambda_{ji}\psi_j(\mathbf{r};\mathbf{R}), \quad i = 1, 2, \ldots, n \tag{1.40}$$

Other standard approximations are necessary used to reduce the computational complexity as follows in the next discussions.

1.7 Hartree–Fock Molecular Dynamics

Consider a (bio)molecule of N electrons and n orbitals. The focus is on finding the minimum energy, or the best electronic wave-function for this (bio)molecular system of many electrons. Furthermore, we suppose that N electrons are assigned to individual n spatial orbitals. Thus, we try to find out how the electronic motions, as the change in the spatial orbital for one electron, influence the behaviour of an electron in another spatial orbital because of the coupling of the electronic motions (Kitaura and Morokuma 1976).

In particular, we will focus on a single-electron in a spatial orbital, $\psi_i(\mathbf{r}; \mathbf{R})$, in the field of other electrons in their fixed spatial orbitals, $\psi_j(\mathbf{r}'; \mathbf{R})$, and in the field of nuclei at their fixed configuration $\mathbf{R}$. Here, $\mathbf{r}'$ gives the position of other electrons relative to the electron occupying $\psi_i(\mathbf{r}; \mathbf{R})$.

Then, the Fock operator of one-electron Hamiltonian occupying orbital i is introduced (see Leach 2001):

$$\hat{f}_i(1) = \hat{h}^{\text{core}}(\mathbf{r}; \mathbf{R}) + \sum_{j=1}^{n} \left[s_j \hat{J}_j(\mathbf{r}; \mathbf{R}) - \hat{K}_j(\mathbf{r}; \mathbf{R}) \right] \tag{1.41}$$

Here, the core Hamiltonian operator, $\hat{h}^{\text{core}}(\mathbf{r}; \mathbf{R})$ is

$$\hat{h}^{\text{core}}(\mathbf{r}; \mathbf{R}) = -\frac{\hbar^2}{2m_e}\nabla_{\mathbf{r}}^2 - k_e e^2 \sum_{\mu=1}^{M} \frac{Z_\mu}{\mid \mathbf{R}_\mu - \mathbf{r} \mid} \tag{1.42}$$

the Coulomb interaction operator, $\hat{J}_j(\mathbf{r}; \mathbf{R})$, corresponding to the average potential due to the electron occupying spatial orbital $\psi_j(\mathbf{r}'; \mathbf{R})$:

$$\hat{J}_j(\mathbf{r}; \mathbf{R}) = k_e e^2 \int \psi_j^*(\mathbf{r}'; \mathbf{R}) \frac{1}{\mid \mathbf{r}' - \mathbf{r} \mid} \psi_j(\mathbf{r}'; \mathbf{R})\, d\mathbf{r}' \tag{1.43}$$

and $\hat{K}_j(\mathbf{r}; \mathbf{R})$ is the exchange operator, which counts for the quantum Pauli exclusion principle:

$$\hat{K}_j(\mathbf{r}; \mathbf{R})\psi_i(\mathbf{r}; \mathbf{R}) = \left[k_e e^2 \int \psi_j^*(\mathbf{r}'; \mathbf{R}) \frac{1}{\mid \mathbf{r}' - \mathbf{r} \mid} \psi_i(\mathbf{r}'; \mathbf{R})\, d\mathbf{r}' \right] \psi_j(\mathbf{r}; \mathbf{R}) \tag{1.44}$$

Thus, the contribution to the energy of the spatial orbital $\psi_i(\mathbf{r}; \mathbf{R})$ is

$$\hat{f}_i(1)\psi_i(\mathbf{r}; \mathbf{R}) = E_i^0 \psi_i(\mathbf{r}; \mathbf{R}) \tag{1.45}$$

where E_i^0 is the energy of a single-electron at orbital i.

The kinetic and potential energy from all N electrons occupying the spatial orbital $\psi_i(\mathbf{r}; \mathbf{R})$ moving in the field of nuclei give the following contribution:

$$H_{ii}^{\text{core}} = \int \psi_i^*(\mathbf{r}; \mathbf{R})\hat{h}^{\text{core}}(\mathbf{r}; \mathbf{R})\psi_i(\mathbf{r}; \mathbf{R})\, d\mathbf{r} \tag{1.46}$$

For N electrons in n spatial orbitals the contribution to the total energy is

$$E_0^{\text{core}} = \sum_{i=1}^{n} s_i H_{ii}^{\text{core}} \tag{1.47}$$

The electron-electron electrostatic interaction is

$$J_{ij} = \int \psi_i^*(\mathbf{r}; \mathbf{R})\hat{J}_j(\mathbf{r}; \mathbf{R})\psi_i(\mathbf{r}; \mathbf{R})\, d\mathbf{r} \tag{1.48}$$

which characterises Coulombic interaction between electrons in spatial orbitals i and j; it's positive (repulsive), and hence unfavourable. The total electrostatic interaction energy between a single electron occupying orbital ψ_i and other $N - 1$ electrons is

$$E_i^{\text{Coul}} = \sum_{j \neq i}^{n} s_j J_{ij} \tag{1.49}$$

Total Coulombic contribution to the electronic energy is

$$E_0^{\text{Coul}} = \sum_{i=1}^{n} s_i E_i^{\text{Coul}} = \sum_{i=1}^{n} \sum_{j=i+1}^{n} s_i s_j J_{ij} \tag{1.50}$$

The third contribution to the energy is the exchange interaction; this term is pure quantum mechanical and it is due to the correlations of motions of electrons with parallel spins. That is, there is a finite probability of finding two paired electrons with opposite spins at the same point in space; however, if the spins are the same, then this probability is zero. This is known as Pauli principle, which says that if two electrons occupy the same region of space and have the same spins they could be considered to have the same set of quantum numbers. Therefore, the same spins avoid each other, by experiencing a lower Coulombic repulsion energy term. The exchange interaction is integral of the form:

$$K_{ij} = \int \psi_i^*(\mathbf{r}; \mathbf{R})\hat{K}_j(\mathbf{r}; \mathbf{R})\psi_j(\mathbf{r}; \mathbf{R})\, d\mathbf{r} \tag{1.51}$$

The integral is non-zero if the spins of the electrons occupying the same spatial orbitals ψ_i and ψ_j are the same. The exchange energy between a single electron in the spatial orbital ψ_i and other $N - 1$ electrons is

$$E_i^{\text{exchange}} = \sum_{j \neq i}^{n} K_{ij} \tag{1.52}$$

Total exchange energy contribution to the electronic energy is given as

$$E_0^{\text{exchange}} = \sum_{i=1}^{n} s_i E_i^{\text{exchange}} = \sum_{i=1}^{n} \sum_{j'=i+1}^{n} s_i K_{ij} \tag{1.53}$$

where j' indicates that the sum is only over electrons j with the same spins as electron i because other terms are zero.

Most of ground state molecules have a closed-shell configuration. Therefore, a closed-shell system with N electrons has $N/2$ orbitals, and hence two spin orbitals associate with a spatial orbital ψ_i^0: ψ_i^α and ψ_i^β. If there is one electron in a spatial orbital ψ_i, then it contributes an energy H_{ii}^{core}. If there are two electrons in a molecular orbital ψ_i, then their energy contribution is $2H_{ii}^{\text{core}}$ (Leach 2001).

Interactions between each pair of spatial orbitals ψ_i^0 and ψ_j^0 include four electrons. There exist four ways that 2 electrons in one orbital interact with 2 electrons in another second orbital. Thus, there is a contribution to energy of $4J_{ij}$. The number of ways we can arrange a pair of electrons is 2, and hence the total exchange contribution is $-2K_{ij}$. Also, there is a contribution arising from the Coulombic interactions between each pair of electrons in the same orbital J_{ii}.

For n orbitals, the total contribution to the energy is

$$E_0(\mathbf{R}) = \sum_{i=1}^{n} s_i H_{ii}^{core} + \sum_{i=1}^{n} \sum_{j=i+1}^{n} (s_i s_j J_{ij} - s_i K_{ij}) + \sum_{i=1}^{n} J_{ii} \tag{1.54}$$

Note that $J_{ii} = K_{ii}$; therefore,

$$E_0(\mathbf{R}) = \sum_{i=1}^{n} s_i H_{ii}^{core} + \sum_{i=1}^{n} \sum_{j=1}^{n} (s_i J_{ij} - K_{ij}) \tag{1.55}$$

The above integrals can further be simplified for computations using the so-called *Roothaan–Hall equations* (Raathaan 1951). For that, we substitute Eq. (1.30) into Eq. (1.45):

$$\hat{f}_i(1) \sum_{\mu=1}^{K} c_{\mu i} \phi_\mu(\mathbf{r}; \mathbf{R}) = E_i^0 \sum_{\mu=1}^{K} c_{\mu i} \phi_\mu(\mathbf{r}; \mathbf{R}) \tag{1.56}$$

We can multiply both sides by $\phi_\nu^*(\mathbf{r}; \mathbf{R})$ and integrate over $\mathbf{r}$:

$$\sum_{\mu=1}^{K} c_{\mu i} \int \phi_\nu^*(\mathbf{r}; \mathbf{R}) \hat{f}_i(1) \phi_\mu(\mathbf{r}; \mathbf{R})\, d\mathbf{r} = E_i^0 \sum_{\mu=1}^{K} c_{\mu i} \int \phi_\nu^*(\mathbf{r}; \mathbf{R}) \phi_\mu(\mathbf{r}; \mathbf{R})\, d\mathbf{r} \tag{1.57}$$

Now, **F** Hartree–Fock matrix is introduced with elements:

$$F_{\nu\mu} = \int \phi_\nu^*(\mathbf{r};\mathbf{R})\hat{f}_i(1)\phi_\mu(\mathbf{r};\mathbf{R})\,d\mathbf{r} \tag{1.58}$$

and the overlapping integral between the basis functions $S_{\nu\mu}$:

$$S_{\nu\mu} = \int \phi_\nu^*(\mathbf{r};\mathbf{R})\phi_\mu(\mathbf{r};\mathbf{R})\,d\mathbf{r} \tag{1.59}$$

Note that, unlike the spatial orbitals ψ_i and ψ_j, the basis functions are not necessary orthonormal, and therefore $S_{\nu\mu} \neq \delta_{\nu\mu}$.

Substituting Eq. (1.41) into Eq. (1.58), we get

$$\begin{aligned} F_{\nu\mu} &= \int \phi_\nu^*(\mathbf{r};\mathbf{R})\hat{h}^{\text{core}}(\mathbf{r};\mathbf{R})\phi_\mu(\mathbf{r};\mathbf{R})\,d\mathbf{r} \\ &\quad + \sum_{j=1}^{n} s_j \int \phi_\nu^*(\mathbf{r};\mathbf{R})\hat{J}_j(\mathbf{r};\mathbf{R})\phi_\mu(\mathbf{r};\mathbf{R})\,d\mathbf{r} \\ &\quad - \sum_{j=1}^{n} \int \phi_\nu^*(\mathbf{r};\mathbf{R})\hat{K}_j(\mathbf{r};\mathbf{R})\phi_\mu(\mathbf{r};\mathbf{R})\,d\mathbf{r} \end{aligned} \tag{1.60}$$

where

$$\hat{J}_j(\mathbf{r};\mathbf{R}) = k_e e^2 \sum_{\alpha=1}^{K} c_{\alpha j} \sum_{\beta=1}^{K} c_{\beta j} \int \phi_\alpha^*(\mathbf{r}';\mathbf{R}) \frac{1}{\mid \mathbf{r}' - \mathbf{r} \mid} \phi_\beta(\mathbf{r}';\mathbf{R})\,d\mathbf{r}' \tag{1.61}$$

and $\hat{K}_j(\mathbf{r};\mathbf{R})$ is the exchange operator, which counts for the quantum Pauli exclusion principle:

$$\begin{aligned} \hat{K}_j(\mathbf{r};\mathbf{R})\psi_i(\mathbf{r};\mathbf{R}) &= \left[k_e e^2 \sum_{\alpha=1}^{K} c_{\alpha j} \sum_{\beta=1}^{K} c_{\beta j} \int \phi_\alpha^*(\mathbf{r}';\mathbf{R}) \frac{1}{\mid \mathbf{r}' - \mathbf{r} \mid} \psi_i(\mathbf{r}';\mathbf{R})\,d\mathbf{r}' \right] \\ &\quad \times \phi_\beta(\mathbf{r}';\mathbf{R}) \end{aligned} \tag{1.62}$$

Therefore, Eq. (1.57) can be written as

$$\sum_{\mu=1}^{K} F_{\nu\mu} c_{\mu i} = \left(\sum_{\mu=1}^{K} S_{\nu\mu} c_{\mu i} \right) E_i^0 \tag{1.63}$$

Alternatively, in a matrix form:

$$\mathbf{FC} = \mathbf{SCE} \tag{1.64}$$

where $\mathbf{E}$ is a $K \times K$ diagonal matrix of the orbital energies E_i^0 (single-electron contribution), $\mathbf{C}$ is a $K \times K$ matrix of coefficients $c_{\mu i}$, and $\mathbf{S}$ is the $K \times K$ overlapping matrix of $S_{\nu\mu}$, and $\mathbf{F}$ is the $K \times K$ Hartree–Fock matrix. Equation (1.64) is known as Roothaan equation (Raathaan 1951).

Note that Eq. (1.64) is not in the standard form of an eigenvalue problem, namely $\mathbf{FC} = \mathbf{CE}$. We can follow method suggested in Leach (2001) to bring it in the form that we can use the standard matrix eigenvalue approaches. For that, we multiply both sides of Eq. (1.64) by $\mathbf{S}^{-1/2}$:

$$\mathbf{S}^{-1/2}\mathbf{FC} = \mathbf{S}^{-1/2}\mathbf{SCE} = \mathbf{S}^{1/2}\mathbf{CE} \tag{1.65}$$

Equation (1.65) can further be written as (since $\mathbf{S}^{-1/2}\mathbf{S}^{1/2} = \mathbf{I}$, where $\mathbf{I}$ is a unitary matrix):

$$\mathbf{S}^{-1/2}\mathbf{F}\left(\mathbf{S}^{-1/2}\mathbf{S}^{1/2}\right)\mathbf{C} = \mathbf{S}^{1/2}\mathbf{CE} \tag{1.66}$$

which is re-arranged as

$$\mathbf{S}^{-1/2}\mathbf{FS}^{-1/2}\left(\mathbf{S}^{1/2}\mathbf{C}\right) = \left(\mathbf{S}^{1/2}\mathbf{C}\right)\mathbf{E} \tag{1.67}$$

Denoting $\mathbf{F}' \equiv \mathbf{S}^{-1/2}\mathbf{FS}^{-1/2}$ and $\mathbf{C}' \equiv \mathbf{S}^{1/2}\mathbf{C}$, we obtain

$$\mathbf{F}'\mathbf{C}' = \mathbf{C}'\mathbf{E} \tag{1.68}$$

which is called here *reduced Roothaan* equation. Equation (1.68) has a non-trivial solution, which is obtained if the following condition is fulfilled (Atkins and Friedman 2005):

$$\det \mid \mathbf{F}' - E_i^0\mathbf{I} \mid = 0 \tag{1.69}$$

The equation can not be solved explicitly because the matrix elements $F_{\nu\mu}$ are expressed as integrals over the exchange and Coulomb operators that depend on the spatial orbital wave-functions. Therefore, in the following, we will present a self-consistent field method.

Introducing the so-called *density matrix* (Leach 2001; Atkins and Friedman 2005):

$$P_{\nu\mu} = \sum_{j=1}^{n} s_j c_{\nu j} c_{\mu j} \tag{1.70}$$

Fock matrix becomes

$$F_{\nu\mu} = H_{\nu\mu}^{\text{core}} + \sum_{\alpha=1}^{K}\sum_{\beta=1}^{K} P_{\alpha\beta}\left[(\nu\mu|\alpha\beta) - \frac{1}{s}(\nu\alpha|\mu\beta)\right] \tag{1.71}$$

where $s = s_j, \ \forall j$ is considered, and

$$H_{\nu\mu}^{\text{core}} = \int \phi_\nu^*(\mathbf{r};\mathbf{R})\hat{h}^{\text{core}}(\mathbf{r};\mathbf{R})\phi_\mu(\mathbf{r};\mathbf{R})\,d\mathbf{r} \tag{1.72}$$

In Eq. (1.71), $(\nu\mu|\alpha\beta)$ denotes

$$(\nu\mu|\alpha\beta) = k_e e^2 \int\int \phi_\nu^*(\mathbf{r};\mathbf{R})\phi_\mu(\mathbf{r};\mathbf{R})\frac{1}{|\mathbf{r}'-\mathbf{r}|}\phi_\alpha^*(\mathbf{r}';\mathbf{R})\phi_\beta(\mathbf{r}';\mathbf{R})\,d\mathbf{r}'\,d\mathbf{r} \tag{1.73}$$

Then, the electronic energy of the (bio)molecule ground state is

$$E_0(\mathbf{R}) = \frac{1}{2}\sum_{\mu=1}^{K}\sum_{\nu=1}^{K} P_{\nu\mu}\left(H_{\nu\mu}^{core} + F_{\nu\mu}\right) \tag{1.74}$$

The Hartree–Fock self-consistent field (HF-SCF) algorithm for solving Roothaan equations is given by Algorithm 1. It can be seen that the algorithm requires to choose a basis set functions and an initial set of trial matrix **C**. The density matrix **P** can either be calculated using the initial trial matrix **C** or guessed, for example, use a zero matrix (that is, initially ignoring all the electron-electron interaction terms, and hence the electrons interact only with nuclei); however, that may have convergence problems (Leach 2001). Better guess can be suggested using a lower level of theory, such as semi-empirical approaches, discussed in the next chapter. Moreover, several criteria can be used to check the convergence; for example, the density matrix **P** can be compared with its value in the previous iteration step. Besides, the change in energy and in the basis set coefficients can be monitored. Note that one-electron matrix elements $H_{\nu\mu}^{\text{core}}$ need to be calculated once because they remain unchanged during the iteration procedure.

Note that the number of two-electron integrals, see Eq. (1.73), to be calculated is of order K^4; therefore, even a small number of basis sets, typical for moderate size of molecules, will require up to millions of two-electron integral calculations (Atkins and Friedman 2005), making these calculations a great challenge in the Hartree–Fock self-consistent field calculations. Nevertheless, the computation efforts may be reduced by using the symmetry since some of non-vanishing two-electron integrals are equal due to the symmetry and some integrals may be very small because the separations become very large. Nevertheless, the computer memory demands for storing such two-electron integrals remain a problem (Atkins and Friedman 2005).

The HF-SCF calculations provide the $K \times K$ matrix of coefficients **C**, and hence the set of n spatial orbitals for each spin. The set of N electrons distributed into these orbitals according to Aufbau principle; that is, two electrons per orbital, starting with the lowest energy orbitals (Leach 2001). If $n > 2N$, then the remaining orbitals are empty, which are known as virtual orbitals. Besides, the Hartree–Fock electronic calculations provide the energy of each single-electron orbital E_i^0 in the ground state.

Algorithm 1: **Hartree–Fock-SCF**

Step 1: Choose a set of basis functions ϕ_μ ($\mu = 1, 2, \cdots, K$)
Step 2: Choose a set of trial coefficients $c_{\mu i}$ ($\mu = 1, 2, \cdots, K, i = 1, 2, \cdots, n$)
Step 3: Calculate the spatial orbitals ψ_i ($i = 1, 2, \cdots, n$) using Eq. (1.30)
Step 4: Calculate the overlap integral elements $S_{\nu\mu}$ ($\nu, \mu = 1, 2, \cdots, K$) using Eq. (1.59)
Step 5: Diagonalise the overlap matrix $\mathbf{S}$
Step 6: Calculate $\mathbf{S}^{-1/2}$
Step 7: Calculate $H^{\text{core}}_{\nu\mu}$ ($\nu, \mu = 1, 2, \cdots, K$) using Eq. (1.72)
Step 8: Calculate or guess the density matrix $P_{\nu\mu}$ ($\nu, \mu = 1, 2, \cdots, K$) using Eq. (1.70)
Step 9: Calculate $\langle\nu\mu|\alpha\beta\rangle$ and $\langle\nu\alpha|\mu\beta\rangle$ ($\nu, \mu, \alpha, \beta = 1, 2, \cdots, K$) using Eq. (1.73)
Step 10: Calculate the Hartree-Fock matrix elements $F_{\nu\mu}$ ($\nu, \mu = 1, 2, \cdots, K$) using Eq. (1.71)
Step 11: Calculate $\mathbf{F}' \equiv \mathbf{S}^{-1/2}\mathbf{F}\mathbf{S}^{-1/2}$
Step 12: Solve secular equation given by Eq. (1.69) to calculate the eigenvalues
 $\mathbf{E}$ (that is, the orbital energies) and eigenvectors $\mathbf{C}'$ by diagonalising $\mathbf{F}'$
Step 13: Calculate spatial orbital coefficients $\mathbf{C}$ using the equation $\mathbf{C} = \mathbf{S}^{-1/2}\mathbf{C}'$
Step 14: Calculate the density matrix $P_{\nu\mu}$ ($\nu, \mu = 1, 2, \cdots, K$) using Eq. (1.70)
Step 15: Check the convergence
Step 16: If converged, STOP; otherwise, goto Step 10.

The energy of an electron in a spatial orbital ψ_i is the sum of core interaction H^{core}_{ii}, the Coulomb interaction E^{Coul}_i, and exchange interactions with other electrons in the set of orbitals E^{exchange}_i:

$$E^0_i = H^{\text{core}}_{ii} + \sum_{j=1}^{n} \left(s_i J_{ij} - K_{ij}\right) \tag{1.75}$$

The total electronic energy of the ground state equals $E_0(\mathbf{R})$ at a fixed nuclei configuration $\mathbf{R}$ as in Eq. (1.55). Therefore, $E_0(\mathbf{R})$ is not equal to the sum of individual orbital energies. Indeed, it is

$$E_0(\mathbf{R}) = \sum_{i=1}^{n} E^0_i - \sum_{i=1}^{n}\sum_{j=1}^{n} \left(s_i J_{ij} - K_{ij}\right) \tag{1.76}$$

Again, the following Hamiltonian function is introduced for the Hartree–Fock approximation:

$$\mathcal{H}_{\text{HF}} = \sum_{\mu=1}^{M} \frac{\mathbf{P}^2_\mu}{2M_\mu} + E_0(\mathbf{R}) + U_{\text{NN}}(\mathbf{R}) + U_{\text{penalty}} \tag{1.77}$$

where the penalty potential energy term is used to ensure the orthonormality of spatial orbitals at all times. Here, $\mathbf{P}_\mu$ (for $\mu = 1, 2, \ldots, M$) are the nuclei momenta.

Then, the dynamics of the nuclei are described classically by the following equations:

$$\dot{\mathbf{R}}_\mu = \frac{\mathbf{P}_\mu}{M_\mu} \tag{1.78}$$
$$\dot{\mathbf{P}}_\mu = -\nabla_\mu \left[E_0(\mathbf{R}) + U_{\mathrm{NN}}(\mathbf{R}) \right], \quad \text{for } \mu = 1, 2, \ldots, M$$

where $\mathbf{R}$ is the nuclei positions vector at the $E_0(\mathbf{R})$ ground state of electrons at the nuclei configuration $\mathbf{R}$ calculated using Eq. (1.76). The equation of motion for every one-electron spatial orbital i (for $i = 1, 2, \ldots, n$) is

$$\hat{f}_i(1)\psi_i(\mathbf{r}; \mathbf{R}) = \sum_{j=1}^{n} \lambda_{ji}\psi_j(\mathbf{r}; \mathbf{R}) \tag{1.79}$$

Note that Eq. (1.79) is not computationally a standard form of the eigenvalue problem. Besides, the set of spatial orbitals are not unique. Therefore, it is possible to form a new set of spatial orbitals as a linear combination of the $\psi_i(\mathbf{r}; \mathbf{R})$:

$$\psi_i'(\mathbf{r}; \mathbf{R}) = \sum_{j=1}^{K'} c_{ij}'\psi_j(\mathbf{r}; \mathbf{R}) \tag{1.80}$$

Furthermore, the new set of orthonormal spatial orbitals has the same minimum energy E_0, the same transformed Fock operator $\hat{f}_i'(1) = \hat{f}_i(1)$, and the matrix of Lagrange multipliers is diagonal with diagonal elements $E_i^{0\prime}$:

$$\hat{f}_i(1)\psi_i'(\mathbf{r}; \mathbf{R}) = E_i^{0\prime}\psi_i'(\mathbf{r}; \mathbf{R}) \tag{1.81}$$

Transformations given by Eq. (1.80) are called canonical transformations, and $\psi_i'(\mathbf{r}; \mathbf{R})$ are called canonical spatial orbitals. Equation (1.81) is the equation of an eigenvalue problem that can be solved using the standard methods, and $E_i^{0\prime}$ and $\psi_i'(\mathbf{r}; \mathbf{R})$ give, respectively, the eigenvalues and eigenstates of Fock's Hamiltonian operator.

The total electronic energy of the ground state is the same as in Eq. (1.76) (because $\sum_{i=1}^{n} E_i^0 = \sum_{i=1}^{n} E_i^{0\prime}$):

$$E_0(\mathbf{R}) = \sum_{i=1}^{n} E_i^{0\prime} - \sum_{i=1}^{n}\sum_{j=1}^{n} \left(s_i J_{ij} - K_{ij} \right) \tag{1.82}$$

1.8 Density Functional Theory (DFT) Molecular Dynamics

The ab-initio MD simulations using the HF-SCF approach considers first solving the HF equations to calculate the spatial orbitals for each spin, used to construct electronic configuration state functions, then the electronic ground state energy for constructing the nuclei configuration using classical approach. The Hartree–Fock SCF method is widely used in quantum mechanics calculations. However, the method is limited to molecules with a small number of atoms and electrons, and furthermore, it is less accurate for large basis set of functions.

Alternatively, density functional theory method is used, which, in contrast to HF-SCF method, is based on the electron probability density (Ziegler 1991; Kohn 1999; Nakajima et al. 2002; Harbola and Banerjee 2003). In particular, DFT considers the electron correlation and it is less demanding from computational viewpoint. Furthermore, it can be used for molecules with larger than 100 of atoms in relatively less time-complexity than HF-SCF method. Using DFT approach, the electronic energy of a quantum system is expressed in terms of the electron probability density ρ (Leach 2001; Atkins and Friedman 2005).

The electron probability density at position $\mathbf{r}$ is

$$\rho(\mathbf{r}) = \sum_{i=1}^{n} s_i \mid \psi_i(\mathbf{r}; \mathbf{R}) \mid^2 \tag{1.83}$$

where $\mid \cdots \mid^2$ indicates the square moduli of the set of one-electron orthonormal spatial orbitals and s_i is the occupation number of the ith orbital, as defined above.

The ground state electronic energy is a functional form of the electron density, $E_0[\rho]$; that is, for a given distribution $\rho(\mathbf{r})$ there is a unique value of energy (Leach 2001; Atkins and Friedman 2005). In DFT, the energy functional is given as the following sum:

$$E_0[\rho(\mathbf{r})] = e \int V_{\text{ext}}(\mathbf{r}; \mathbf{R})\rho(\mathbf{r})\, d\mathbf{r} + E_{\text{KE}}[\rho(\mathbf{r})] + E_{\text{H}}[\rho(\mathbf{r})] + E_{\text{XC}}[\rho(\mathbf{r})] \tag{1.84}$$

In Eq. (1.84), the first term gives the interactions of the electrons with an external potential $V_{\text{ext}}(\mathbf{r})$, such as the electrostatic potential of the nuclei with a fixed configuration $\mathbf{R}$ at position $\mathbf{r}$:

$$V_{\text{ext}}(\mathbf{r}; \mathbf{R}) = -k_e \sum_{\mu=1}^{M} \frac{eZ_\mu}{\mid \mathbf{r} - \mathbf{R}_\mu \mid} \tag{1.85}$$

The second term denotes the kinetic energy of system of non-interacting electrons with the same density $\rho(\mathbf{r})$ as the real system:

$$E_{\text{KE}}[\rho(\mathbf{r})] = \sum_{i=1}^{n} s_i \int \psi_i^*(\mathbf{r}; \mathbf{R}) \left(-\frac{\hbar^2}{2m_e}\nabla_{\mathbf{r}}^2\right) \psi_i(\mathbf{r}; \mathbf{R})\, d\mathbf{r} \tag{1.86}$$

The third term is the electron-electron Coulombic interactions energy, known as the Hartree electrostatic energy, which arises from the classical interactions between the two charge densities summed over all possible pairwise interactions (Leach 2001; Atkins and Friedman 2005):

$$E_{\mathrm{H}}[\rho(\mathbf{r})] = \frac{1}{2}k_e e^2 \int\int \frac{\rho(\mathbf{r}_1)\rho(\mathbf{r}_2)}{\mid \mathbf{r}_1 - \mathbf{r}_2 \mid}\, d\mathbf{r}_1\, d\mathbf{r}_2 \tag{1.87}$$

The fourth term $E_{\mathrm{XC}}[\rho(\mathbf{r})]$ denotes the contributions from exchange and correlations, including the contribution due to the difference between the true kinetic energy of electrons and $E_{\mathrm{KE}}[\rho(\mathbf{r})]$ (Leach 2001).

Next, we introduce the one-electron Kohn–Sham Hamiltonian operator as (Leach 2001):

$$\begin{aligned} \hat{h}_i^{\mathrm{KS}}(1) &= -\frac{\hbar^2}{2m_e}\nabla_{\mathbf{r}}^2 - k_e e^2 \sum_{\mu=1}^{M} \frac{Z_\mu}{\mid \mathbf{R}_\mu - \mathbf{r} \mid} + k_e e^2 \int \frac{\rho(\mathbf{r}')}{\mid \mathbf{r}' - \mathbf{r} \mid}\, d\mathbf{r}' + \hat{V}_i^{\mathrm{XC}}[\mathbf{r}] \\ &\equiv \hat{h}^{\mathrm{core}}(\mathbf{r}; \mathbf{R}) + \hat{V}_i^{\mathrm{Coul}}(\mathbf{r}; \mathbf{R}) + \hat{V}_i^{\mathrm{XC}}[\mathbf{r}] \end{aligned} \tag{1.88}$$

where $\hat{V}_i^{\mathrm{XC}}[\mathbf{r}]$ denotes the one-electron exchange-correlation operator functional that relates to the exchange-correlation energy as (Leach 2001):

$$\hat{V}_i^{\mathrm{XC}}[\mathbf{r}] = \frac{\delta E_{\mathrm{XC}}[\rho(\mathbf{r})]}{\delta \rho(\mathbf{r})} \tag{1.89}$$

In Eq. (1.88), $\hat{h}^{\mathrm{core}}(\mathbf{r}; \mathbf{R})$ is the core Hamiltonian operator (as in the HF approach) and $\hat{V}_i^{\mathrm{Coul}}(\mathbf{r}; \mathbf{R})$ is the Coulombic one-electron operator given as

$$\hat{V}_i^{\mathrm{Coul}}(\mathbf{r}; \mathbf{R}) = k_e e^2 \int \frac{\rho(\mathbf{r}')}{\mid \mathbf{r}' - \mathbf{r} \mid}\, d\mathbf{r}' \tag{1.90}$$

Then, the one-electron Kohn–Sham equations are written as (Hohenberg and Kohn 1964; Kohn and Sham 1965)

$$\hat{h}_i^{\mathrm{KS}}(1)\psi_i(\mathbf{r}; \mathbf{R}) = E_i^0(\mathbf{r}; \mathbf{R})\psi_i(\mathbf{r}; \mathbf{R}) \tag{1.91}$$

where $E_i^0(\mathbf{r}; \mathbf{R})$ is the one-electron orbital i ground state energy at configuration $\mathbf{R}$ of nuclei. The total electronic ground state energy is given by Eq. (1.84):

$$\begin{aligned} E_0(\mathbf{R}) &= \sum_{i=1}^{n} s_i \int \psi_i^*(\mathbf{r}; \mathbf{R}) \left[\hat{h}^{\mathrm{core}}(\mathbf{r}; \mathbf{R}) + \hat{V}_i^{\mathrm{Coul}}(\mathbf{r}; \mathbf{R}) + \hat{V}_i^{\mathrm{XC}}[\mathbf{r}]\right] \psi_i(\mathbf{r}; \mathbf{R})\, d\mathbf{r} \\ &= \int E_i^0(\mathbf{r}; \mathbf{R})\rho(\mathbf{r})\, d\mathbf{r} \end{aligned} \tag{1.92}$$

A self-consistent field approach is suggested (Kohn and Sham 1965), as discussed in the following, to solve Kohn–Sham equations. In DFT approach, the exchange-correlation functional are treated using the so-called local density approximation (LDA), based on the model of a uniform electronic gas with constant electron density. In that approach, the total exchange-correlation energy of the system is

$$E_{XC}[\rho(\mathbf{r})] = \int \rho(\mathbf{r})\varepsilon_{XC}(\rho(\mathbf{r}))\,d\mathbf{r} \tag{1.93}$$

where $\varepsilon_{XC}(\rho(\mathbf{r}))$ is the exchange-correlation energy of one electron, which is a function of density of electrons in the uniform gas. Then, the exchange-correlation functional can be obtained as (Leach 2001)

$$\hat{V}_i^{XC}[\mathbf{r}] = \rho(\mathbf{r})\frac{d\varepsilon_{XC}(\rho(\mathbf{r}))}{d\rho(\mathbf{r})} + \varepsilon_{XC}(\rho(\mathbf{r})) \tag{1.94}$$

Some analytical expressions of the exchange-correlation energy per electron are given (Gunnarsson and Lundqvist 1976)

$$\begin{aligned} \varepsilon_{XC}(\rho(\mathbf{r})) &= -\frac{0.458}{r_0} - 0.0333\left[\left(1+\frac{r_0}{11.4}\right)\log\left(1+\frac{11.4}{r_0}\right)\right. \\ &\quad \left. -\left(\frac{r_0}{11.4}\right)^2 + \frac{r_0}{22.8} - \frac{1}{3}\right] \\ r_0 &= \left(\frac{3}{4\pi\rho(\mathbf{r})}\right)^{1/3} \end{aligned} \tag{1.95}$$

For the exchange energy term only, based on the local density approximation, the following expression is obtained (Gritsenko et al. 1994) (and the references therein):

$$E_X[\rho_\uparrow(\mathbf{r}), \rho_\downarrow(\mathbf{r}))] = -\frac{3}{2}\left(\frac{3}{4\pi}\right)^{1/3}\int \left[(\rho_\uparrow(\mathbf{r}))^{4/3} + (\rho_\downarrow(\mathbf{r}))^{4/3}\right] d\mathbf{r} \tag{1.96}$$

where $\uparrow\downarrow$ represent up and down spins. The following expression is also suggested for correlation energy per electron if the number of spins up and down are the same (Leach 2001)

$$\varepsilon_C(\rho(\mathbf{r})) = \begin{cases} -\dfrac{0.1423}{1 + 1.9529 r_0^{1/2} + 0.3334 r_0}, & r_0 \geq 1 \\ -0.0480 + 0.0311\ln r_0 - 0.0116 r_0 + 0.0020 r_0 \ln r_0 & r_0 < 1 \end{cases} \tag{1.97}$$

Equation (1.94) can then be used to obtain an expression for $\hat{V}_i^{XC}[\rho(\mathbf{r}]$ to be used for solving the Kohn–Sham equations.

The Kohn–Sham equations can be solved using the same strategy as using HF-SCF approach. For that, Eq. (1.30) can be used to expand the Kohn–Sham orbitals as a linear combinations of the atomic-centred basis functions. Substituting Eq. (1.30) into Eq. (1.91):

$$\hat{h}_i^{\mathrm{KS}}(1) \sum_{\mu=1}^{K} c_{\mu i} \phi_\mu(\mathbf{r}; \mathbf{R}) = E_i^0 \sum_{\mu=1}^{K} c_{\mu i} \phi_\mu(\mathbf{r}; \mathbf{R}) \tag{1.98}$$

We can multiply both sides by $\phi_\nu^*(\mathbf{r}; \mathbf{R})$ and integrate over $\mathbf{r}$:

$$\sum_{\mu=1}^{K} c_{\mu i} \int \phi_\nu^*(\mathbf{r}; \mathbf{R}) \hat{h}_i^{\mathrm{KS}}(\mathbf{r}; \mathbf{R}) \phi_\mu(\mathbf{r}; \mathbf{R})\, d\mathbf{r} = E_i^0 \sum_{\mu=1}^{K} c_{\mu i} \int \phi_\nu^*(\mathbf{r}; \mathbf{R}) \phi_\mu(\mathbf{r}; \mathbf{R})\, d\mathbf{r} \tag{1.99}$$

Alternatively,

$$\sum_{\mu=1}^{K} H_{\nu\mu} c_{\mu i} = \left(\sum_{\mu=1}^{K} S_{\nu\mu} c_{\mu i} \right) E_i^0 \tag{1.100}$$

where $S_{\nu\mu}$ is the above defined overlap integral matrix, and $H_{\nu\mu}$ are the elements of the Kohn–Sham matrix, defined as

$$\begin{aligned} H_{\nu\mu} &\equiv \int \phi_\nu^*(\mathbf{r}; \mathbf{R}) \hat{h}_i^{\mathrm{KS}}(\mathbf{r}; \mathbf{R}) \phi_\mu(\mathbf{r}; \mathbf{R})\, d\mathbf{r} \\ &= \int \phi_\nu^*(\mathbf{r}; \mathbf{R}) \left[-\frac{\hbar^2}{2m_e} \nabla_{\mathbf{r}}^2 - k_e e^2 \sum_{\mu=1}^{M} \frac{Z_\mu}{\mid \mathbf{R}_\mu - \mathbf{r} \mid} \right] \phi_\mu(\mathbf{r}; \mathbf{R})\, d\mathbf{r} \\ &\quad + \int \phi_\nu^*(\mathbf{r}; \mathbf{R}) \left[k_e e^2 \int \frac{\rho(\mathbf{r}')}{\mid \mathbf{r}' - \mathbf{r} \mid} d\mathbf{r}' + \hat{V}_i^{\mathrm{XC}}[\mathbf{r}] \right] \phi_\mu(\mathbf{r}; \mathbf{R})\, d\mathbf{r} \\ &\equiv H_{\nu\mu}^{\mathrm{core}} + \int \phi_\nu^*(\mathbf{r}; \mathbf{R}) \left[k_e e^2 \int \frac{\rho(\mathbf{r}')}{\mid \mathbf{r}' - \mathbf{r} \mid} d\mathbf{r}' + \hat{V}_i^{\mathrm{XC}}[\mathbf{r}] \right] \phi_\mu(\mathbf{r}; \mathbf{R})\, d\mathbf{r} \end{aligned} \tag{1.101}$$

Equation (1.100) can also be written in the matrix form as the Roothaan–Hall equations:

$$\mathbf{HC} = \mathbf{SCE} \tag{1.102}$$

In Eq. (1.101), the second term, namely the Coulombic repulsion term, can be simplified as

$$H^{\text{Coul}}_{\nu\mu} = k_e e^2 \int\int \phi^*_\nu(\mathbf{r};\mathbf{R})\frac{\rho(\mathbf{r}')}{\mid \mathbf{r}'-\mathbf{r}\mid}\phi_\mu(\mathbf{r};\mathbf{R})\,d\mathbf{r}\,d\mathbf{r}' \\ = e\int \phi^*_\nu(\mathbf{r};\mathbf{R})\hat{V}^{\text{elec}}_i(\mathbf{r})\phi_\mu(\mathbf{r};\mathbf{R})\,d\mathbf{r} \tag{1.103}$$

where $\hat{V}^{\text{elec}}_i(\mathbf{r})$ is an electric potential of the charge density $\rho(\mathbf{r})$:

$$\hat{V}^{\text{elec}}_i(\mathbf{r}) = k_e e\int \frac{\rho(\mathbf{r}')}{\mid \mathbf{r}'-\mathbf{r}\mid}\,d\mathbf{r}' \tag{1.104}$$

The last term in Eq. (1.101), is the exchange-correlation term, which is written as

$$H^{\text{XC}}_{\nu\mu} = \int \phi^*_\nu(\mathbf{r};\mathbf{R})\hat{V}^{\text{XC}}_i[\mathbf{r}]\phi_\mu(\mathbf{r};\mathbf{R})\,d\mathbf{r} \tag{1.105}$$

Similar to HF-SCF approach, we multiply both sides of Eq. (1.102) by $\mathbf{S}^{-1/2}$:

$$\mathbf{S}^{-1/2}\mathbf{HC} = \mathbf{S}^{-1/2}\mathbf{SCE} = \mathbf{S}^{1/2}\mathbf{CE} \tag{1.106}$$

Since $\mathbf{S}^{-1/2}\mathbf{S}^{1/2} = \mathbf{I}$, we can write:

$$\mathbf{S}^{-1/2}\mathbf{H}\left(\mathbf{S}^{-1/2}\mathbf{S}^{1/2}\right)\mathbf{C} = \mathbf{S}^{1/2}\mathbf{CE} \tag{1.107}$$

or,

$$\mathbf{S}^{-1/2}\mathbf{H}\mathbf{S}^{-1/2}\left(\mathbf{S}^{1/2}\mathbf{C}\right) = \left(\mathbf{S}^{1/2}\mathbf{C}\right)\mathbf{E} \tag{1.108}$$

Again denoting $\mathbf{H}' \equiv \mathbf{S}^{-1/2}\mathbf{H}\mathbf{S}^{-1/2}$ and $\mathbf{C}' \equiv \mathbf{S}^{1/2}\mathbf{C}$, we obtain

$$\mathbf{H}'\mathbf{C}' = \mathbf{C}'\mathbf{E} \tag{1.109}$$

From Eq. (1.109) we obtain the following secular equation:

$$\det \mid \mathbf{H}' - E^0_i\mathbf{I} \mid = 0 \tag{1.110}$$

Algorithm 2 presents a self-consistent field method for Kohn–Sham equations.

The total electronic energy of the ground state equals $E_0(\mathbf{R})$ at a fixed nuclei configuration $\mathbf{R}$ calculated from Eq. (1.92).

Then, the following classical Hamiltonian function is introduced:

$$\mathcal{H}_{\text{KS}} = \sum_{\mu=1}^{M}\frac{\mathbf{P}^2_\mu}{2M_\mu} + E_0(\mathbf{R}) + U_{\text{NN}}(\mathbf{R}) + U_{\text{penalty}} \tag{1.111}$$

The dynamics of the nuclei are described classically by the following equations:

Algorithm 2: **Kohn–Sham-SCF**
Step 1: Choose a set of basis functions ϕ_μ $(\mu = 1, 2, \cdots, K)$
Step 2: Choose a set of trial coefficients $c_{\mu i}$ $(\mu = 1, 2, \cdots, K, i = 1, 2, \cdots, n)$
Step 3: Calculate the spatial orbitals ψ_i $(i = 1, 2, \cdots, n)$ using Eq. (1.30)
Step 4: Calculate the overlap integral elements $S_{\nu\mu}$ $(\nu, \mu = 1, 2, \cdots, K)$ using Eq. (1.59)
Step 5: Diagonalise the overlap matrix $\mathbf{S}$
Step 6: Calculate $\mathbf{S}^{-1/2}$
Step 7: Calculate $H^{\text{core}}_{\nu\mu}$ $(\nu, \mu = 1, 2, \cdots, K)$ using Eq. (1.101)
Step 8: Calculate electron density $\rho(\mathbf{r})$ using Eq. (1.83)
Step 9: Calculate $H^{\text{Coul}}_{\nu\mu}$ $(\nu, \mu = 1, 2, \cdots, K)$ using Eq. (1.103)
Step 10: Calculate $H^{\text{XC}}_{\nu\mu}$ $(\nu, \mu = 1, 2, \cdots, K)$ using Eq. (1.105)
Step 11: Calculate the Kohn-Sham matrix elements $H_{\nu\mu}$ $(\nu, \mu = 1, 2, \cdots, K)$ using Eq. (1.101)
Step 12: Calculate $\mathbf{H}' \equiv \mathbf{S}^{-1/2}\mathbf{H}\mathbf{S}^{-1/2}$
Step 13: Solve secular equation given by Eq. (1.110) to calculate the eigenvalues $\mathbf{E}$ (that is, the orbital energies) and eigenvectors $\mathbf{C}'$ by diagonalising $\mathbf{H}'$
Step 14: Calculate spatial orbital coefficients $\mathbf{C}$ using the equation $\mathbf{C} = \mathbf{S}^{-1/2}\mathbf{C}'$
Step 15: Check the convergence
Step 16: If converged, STOP; otherwise, goto Step 8.

$$\begin{aligned}
\dot{\mathbf{R}}_\mu &= \frac{\mathbf{P}_\mu}{M_\mu} \\
\dot{\mathbf{P}}_\mu &= -\nabla_\mu \left[E_0(\mathbf{R}) + U_{\text{NN}}(\mathbf{R})\right], \quad \text{for } \mu = 1, 2, \ldots, M
\end{aligned} \tag{1.112}$$

where $\mathbf{R}$ is the nuclei positions vector at the $E_0(\mathbf{R})$ ground state of electrons at the nuclei configuration $\mathbf{R}$ calculated using Eq. (1.92). The equation of motion for an one-electron spatial orbital i is given as follows:

$$\hat{h}_i^{\text{KS}}(1)\psi_i(\mathbf{r}; \mathbf{R}) = \sum_{j=1}^{n} \lambda_{ji}\psi_j(\mathbf{r}; \mathbf{R}), \quad i = 1, 2, \ldots, n \tag{1.113}$$

Standard QM simulations are computationally costly solving electronic wavefunctions in terms of Hartree–Fock or Kohn–Sham operators; typically, the computational time complexity is of order of $O(K^4)$ with K being the number of orbitals (Goedecker 1999; Gogonea et al. 2001). Sparsity of the Fock matrix allows development of algorithms using linear algebra reducing the time complexity to $O(K \log K)$ or even $O(K)$. The divide-and-conquer algorithms are used to localise electronic calculations. Besides, different treatment of strong intramolecular and weak intermolecular interactions are also suggested.

1.9 Car–Parrinello Molecular Dynamics

Car–Parrinello molecular dynamics approach involves the dynamics of the electrons into equations of motion. For that, the electronic degrees of freedom are included into the classical dynamics by the fictitious electronic kinetic energy term. Besides, the Kohn–Sham spatial orbitals are considered time-dependent; that is, $\psi_i(\mathbf{r}, t; \mathbf{R})$. The dynamics of the orbitals propagates an initially minimised set of spatial orbitals for each spin via the electronic calculations to the subsequent minima corresponding to the new nuclear configuration $\mathbf{R}$ (Car and Parrinello 1985; Iftime et al. 2005). During the time evolution of the spatial orbitals, they are coupled to a thermostat to be maintained at a fixed fictitious temperature T_ψ, which is assumed to be much smaller then the temperature T of the nuclei $T_\psi \ll T$ (Iftime et al. 2005); however, they are still allowed to relax quickly in response to the nuclear motion.

The extended classical Hamiltonian function is introduced:

$$\mathcal{H}_{\text{CP}} = \sum_{\mu=1}^{M} \frac{\mathbf{P}_\mu^2}{2M_\mu} + \sum_{i=1}^{n} \frac{P_{\psi_i}^2}{2m_\psi} + E_0(\{\psi_i(t)\}, \mathbf{R}) + U_{\text{NN}}(\mathbf{R}) + U_{\text{penalty}} \tag{1.114}$$

where P_{ψ_i} is the conjugated momentum associated with spatial orbital ψ_i and $\mathbf{P}_\mu$ (for $\mu = 1, 2, \ldots, M$) are the usual conjugated momenta associated with nuclei coordinates $\mathbf{R}$. Again, the penalty potential energy term ensures the orthonormality of spatial orbitals at all times. Here, m_ψ is a mass-like parameter with units of energy$\times$time^2. The equations of motion are:

$$\begin{aligned}
\dot{R}_\mu &= \frac{P_\mu}{M_\mu} \\
\dot{P}_\mu &= -\nabla_\mu \left[E_0(\{\psi_i(t)\}, \mathbf{R}) + U_{\text{NN}}(\mathbf{R})\right], \quad \mu = 1, 2, \ldots, 3M \\
\dot{\psi}_i &= \frac{P_{\psi_i}}{m_\psi} \\
\dot{P}_{\psi_i} &= -\hat{h}_i^{\text{KS}}(1)\psi_i(\mathbf{r}, t; \mathbf{R}) + \sum_{j=1}^{n} \lambda_{ji}\psi_j(\mathbf{r}, t; \mathbf{R}), \quad i = 1, 2, \ldots, n
\end{aligned} \tag{1.115}$$

In Eq. (1.115), the fictitious temperature T_ψ is adjusted initially with the choice of the fictitious momenta P_{ψ_i} and the adiabatic decoupling from nuclear dynamics is controlled by adjusting the value of m_ψ. Typically, small values of m_ψ are suggested (Iftime et al. 2005). Furthermore, the thermostats can be used to increase the value of m_ψ, as suggested in Iftime et al. (2005) (and the references therein).

In the case of the massive chain thermostatting approach of length L_e, the following electronic motions are suggested for every spatial orbital $i = 1, 2, \ldots, n$:

$$
\begin{aligned}
\dot{\psi}_i &= \frac{P_{\psi_i}}{m_\psi} \\
\dot{P}_{\psi_i} &= -\hat{h}_i^{\mathrm{KS}}(1)\psi_i(\mathbf{r}, t; \mathbf{R}) + \sum_{j=1}^{n} \lambda_{ji}\psi_j(\mathbf{r}, t; \mathbf{R}) - \frac{\pi_{1,i}^e}{Q_{1,i}^e} P_{\psi_i} \\
\dot{\pi}_{1,i}^e &= \frac{P_{\psi_i}^2}{m_\psi} - k_B T_\psi - \frac{\pi_{2,i}^e}{Q_{2,i}^e}\pi_{1,i}^e \\
\dot{\pi}_{k,i}^e &= \frac{(\pi_{k-1,i}^e)^2}{Q_{k-1,i}^e} - k_B T_\psi - \frac{\pi_{k+1,i}^e}{Q_{k+1,i}^e}\pi_{k,i}^e, \quad k = 2, \ldots, L_e - 1 \\
\dot{\pi}_{L_e,i}^e &= \frac{(\pi_{L_e-1,i}^e)^2}{Q_{L_e-1,i}^e} - k_B T_\psi \\
\dot{\eta}_{k,i}^e &= \frac{\pi_{k,i}^e}{Q_{k,i}^e}, \quad k = 1, \ldots, L_e
\end{aligned}
\tag{1.116}
$$

and for the nuclear degrees of freedom, the equations of motion coupled to another massive chain of thermostats of length L_n each are given as (for $\mu = 1, 2, \ldots, 3M$):

$$
\begin{aligned}
\dot{R}_\mu &= \frac{P_\mu}{M_\mu} \\
\dot{P}_\mu &= -\nabla_\mu \left[E_0(\{\psi_i(t)\}, \mathbf{R}) + U_{\mathrm{NN}}(\mathbf{R})\right] - \frac{\pi_{1,\mu}^n}{Q_{1,\mu}^n} P_\mu \\
\dot{\pi}_{1,\mu}^n &= \frac{P_\mu^2}{M_\mu} - k_B T - \frac{\pi_{2,\mu}^n}{Q_{2,\mu}^n}\pi_{1,\mu}^n \\
\dot{\pi}_{k,\mu}^n &= \frac{(\pi_{k-1,\mu}^n)^2}{Q_{k-1,\mu}^n} - k_B T - \frac{\pi_{k+1,\mu}^n}{Q_{k+1,\mu}^n}\pi_{k,\mu}^n, \quad k = 2, \ldots, L_n - 1 \\
\dot{\pi}_{L_n,\mu}^n &= \frac{(\pi_{L_n-1,\mu}^n)^2}{Q_{L_n-1,\mu}^n} - k_B T \\
\dot{\eta}_{k,\mu}^n &= \frac{\pi_{k,\mu}^n}{Q_{k,\mu}^n}, \quad k = 1, \ldots, L_n
\end{aligned}
\tag{1.117}
$$

Here, $(\eta_{k,i}^e, \pi_{k,i}^e)$, $k = 1, 2, \ldots, L_e$, denote the thermostat variables and their conjugated momenta coupled to electronic degrees of freedom and $(\eta_{k,\mu}^n, \pi_{k,\mu}^n)$, $k = 1, 2, \ldots, L_n$, denote the thermostat variables and their conjugated momenta coupled to nuclear degrees of freedom. $Q_{k,i}^e$ (or $Q_{k,\mu}^n$) represent the thermostat masses, optimised as in Kamberaj (2020). In Eq. (1.117), T is the temperature of thermostat bath coupled to nuclear degrees of freedom and k_B is Boltzmann's constant.

1.10 Quantum Electronic Friction Molecular Dynamics

In Born–Oppenheimer approximation, it is assumed that electrons respond immediately to the changes on the nuclear configurations $\mathbf{R}$. However, the slow motion of heavy nucleus disrupts the equilibrium structure of the fast moving electrons, which, in turn, exert of force upon the nucleus (Dou et al. 2017; Dou and Subotnik 2018). Therefore, the Born–Oppenheimer approach can strictly apply when the nuclear dynamical motion is infinitesimally slow compare to the motion of electrons. There exist cases when the Born–Oppenheimer approximation fails, such as the studies of electrons at the surface of metals (Huang et al. 2000; Bünermann et al. 2015) and the water flow fluid though nanoscale carbon pipes (Whitby et al. 2008; Bocquet 2020).

The force that the electrons apply on the nuclei as they move close or within the manifold of electrons is modelled as a sum of a friction force and a random force (Dou et al. 2017):

$$M_\mu \ddot{\mathbf{R}}_\mu = -\nabla_\mu \left[E_0(\mathbf{R}) + U_{\text{NN}}(\mathbf{R})\right] - \sum_{\nu=1}^{M} \gamma_{\mu\nu} \dot{\mathbf{R}}_\nu + \mathbf{B}_\mu(t), \quad \text{for } \mu = 1, 2, \ldots, M \tag{1.118}$$

where $\mathbf{B}_\mu(t)$ is the random force such that at equilibrium it obeys to the second fluctuation-dissipation theorem:

$$\begin{aligned} &\frac{1}{2}\left(\langle \mathbf{B}_\mu(t)\mathbf{B}_\nu(t')\rangle + \langle \mathbf{B}_\nu(t)\mathbf{B}_\mu(t')\rangle\right) = k_B T \gamma_{\mu\nu} \delta(t - t') \\ &\langle \mathbf{B}_\mu(t)\rangle = 0 \end{aligned} \tag{1.119}$$

$\gamma_{\mu\nu}$ is the electronic friction matrix defined as follows in the Markovian limit (Dou et al. 2017; Dou and Subotnik 2018):

$$\gamma_{\mu\nu} = -\int_0^\infty \left[\int \nabla_\mu \hat{h}_i^{\text{KS}}(1) e^{-\hat{\mathcal{L}}t} \nabla_\nu \rho(\mathbf{r}; \mathbf{R})\, d\mathbf{r}\right] dt \tag{1.120}$$

where $-\hat{\mathcal{L}}(\cdot) = -\frac{i}{\hbar}[\hat{h}_i^{\text{KS}}, \cdot]$ is the Liouvillian operator. Here, $\gamma_{\mu\nu}$ is real and symmetric, $\gamma_{\mu\nu} = \gamma_{\nu\mu}$. The idea behind is that the time scale of the electronic density motion can be divided into the time scale for electronic relaxation and time scale for nuclear motion; if we consider that the nuclei are fixed during the fast electronic motion, as in Born–Oppenheimer approximation, then $\dot{\mathbf{R}}_\nu = 0$ and electrons have a steady state density matrix (that is, $-\hat{\mathcal{L}}\rho = 0$, and thus $\partial\rho/\partial t = 0$). The time scale of electronic relaxation τ is given by the condition $e^{-\hat{\mathcal{L}}\tau}\rho' \approx \rho$, for any total electronic density ρ':

$$\rho'(\mathbf{r}; \mathbf{R}) = \rho(\mathbf{r}; \mathbf{R}) + \Delta\rho(\mathbf{r}; \mathbf{R}) \tag{1.121}$$

where $\Delta\rho(\mathbf{r};\mathbf{R})$ is a correction term characterising the electronic excitation due to nuclear motion (Race et al. 2012).

If the nuclear motion is present, then (Dou et al. 2017)

$$\frac{\partial\Delta\rho}{\partial t} = -\hat{\mathcal{L}}\Delta\rho - \sum_{\nu=1}^{M}\dot{\mathbf{R}}_\nu\nabla_\nu\rho' \tag{1.122}$$

Applying the adiabatic approximation (that is, the nuclear motion is much slower than the electronic motion), $\Delta\rho$ is small, and hence $\nabla_\nu\rho' \approx \nabla_\nu\rho$. Thus,

$$\frac{\partial\Delta\rho}{\partial t} \approx -\hat{\mathcal{L}}\Delta\rho - \sum_{\nu=1}^{M}\dot{\mathbf{R}}_\nu\nabla_\nu\rho \tag{1.123}$$

Solving Eq. (1.123) for $\Delta\rho$, using the initial condition $\Delta\rho = 0$ at $t = 0$, we obtain

$$\Delta\rho = -\sum_{\nu=1}^{M}\int_0^t \dot{\mathbf{R}}_\nu e^{-\hat{\mathcal{L}}(t-t')}\nabla_\nu\rho\, dt' \tag{1.124}$$

Therefore, the friction force on the nucleus is given as (Dou et al. 2017)

$$\mathbf{F}_\mu^{\text{friction}} = -\sum_{\nu=1}^{M}\int_0^t \Gamma_{\mu\nu}(t,t')\dot{\mathbf{R}}_\nu(t')\, dt' \tag{1.125}$$

where $\Gamma_{\mu\nu}(t,t')$ is the kernel function:

$$\Gamma_{\mu\nu}(t,t') = -\int \nabla_\mu\hat{h}_i^{\text{KS}}(1)e^{-\hat{\mathcal{L}}(t-t')}\nabla_\nu\rho\, d\mathbf{r} \tag{1.126}$$

In the case of the Markovian limit for which the memory kernel is faster than the nuclei dynamics, we have:

$$\int_0^t \Gamma_{\mu\nu}(t,t')\dot{\mathbf{R}}_\nu(t')\, dt' \approx \gamma_{\mu\nu}\dot{\mathbf{R}}_\nu(t) \tag{1.127}$$

Thus, the friction force exerted by the electrons on the nucleus is:

$$\mathbf{F}_\mu^{\text{friction}} = -\sum_{\nu=1}^{M}\gamma_{\mu\nu}\dot{\mathbf{R}}_\nu(t) \tag{1.128}$$

Note that the Markovian limit is consistent with the adiabatic approximation.

1.11 Summary

We introduced the quantum mechanics molecular models and quantum mechanics molecular dynamics simulations of molecules and biomolecular systems. Different numerical aspects of the non-relativistic electronic structure of molecular systems were also discussed. In particular, we described the HF-SCF and DFT approaches for electronic ground state calculations.

Furthermore, the Born–Oppenheimer quantum mechanics MD simulations were introduced, including Car–Parrinello and quantum electronic friction MD simulation approaches.

References

P. Atkins, R. Friedman, *Molecular Quantum Mechanics*, 4th edn. (Oxford University Press, 2005)

L. Bocquet, Nanofluidics coming of age. Nat. Mater. **19**, 254–256 (2020)

M. Born, R. Oppenheimer, Zur Quantentheorie der Molekeln. Ann. Phys. **389**, 457–484 (1927)

O. Bünermann, H. Jiang, Y. Dorenkamp, A. Kandratsenka, S.M. Janke, D.J. Auerbach, A.M. Wodtke, Electron-hole pair excitation determines the mechanism of hydrogen atom adsorption. Science **350**, 1346–1349 (2015)

R. Car, M. Parrinello, Unified approach for molecular dynamics and density-functional theory. Phys. Rev. Lett. **55**, 2471–2474 (1985)

F. Ding, J.J. Goings, H. Liu, D.B. Lingerfelt, X. Li, Ab initio two-component Ehrenfest dynamics. J. Chem. Phys. **143**, 114105 (2015)

W. Dou, J.E. Subotnik, Perspective: how to understand electronic friction. J. Chem. Phys. **148**, 230901 (2018)

W. Dou, G. Miao, J.E. Subotnik, Born-Oppenheimer dynamics, electronic friction, and the inclusion of electron-electron interactions. Phys. Rev. Lett. **119**, 046001–6 (2017)

S. Goedecker, Linear scaling electronic structure methods. Rev. Mod. Phys. **71**, 1085–1123 (1999)

V. Gogonea, D. Suárex, A. van der Vaart, K.M. Merz, New developments in applying quantum mechanics to proteins. Curr. Opin. Struct. Biol. **11**, 217–223 (2001)

O.V. Gritsenko, A. Rubio, L.C. Balbas, J.A. Alonso, Density approximation to the average Hartree-Fock exchange potential for atoms. Proc. Indian Acad. Sci. **106**, 91–102 (1994)

O. Gunnarsson, B.I. Lundqvist, Exchange and correlation in atoms, molecules, and solids by spin-density-functional formalism. Phys. Rev. B 13:4274 (1976). With erratum in Ref. Gunnarsson and Lundqvist (1977)

B. Hall, *Quantum Theory for Mathematicians*. Graduate Texts in Mathematics (Springer, 2013)

M.K. Harbola, A. Banerjee, Many-electron problem in terms of the density: from Thomas-Fermi to modern density-functional theory. J. Theor. Comput. Chem. **2**, 301–322 (2003)

W.J. Hehre, L. Radom, P.v.R. Schleyer, J.A. Pople, *Ab Initio Molecular Orbital Theory* (Wiley, New York, 1986)

P. Hohenberg, W. Kohn, Inhomogeneous electron gas. Phys. Rev. **136**, B864 (1964)

Y. Huang, C.T. Rettner, D.J. Auerbach, A.M. Wodtke, Vibrational promotion of electron transfer. Science **290**, 111–114 (2000)

R. Iftime, P. Minary, M.E. Tuckerman, Ab initio molecular dynamics: concepts, recent developments, and future trends. Proc. Natl. Acad. Sci. USA **102**, 6654–6659 (2005)

F. Jensen, *Introduction to Computational Chemistry* (Wiley, Chichester, UK, 2017)

H. Kamberaj, *Molecular Dynamics Simulations in Statistical Physics: Theory and Applications*. Computational Series (Springer Nature, Switzerland, 2020)

K. Kitaura, K. Morokuma, A new energy decomposition scheme for molecular interactions within the Hartree-Fock approximation. Int. J. Quantum Chem. **10**, 325–340 (1976)
W. Kohn, Nobel lecture: Electronic structure of matter-wave functions and density functionals. Rev. Mod. Phys. **71**, 1253–1266 (1999)
W. Kohn, L.J. Sham, Self-consistent equations including exchange and correlation effects. Phys. Rev. **1133**, A140 (1965)
A.R. Leach, Molecular Modelling, in *Principles and Applications* 2nd edn. (Prentice Hall, Pearson Education Limited, Edingburgh Gate, 2001)
S. Lehtola, A review on non-relativistic, fully numerical electronic structure calculations on atoms and diatomic molecules. Int. J. Quantum Chem. **119**, e25968-31 (2019)
T. Nakajima, T. Tsuneda, H. Nakano, K. Hirao, Recent advances in electronic structure theory. J. Theor. Comput. Chem. **1**, 109–136 (2002)
E. Paquet, H.L. Viktor, Computational methods for ab initio molecular dynamics. Adv. Chem. **2018**, 1–14 (2018)
C. Raathaan, New developments in molecular orbital theory. Rev. Mod. Phys. **23**, 69–89 (1951)
C.P. Race, D.R. Mason, A.P. Sutton, A new directional model for the electronic frictional forces in molecular dynamics simulations of radiation damage in metals. J. Nucl. Mat. **425**, 33–40 (2012)
A. Reyes, F. Moncada, J. Charry, The any particle molecular orbital approach: a short review of the theory and applications. Int. J. Quantum Chem. **119**, e25705-13 (2019)
K. Schwarz, P. Blaha, DFT calculations for real solids, in *Handbook of Solid State Chemistry. Theoretical Description* ed. by R. Dronkowski, S. Kikkawa, A. Stein, vol. 5 (Wiley, 2017)
M. Whitby, L. Cagnon, M. Thanon, N. Quirke, Enhanced fluid flow though nanoscale carbon pipes. Nano Lett. **8**, 2632–2637 (2008)
T. Ziegler, Approximate density functional theory as a practical tool in molecular energetics and dynamics. Chem. Rev. **91**, 651–667 (1991)

Chapter 2
Basis Set Functions

The chapter aims to introduce the basis set functions for quantum mechanics computer simulations and numerical electronic calculations.

In this chapter, we discuss the basis set functions for standard molecular quantum mechanics calculations. Furthermore, the basis sets for quantum periodic systems, such as solids and metals, will be described.

Furthermore, we introduce an interactive Python graphical user interface software for basis set exchange and quantum mechanics computer simulations. The focus is on implementing all the theoretical aspects of quantum mechanics calculations described in Chap. 1 in computer simulations of single (bio)molecular system and periodic systems.

2.1 Introduction

The electronic structure calculations depend on the specified degrees of freedom for the one-particle states, or the so-called orbitals of a many-electron wave function. For that, the basis set functions were introduced. A common approach is using the linear combination of atomic orbitals (LCAOs) strategy for the basis set functions constructed from atomic orbitals (AOs) (Heitler and London 1927; Lennard-Jones 1929) with a more recent review in Lehtola (2019b). According to LCAO approach, the single-electron molecular spatial orbital of the spin α, $\psi_i^\alpha(\mathbf{r})$, is expressed in terms of AO basis set functions, ϕ_μ, as:

$$\psi_i^\alpha(\mathbf{r}) = \sum_\mu c_{\mu i}^\alpha \phi_\mu(\mathbf{r} - \mathbf{R}_\mu) \tag{2.1}$$

H. Kamberaj, *Computer Simulations in Molecular Biology*, Scientific Computation,
https://doi.org/10.1007/978-3-031-34839-6_2

where the sum runs over all shells of the atoms in the (bio)molecular system and $c^{\alpha}_{\mu i}$ denote the molecular orbital coefficients. From Eq. (2.1), the AO basis set functions ϕ_{μ} are calculated relative to the coordinate system with origin at the nucleus's position $\mathbf{R}_{\mu}$. In the following, the superscript α is dropped out. The AO basis set functions are taken as the following (Leach 2001; Atkins and Friedman 2005):

$$\phi^{(nlm)}_{\mu}(\mathbf{r}; \mathbf{R}_{\mu}) = R^{\mu}_{nl}(r) Y^{\mu}_{lm}(\theta, \phi) \tag{2.2}$$

where $n = 1, 2, \ldots, l = 0, 1, \ldots, n-1$, and $m = -l, -(l-1), \ldots, l$ denote the principal, azimuthal, and magnetic quantum numbers, respectively. $R^{\mu}_{nl}(r)$ is the radial function (where $r \equiv | \mathbf{r} - \mathbf{R}_{\mu} |$) and $Y^{\mu}_{lm}(\theta, \phi)$ are the spherical harmonics of the AO μ, which are chosen to be real. In principle, an infinite number of basis set functions would be needed to calculate exactly the Hartree–Fock energy, which is the limiting energy, known as the Hartree–Fock limit. In fact, the HF limit is not precisely the ground-state energy of a molecule because the Hartree–Fock approximation ignores the electron correlation energy. Furthermore, an infinite basis set functions is practically impossible, and thus a finite basis set functions is chosen, K, which is known as basis-set truncation error. The error of this truncation is measured as the difference between the Hartree–Fock limit and the calculated energy using HF-SCF approach. Therefore, the main goal is to minimise the number of the basis set functions (and hence minimise the number of two-electron integral computations) and chose these basis functions such that to minimise the computational complexity of the integrals; however, the basis-set truncation error should be small.

Then, Eq. (2.1) is written as:

$$\psi_i(\mathbf{r}) = \sum_{\mu=1}^{K} c_{\mu i} \phi^{(nlm)}_{\mu}(\mathbf{r}; \mathbf{R}_{\mu}) \tag{2.3}$$

If an electron is represented by the wave-function $\psi_i(\mathbf{r})$, then the electron occupies the orbital i.

The LCAO allows the electrons to be explicitly included in calculations. Furthermore, for heavy atoms, that does not result in significant extra computation cost. Moreover, using AO as the basis set functions, the basis-set truncation errors are only systematic; that is, these errors are similar at different geometries or spin states, which allow for adjusting the computational speed and accuracy with the size of the basis sets (Lehtola 2019b). For example, single-ζ basis sets are suggested at a qualitative level of accuracy for preliminary studies; polarised double-ζ or triple-ζ basis sets work well for quantitative studies with DFT approach; quadruple-ζ or high basis sets are often used for benchmark studies with DFT calculations of the molecular properties, which require large basis sets to produce reliable results (Lehtola 2019b).

Three type of LCAO basis set functions are often used (Leach 2001; Atkins and Friedman 2005; Lehtola 2019b), as described in the following.

2.1.1 Slater-Type Orbitals

The Slater-type orbitals (STOs) (Zener 1930; Slater 1930; Eckart 1930) are also known as exponential type orbitals:

$$\phi_\mu^{(nlm)} = -\left[\left(\frac{2Z}{na}\right)^3 \frac{(n-l-1)!}{2n((n+l)!)^3}\right]^{1/2} \rho^l L_{n+1}^{2l+1}(\rho) \exp\left(-\zeta_n \frac{r}{a}\right) Y_{lm}(\theta,\phi) \quad (2.4)$$

where (n, l, m) are the quantum numbers specifying a nucleus of an atom with atomic number Z, L_{n+1}^{2l+1} is the Laguerre polynomial special function, $\rho = 2\zeta_n r/a$, and the orbital exponents $\zeta_n = Z/(na)$. Here, a is:

$$a = 4\pi\varepsilon_0 \frac{\hbar^2}{m_r e^2} = 5.29177 \times 10^{-11} \text{ m} \quad (2.5)$$

where $m_r = m_e M_\mu / \left(m_e + M_\mu\right)$ is the reduced mass (m_e is the mass of electron and M_μ is the mass of a nucleus), which for infinitely heavy nucleus equals the Bohr radius:

$$a = a_0 = 4\pi\varepsilon_0 \frac{\hbar^2}{m_e e^2} = 5.29177 \times 10^{-11} \text{ m} \quad (2.6)$$

In Eq. (2.2), the radial function $R_{nl}(r)$ is also written as

$$R_{nl}(r) = -2\left[(\zeta_n)^3 \frac{(n-l-1)!}{n((n+l)!)^3}\right]^{1/2} \rho^l L_{n+1}^{2l+1}(\rho) \exp\left(-\zeta_n r\right) \quad (2.7)$$

and the angular part of the AO is given as product of these two functions:

$$Y_{lm}(\theta,\phi) = \Theta_{lm}(\theta)\Phi_m(\phi) \quad (2.8)$$

where

$$\begin{aligned} \Theta_{lm}(\theta) &= \left[\frac{2l+1}{2}\frac{(l-|m|)!}{(l+|m|)!}\right]^{1/2} P_l^{|m|}(\cos\theta) \\ \Phi_m(\phi) &= \frac{1}{\sqrt{2\pi}} \exp(im\phi) \end{aligned} \quad (2.9)$$

where $(\cdots)!$ denotes the factorial and $P_l^{|m|}(\cos\theta)$ are the associated Legendre polynomial functions (which for $m = 0$ give the Legendre polynomials $P_l(\cos\theta)$):

$$P_l^m(x) = (-1)^m \left(1 - x^2\right)^{m/2} \frac{d^m}{dx^m} P_l(x) \tag{2.10}$$

$$P_l(x) = \frac{1}{2^l l!} \frac{d^l}{dx^l} (x^2 - 1)^l$$

Here, the atomic orbitals with $l = 0$ are called s-type orbitals, those with $l = 1$ are called p-type orbitals, those with $l = 2$ are called d-type orbitals, and those with $l = 3$ are called f-type orbitals. If the electron occupies an s-type orbital, it is called an s-electron. Similarly, there exist p-, d-, and f-electrons for the electrons occupying, respectively, p-, d-, and f-type orbitals.

The analytical expressions for the one-electron orbitals centred at the nucleus coordinate $\mathbf{R}_\mu = (0, 0, 0)$ are given as (Leach 2001):

$$\begin{aligned}
R_{nl}(r) &= 2\,(\zeta_1)^{3/2} \exp\left(-\zeta_1 r\right), \quad n = 1, l = 0 \\
R_{nl}(r) &= 2\,(\zeta_2)^{3/2} \left(1 - \zeta_2 r\right) \exp\left(-\zeta_2 r\right), \quad n = 2, l = 0 \\
R_{nl}(r) &= \left(\frac{4}{3}\right)^{1/2} (\zeta_2)^{5/2}\, r \exp\left(-\zeta_2 r\right), \quad n = 2, l = 1 \\
R_{nl}(r) &= \left(\frac{2}{3}\right)^{1/2} (\zeta_3)^{3/2} \left(3 - 6\zeta_3 r + 2(\zeta_3 r)^2\right) \exp\left(-\zeta_3 r\right), \quad n = 3, l = 0 \\
R_{nl}(r) &= \left(\frac{8}{9}\right)^{1/2} (\zeta_3)^{5/2} \left(2 - \zeta_3 r\right) r \exp\left(-\zeta_3 r\right), \quad n = 3, l = 1 \\
R_{nl}(r) &= \left(\frac{8}{45}\right)^{1/2} (\zeta_3)^{7/2}\, r^2 \exp\left(-\zeta_3 r\right), \quad n = 3, l = 2
\end{aligned} \tag{2.11}$$

In Fig. 2.1a, it is plotted the radial function $R_{nl}(r)$ for $n = 1$ and $l = 0$ (1s orbital); in Fig. 2.1b, there are shown the radial functions $R_{nl}(r)$ for $n = 2$ and $l = 0, 1$ (2s and 2p orbitals); and Fig. 2.1c presents the radial functions $R_{nl}(r)$ for $n = 3$ and $l = 0, 1, 2$ (3s, 3p, and 3d orbitals). The atom of nitrogen (N) with $Z = 7$ was considered and the lengths are expressed in the units of Bohr radius a_0.

For $n = 1$ and $l = 0$, the spherical harmonic is ($m = 0$):

$$Y_{00} = \Theta_{00}(\theta)\Phi_0(\phi) = \frac{1}{\sqrt{4\pi}} \tag{2.12}$$

Thus, the atomic 1s orbital is:

$$\phi_\mu^{(1s)}(\mathbf{r}) = \frac{1}{\sqrt{\pi}} (\zeta_1)^{3/2} \exp\left(-\zeta_1 r\right), \quad \zeta_1 = \frac{Z}{a_0} \tag{2.13}$$

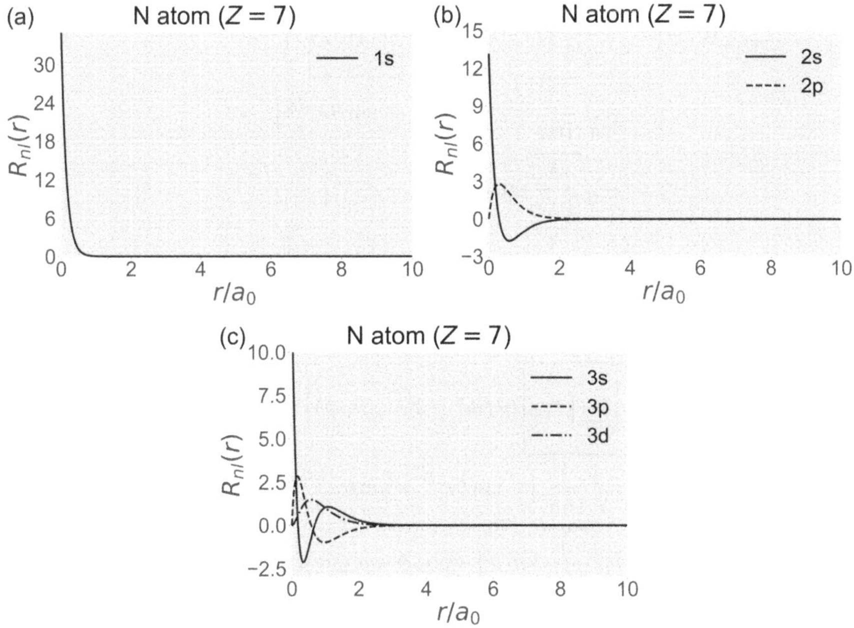

Fig. 2.1 **a** The radial function $R_{nl}(r)$ for $n = 1$ and $l = 0$ (1s orbital); **b** The radial function $R_{nl}(r)$ for $n = 2$ and $l = 0, 1$ (2s and 2p orbitals); **c** The radial function $R_{nl}(r)$ for $n = 3$ and $l = 0, 1, 2$ (3s, 3p, and 3d orbitals). The atom of nitrogen (N) with $Z = 7$ was considered and the lengths are expressed in the units of Bohr radius a_0

For $n = 2$ and $l = 0$ (2s orbital), the spherical harmonic (for also $m = 0$) is the same as in Eq. (2.12), and the atomic 2s orbital is:

$$\phi_{\mu}^{(2s)}(\mathbf{r}) = \frac{1}{\sqrt{\pi}} (\zeta_2)^{3/2} (1 - \zeta_2 r) \exp(-\zeta_2 r), \quad \zeta_2 = \frac{Z}{2a_0} \tag{2.14}$$

For $n = 2$, $l = 1$, then $m = 0, \pm 1$, and thus the spherical harmonics are:

$$\begin{aligned} Y_{lm}(\theta, \phi) &= \Theta_{lm}(\theta)\Phi_m(\phi) = \sqrt{\frac{3}{4\pi}} \cos\theta, \quad m = 0 \\ Y_{lm}(\theta, \phi) &= \Theta_{lm}(\theta)\Phi_m(\phi) = -\sqrt{\frac{3}{8\pi}} \sin\theta \exp(\pm i\phi), \quad m = \pm 1 \end{aligned} \tag{2.15}$$

There are three real spherical harmonics, corresponding to $2p_x$, $2p_y$ and $2p_z$, which are taken by the following combinations:

$$
\begin{aligned}
Y_{lm}^{(2\mathrm{p}_x)}(\theta,\phi) &= -\frac{1}{\sqrt{2}}\left[Y_{1+1}(\theta,\phi)+Y_{1-1}(\theta,\phi)\right] \\
&= -\sqrt{\frac{3}{16\pi}}\left[-\sin\theta e^{i\phi}-\sin\theta e^{-i\phi}\right] \\
&= -\sqrt{\frac{3}{16\pi}}\sin\theta\left[-e^{i\phi}-e^{-i\phi}\right] \\
&= -\sqrt{\frac{3}{16\pi}}\sin\theta\left[-\cos\phi-i\sin\phi-\cos\phi+i\sin\phi\right] \\
&= \sqrt{\frac{3}{4\pi}}\sin\theta\cos\phi \\
Y_{lm}^{(2\mathrm{p}_y)}(\theta,\phi) &= \frac{i}{\sqrt{2}}\left[Y_{1+1}(\theta,\phi)-Y_{1-1}(\theta,\phi)\right] \\
&= i\sqrt{\frac{3}{16\pi}}\left[-\sin\theta e^{i\phi}+\sin\theta e^{-i\phi}\right] \\
&= i\sqrt{\frac{3}{16\pi}}\sin\theta\left[-\cos\phi-i\sin\phi+\cos\phi-i\sin\phi\right] \\
&= \sqrt{\frac{3}{4\pi}}\sin\theta\sin\phi \\
Y_{lm}^{(2\mathrm{p}_z)}(\theta,\phi) &= Y_{10}(\theta,\phi)=\sqrt{\frac{3}{4\pi}}\cos\theta
\end{aligned}
\tag{2.16}
$$

where the mathematical formulas $e^{\pm i\phi}=\cos\phi\pm i\sin\phi$ and $i^2=-1$ are used.

Thus, the three atomic 2p orbitals (namely $2\mathrm{p}_x$, $2\mathrm{p}_y$, and $2\mathrm{p}_z$) are:

$$
\begin{aligned}
\phi_\mu^{(2\mathrm{p}_x)}(\mathbf{r}) &= \frac{\sqrt{3}}{2\pi}(\zeta_2)^{3/2}(1-\zeta_2 r)\exp(-\zeta_2 r)\sin\theta\cos\phi, \quad \zeta_2=\frac{Z}{2a_0} \\
\phi_\mu^{(2\mathrm{p}_y)}(\mathbf{r}) &= \frac{\sqrt{3}}{2\pi}(\zeta_2)^{3/2}(1-\zeta_2 r)\exp(-\zeta_2 r)\sin\theta\sin\phi, \\
\phi_\mu^{(2\mathrm{p}_z)}(\mathbf{r}) &= \frac{\sqrt{3}}{2\pi}(\zeta_2)^{3/2}(1-\zeta_2 r)\exp(-\zeta_2 r)\cos\theta
\end{aligned}
\tag{2.17}
$$

For $n=3, l=0$, and $m=0$, the spherical harmonic is again given by Eq. (2.12), and hence the atomic 3s orbital is:

$$
\phi_\mu^{(3\mathrm{s})}(\mathbf{r}) = \sqrt{\frac{1}{6\pi}}(\zeta_3)^{3/2}\left(3-6\zeta_3 r+2(\zeta_3 r)^2\right)\exp(-\zeta_3 r), \quad \zeta_3=\frac{Z}{3a_0}
\tag{2.18}
$$

For $n=3$, $l=1$, and $m=0,\pm1$, the following $3\mathrm{p}_x$, $3\mathrm{p}_y$, and $3\mathrm{p}_z$ atomic orbitals are obtained:

$$\phi_{\mu}^{(3\mathrm{p}_x)}(\mathbf{r}) = \sqrt{\frac{2}{3\pi}}\,(\zeta_3)^{5/2}\,(2-\zeta_3 r)\, r\exp(-\zeta_3 r)\sin\theta\cos\phi, \quad \zeta_3 = \frac{Z}{3a_0} \tag{2.19}$$

$$\phi_{\mu}^{(3\mathrm{p}_y)}(\mathbf{r}) = \sqrt{\frac{2}{3\pi}}\,(\zeta_3)^{5/2}\,(2-\zeta_3 r)\, r\exp(-\zeta_3 r)\sin\theta\sin\phi,$$

$$\phi_{\mu}^{(3\mathrm{p}_z)}(\mathbf{r}) = \sqrt{\frac{2}{3\pi}}\,(\zeta_3)^{5/2}\,(2-\zeta_3 r)\, r\exp(-\zeta_3 r)\cos\theta$$

For $n = 3, l = 2, m = 0, \pm 1, \pm 2$, the spherical harmonics are:

$$Y_{lm}(\theta,\phi) = \Theta_{lm}(\theta)\Phi_m(\phi) = \sqrt{\frac{5}{16\pi}}\left(3\cos^2\theta - 1\right), \quad m = 0 \tag{2.20}$$

$$Y_{lm}(\theta,\phi) = \Theta_{lm}(\theta)\Phi_m(\phi) = -\sqrt{\frac{15}{8\pi}}\sin\theta\cos\theta\exp(\pm i\phi), \quad m = \pm 1$$

$$Y_{lm}(\theta,\phi) = \Theta_{lm}(\theta)\Phi_m(\phi) = \sqrt{\frac{15}{32\pi}}\sin^2\theta\exp(\pm i2\phi), \quad m = \pm 2$$

The following combinations can be taken:

$$\begin{aligned}
Y_{lm}^{(3\mathrm{d}_{xz})}(\theta,\phi) &= -\frac{1}{\sqrt{2}}\left[Y_{2+1}(\theta,\phi) + Y_{2-1}(\theta,\phi)\right] \\
&= \sqrt{\frac{15}{16\pi}}\sin\theta\cos\theta\,[2\cos\phi] \\
&= \sqrt{\frac{15}{4\pi}}\sin\theta\cos\theta\cos\phi \\
Y_{lm}^{(3\mathrm{d}_{yz})}(\theta,\phi) &= \frac{i}{\sqrt{2}}\left[Y_{2+1}(\theta,\phi) - Y_{2-1}(\theta,\phi)\right] \\
&= -i\sqrt{\frac{15}{16\pi}}\sin\theta\cos\theta\,[2i\sin\phi] \\
&= \sqrt{\frac{15}{4\pi}}\sin\theta\cos\theta\sin\phi \\
Y_{lm}^{(3\mathrm{d}_{z^2})}(\theta,\phi) &= Y_{20}(\theta,\phi) = \sqrt{\frac{5}{16\pi}}\left(3\cos^2\theta - 1\right) \\
Y_{lm}^{(3\mathrm{d}_{x^2-y^2})}(\theta,\phi) &= \frac{1}{\sqrt{2}}\left[Y_{2+2}(\theta,\phi) + Y_{2-2}(\theta,\phi)\right] \\
&= \sqrt{\frac{15}{64\pi}}\sin^2\theta\,[2\cos(2\phi)] \\
&= \sqrt{\frac{15}{16\pi}}\sin^2\theta\cos(2\phi)
\end{aligned} \tag{2.21}$$

$$\begin{aligned}
Y_{lm}^{(3d_{xy})}(\theta,\phi) &= -\frac{i}{\sqrt{2}}\left[Y_{2+2}(\theta,\phi) - Y_{2-2}(\theta,\phi)\right] \\
&= -i\sqrt{\frac{15}{64\pi}}\sin^2\theta\,[2i\sin(2\phi)] \\
&= \sqrt{\frac{15}{16\pi}}\sin^2\theta\sin(2\phi)
\end{aligned}$$

Then, the atomic 3d orbitals are given as:

$$\begin{aligned}
\phi_\mu^{(3d_{xy})}(\mathbf{r}) &= \sqrt{\frac{1}{6\pi}}\,(\zeta_3)^{7/2}\,r^2\exp(-\zeta_3 r)\sin^2\theta\sin(2\phi) \qquad (2.22)\\
\phi_\mu^{(3d_{xz})}(\mathbf{r}) &= \sqrt{\frac{2}{3\pi}}\,(\zeta_3)^{7/2}\,r^2\exp(-\zeta_3 r)\sin\theta\cos\phi\cos\theta \\
\phi_\mu^{(3d_{yz})}(\mathbf{r}) &= \sqrt{\frac{2}{3\pi}}\,(\zeta_3)^{7/2}\,r^2\exp(-\zeta_3 r)\sin\theta\sin\phi\cos\theta \\
\phi_\mu^{(3d_{z^2})}(\mathbf{r}) &= \sqrt{\frac{1}{18\pi}}\,(\zeta_3)^{7/2}\,r^2\exp(-\zeta_3 r)\left(3\cos^2\theta - 1\right) \\
\phi_\mu^{(3d_{x^2-y^2})}(\mathbf{r}) &= \sqrt{\frac{1}{6\pi}}\,(\zeta_3)^{7/2}\,r^2\exp(-\zeta_3 r)\sin^2\theta\cos(2\phi)
\end{aligned}$$

For plotting, the atomic 1s- 2s-, and 3s-type orbitals are written in Cartesian coordinates as:

$$\begin{aligned}
\phi_\mu^{(1s)}(x,y,z) &= \frac{1}{\sqrt{\pi}}\,(\zeta_1)^{3/2}\exp\left(-\zeta_1\sqrt{x^2+y^2+z^2}\right) \qquad (2.23)\\
\phi_\mu^{(2s)}(x,y,z) &= \frac{1}{\sqrt{\pi}}\,(\zeta_2)^{3/2}\left[1 - \zeta_2\sqrt{x^2+y^2+z^2}\right]\exp\left(-\zeta_2\sqrt{x^2+y^2+z^2}\right) \\
\phi_\mu^{(3s)}(x,y,z) &= \sqrt{\frac{1}{6\pi}}\,(\zeta_3)^{3/2}\left[3 - 6\zeta_3\sqrt{x^2+y^2+z^2} + 2(\zeta_3)^2(x^2+y^2+z^2)\right] \\
&\quad\times\exp\left(-\zeta_3\sqrt{x^2+y^2+z^2}\right)
\end{aligned}$$

Next, a cubic volume is created with a side-length of 8 units of a_0, which was split into 50 pixels each side (that is, $50 \times 50 \times 50$). Then, to each pixel was assigned a density value equal to the atomic orbital calculated at the centre of that pixel. The density is represented by the colour as the fourth coordinate in a volumetric plot. These plots are shown in Fig. 2.2 for the 1s-, 2s-, and 3s-type atomic orbitals. It can be seen that 2s- and 3s-orbitals show a cusp at the centre of the atomic nucleus, which is not present for 1s-type atomic orbital.

Similarly, the three atomic 2p orbitals (such as $2p_x$, $2p_y$, and $2p_z$) in Cartesian coordinates are written as:

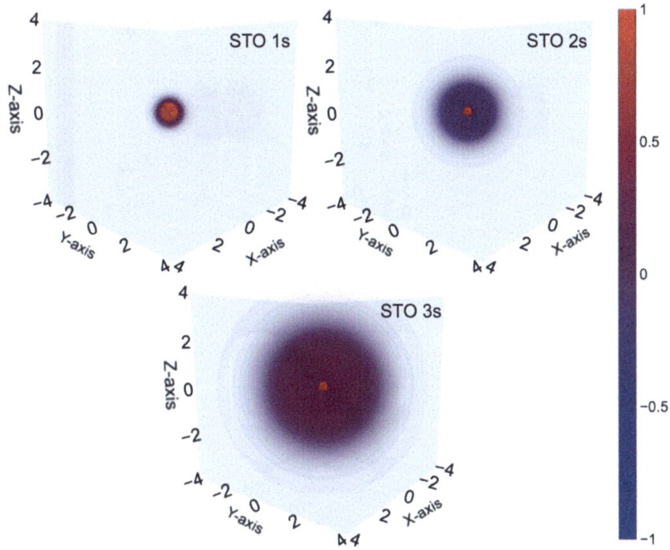

Fig. 2.2 STO 1s-, 2s-, and 3s-type atomic orbitals density functions plotted as a three-dimension volumetric data points. The colour bar represents the density scale, which is the fourth dimension

$$\phi_{\mu}^{(2p_x)}(x,y,z) = \frac{\sqrt{3}}{2\pi}(\zeta_2)^{3/2}\frac{\left(1-\zeta_2\sqrt{x^2+y^2+z^2}\right)x}{\sqrt{x^2+y^2+z^2}}\exp\left(-\zeta_2\sqrt{x^2+y^2+z^2}\right) \tag{2.24}$$

$$\phi_{\mu}^{(2p_y)}(x,y,z) = \frac{\sqrt{3}}{2\pi}(\zeta_2)^{3/2}\frac{\left(1-\zeta_2\sqrt{x^2+y^2+z^2}\right)y}{\sqrt{x^2+y^2+z^2}}\exp\left(-\zeta_2\sqrt{x^2+y^2+z^2}\right)$$

$$\phi_{\mu}^{(2p_z)}(x,y,z) = \frac{\sqrt{3}}{2\pi}(\zeta_2)^{3/2}\frac{\left(1-\zeta_2\sqrt{x^2+y^2+z^2}\right)z}{\sqrt{x^2+y^2+z^2}}\exp\left(-\zeta_2\sqrt{x^2+y^2+z^2}\right)$$

Figure 2.3 presents graphically the $2p_x$-, $2p_y$-, and $2p_z$-type atomic orbitals density functions plotted as a three-dimension volume. The colour bar represents the density scale.

The atomic 3d orbitals in Cartesian coordinates are given as the following:

$$\phi_{\mu}^{(3d_{xy})}(x,y,z) = \sqrt{\frac{2}{3\pi}}(\zeta_3)^{7/2}xy\exp\left(-\zeta_3\sqrt{x^2+y^2+z^2}\right) \tag{2.25}$$

$$\phi_{\mu}^{(3d_{xz})}(x,y,z) = \sqrt{\frac{2}{3\pi}}(\zeta_3)^{7/2}xz\exp\left(-\zeta_3\sqrt{x^2+y^2+z^2}\right)$$

$$\phi_{\mu}^{(3d_{yz})}(x,y,z) = \sqrt{\frac{2}{3\pi}}(\zeta_3)^{7/2}yz\exp\left(-\zeta_3\sqrt{x^2+y^2+z^2}\right)$$

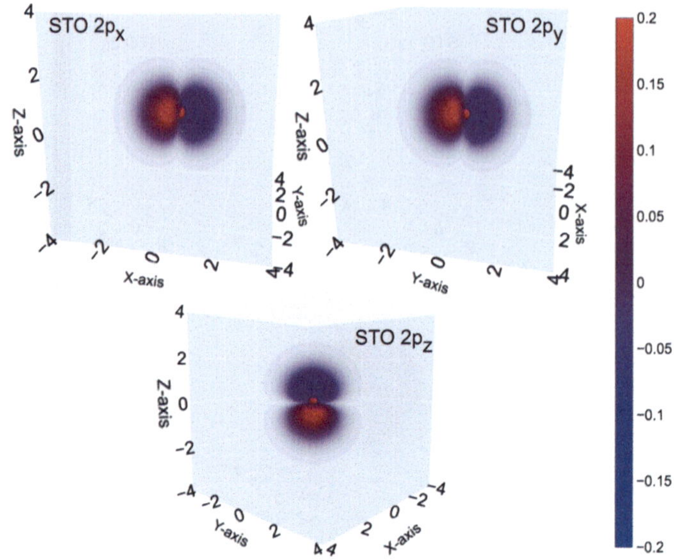

Fig. 2.3 STO $2p_x$-, $2p_y$-, and $2p_z$-type atomic orbitals density functions plotted as a three-dimension volume. The colour bar represents the density scale

$$\phi_{\mu}^{(3d_{z^2})}(x, y, z) = \sqrt{\frac{1}{18\pi}} (\zeta_3)^{7/2} \left(2z^2 - (x^2 + y^2)\right) \exp\left(-\zeta_3\sqrt{x^2 + y^2 + z^2}\right)$$

$$\phi_{\mu}^{(3d_{x^2-y^2})}(x, y, z) = \sqrt{\frac{1}{6\pi}} (\zeta_3)^{7/2} \left(x^2 - y^2\right) \exp\left(-\zeta_3\sqrt{x^2 + y^2 + z^2}\right)$$

A drawback of STOs is that ns-type orbitals for $n > 1$ have zero amplitude at the nucleus position, as it can be depicted from Fig. 2.2. Furthermore, the STOs are difficult from the computation viewpoint; some integrals, in particular when the atomic orbitals are centred at different nuclei, are very difficult to evaluate. However, it is relatively easy to compute the integrals including one or two centres, such as $(\nu\nu|\mu\mu)$, $(\nu\mu|\mu\mu)$, and $(\nu\mu|\nu\mu)$. Moreover, three-centre and four-centre integrals can also be relatively straightforward to be computed with STOs, if the atomic orbitals belong to the same atom; however, if they are located at different atoms, then their computation is very difficult (if not impossible) (Fig. 2.4).

2.1.2 Gaussian-Type Orbitals

Gaussian-type orbitals (GTOs) are introduced by Boys (1950) and reviewed in Jalbout et al. (2004) and Lehtola (2019b). In Cartesian coordinates, the functional form of the Gaussians is

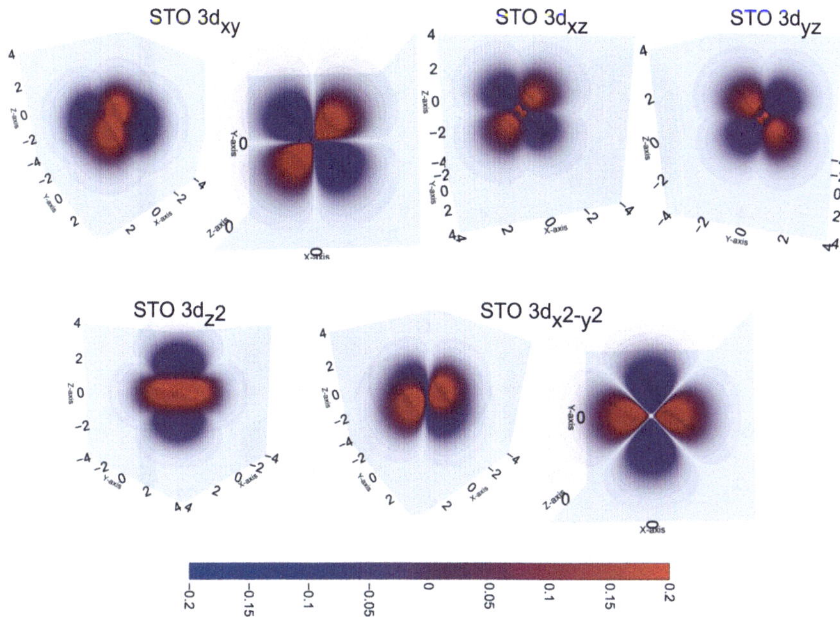

Fig. 2.4 STO 3d-type atomic orbitals density functions plotted as a three-dimension volume. The colour bar represents the density scale

$$\Theta_{ijk}(\mathbf{r}-\mathbf{R}_\mu) = (x-X_\mu)^i(y-Y_\mu)^j(z-Z_\mu)^k \exp\left(-\alpha \mid \mathbf{r}-\mathbf{R}_\mu \mid^2\right) \tag{2.26}$$

where $\left(X_\mu, Y_\mu, Z_\mu\right)$ are the Cartesian coordinates of the centre of the Gaussian function at $\mathbf{R}_\mu$, and (x, y, z) are the Cartesian coordinates of the centre of the electron at $\mathbf{r}$. Here, (i, j, k) are non-negative integers and α is a positive exponent. By definition, for $i+j+k=0$, the Cartesian Gaussian is an s-type Gaussian, and there is only one Gaussian function for s-type, namely Θ_{000}:

$$\Theta_{000}(\mathbf{r}-\mathbf{R}_\mu) = \exp\left(-\alpha \mid \mathbf{r}-\mathbf{R}_\mu \mid^2\right) \tag{2.27}$$

For $i+j+k=1$, it is a p-type Gaussian; there are three Gaussians of p-type as given in the following:

$$\begin{aligned}
\Theta_{100}(\mathbf{r}-\mathbf{R}_\mu) &= (x-X_\mu)\exp\left(-\alpha \mid \mathbf{r}-\mathbf{R}_\mu \mid^2\right) \\
\Theta_{010}(\mathbf{r}-\mathbf{R}_\mu) &= (y-Y_\mu)\exp\left(-\alpha \mid \mathbf{r}-\mathbf{R}_\mu \mid^2\right) \\
\Theta_{001}(\mathbf{r}-\mathbf{R}_\mu) &= (z-Z_\mu)\exp\left(-\alpha \mid \mathbf{r}-\mathbf{R}_\mu \mid^2\right)
\end{aligned} \tag{2.28}$$

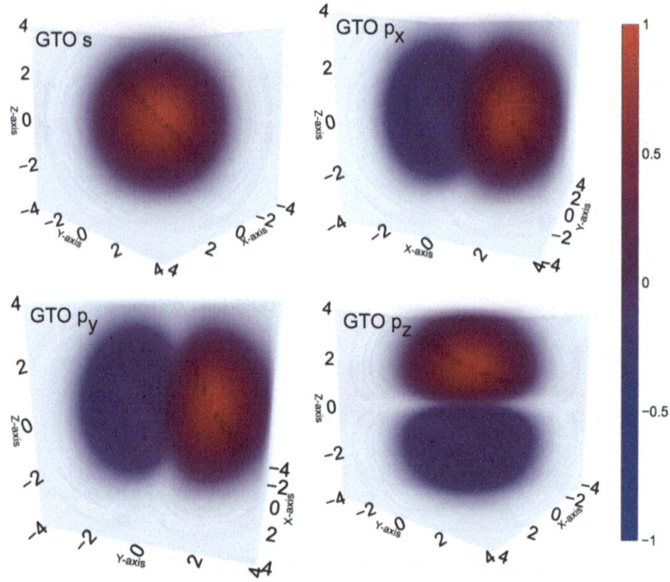

Fig. 2.5 The s- and p-type Gaussian orbitals

For $i + j + k = 2$, it is a d-type Gaussian. There are six d-type Gaussians, namely:

$$\begin{aligned}\Theta_{110}(\mathbf{r}-\mathbf{R}_\mu) &= (x-X_\mu)(y-Y_\mu)\exp\left(-\alpha \mid \mathbf{r}-\mathbf{R}_\mu \mid^2\right)\\ \Theta_{101}(\mathbf{r}-\mathbf{R}_\mu) &= (x-X_\mu)(z-Z_\mu)\exp\left(-\alpha \mid \mathbf{r}-\mathbf{R}_\mu \mid^2\right)\\ \Theta_{011}(\mathbf{r}-\mathbf{R}_\mu) &= (y-Y_\mu)(z-Z_\mu)\exp\left(-\alpha \mid \mathbf{r}-\mathbf{R}_\mu \mid^2\right)\\ \Theta_{200}(\mathbf{r}-\mathbf{R}_\mu) &= (x-X_\mu)^2\exp\left(-\alpha \mid \mathbf{r}-\mathbf{R}_\mu \mid^2\right)\\ \Theta_{020}(\mathbf{r}-\mathbf{R}_\mu) &= (y-Y_\mu)^2\exp\left(-\alpha \mid \mathbf{r}-\mathbf{R}_\mu \mid^2\right)\\ \Theta_{002}(\mathbf{r}-\mathbf{R}_\mu) &= (z-Z_\mu)^2\exp\left(-\alpha \mid \mathbf{r}-\mathbf{R}_\mu \mid^2\right)\end{aligned} \tag{2.29}$$

Figure 2.5 shows a density plot of the s- and p-type Gaussian orbitals.

Figure 2.6 shows a density plot of the d-type Gaussian orbitals.

The linear combinations of d-Gaussian orbitals are also used; thus, six linear combinations of the d-type Gaussians can be created, where five of them have the angular behaviour of the five real 3d-hydrogen type orbitals and the other has a spherical symmetry like an s-type function, and hence it is often ignored from the basis set; however, it is not necessary to be eliminated because the basis set is not necessarily orthogonal.

Furthermore, spherical Gaussians can also be used where the factors $(x - X_\mu)$ (or, $(y - Y_\mu)$ and $(z - Z_\mu)$) are replaced by spherical harmonics. The main advantage of the GTOs is the Gaussian product theorem, which in particular is useful in

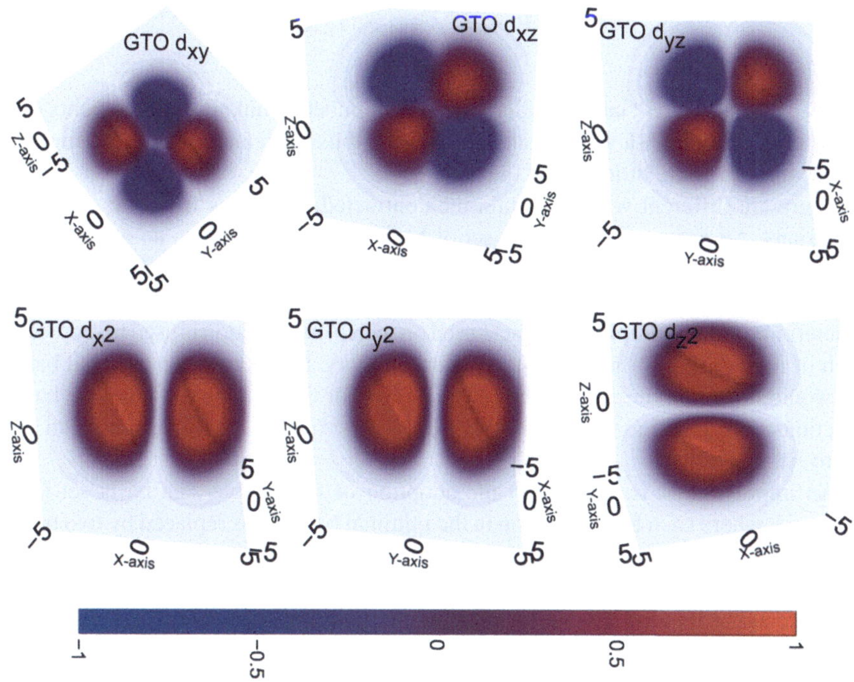

Fig. 2.6 The d-type Gaussian orbitals

computation of the two-electron integrals (Boys 1950; McMurchie and Davidson 1978; Obara and Saika 1986; Gill et al. 1991). According to that theorem, the product of two Gaussians at different centres yields a single-Gaussian function centred at a point between these two centres. Thus, the two-electron integrals on three and four centres are equivalent to integrals over two different centres that are computationally more efficient. However, there is a problem identified for GTO, which has a cusp for 1s hydrogen atomic orbitals at the nucleus position ($r \to 0$); therefore, larger basis sets are used to have the same accuracy as obtained from STO. In contrast, STO for $n = 1$ has a cusp at $r = 0$, but not GTO.

Several GTOs are added together to avoid the disadvantages obtained by a single-function GTO, known as contracted Gaussian functions. Thus, each Gaussian is contracted to produce g_μ, which is a linear combination of the primitive Gaussian functions $\Theta_m^{(ijk)}$ centred at the same nucleus position of the atom:

$$g_\mu(\mathbf{r}; \mathbf{R}_\mu) = \sum_m d_{\mu m} \Theta_m^{(ijk)}(\mathbf{r}_m; \mathbf{R}_\mu) \tag{2.30}$$

where $d_{\mu m}$ are the contraction coefficients. Then, the spatial atomic orbitals are obtained as a linear combination of the those contracted Gaussians:

$$\phi_{\mu}(\mathbf{r};\mathbf{R}_{\mu}) = \sum_{\nu} c_{\nu\mu} g_{\nu}(\mathbf{r};\mathbf{R}_{\mu}) \tag{2.31}$$

The use of contracted Gaussians decreases the number of unknown coefficients $c_{\nu\mu}$ determined from high-level HF calculations, and thus the reduction of the computation costs with some little lose of the accuracy.

There are different ways to obtain the contracted Gaussians from the primitive Gaussians. An approach includes a set of basis functions and then an atomic SCF calculations to obtain the optimised set of exponents α (see Eq. (2.26)) for the basis functions that are used in molecular structure calculations. Then, a minimal basis set is determined as the simplest basis set based on which one function is used to represent each of the orbitals of the elementary valence theory. For example, a minimal basis set would have one function each for H and He (for the 1s-type orbital); five basis functions each for Li to Ar (for the 1s, 2s, and 2p orbitals); nine functions each for Na to Ar, and others.

An improvement is considered the adoption of the double-zeta basis set (DZ basis set), where each basis function in the minimal basis set is replaced by two basis functions. The number of basis set functions is doubled compared with minimal basis set. Another basis set is the so-called triple-zeta basis set (TZ basis set), where three basis functions are used to represent each of the orbitals presented in the elementary valence theory.

As a compromise between disadvantage of the minimal basis set and the computational cost of DZ and TZ basis sets, the so-called split-valence (SV) basis set is introduced (Weigend and Ahlrichs 2005). According to this approach, each valence atomic orbital is described by two basis functions whereas the inner-shell atomic orbital is described by one basis function each.

Note that there is also contribution from basis functions of orbitals with quantum number l larger than the maximum one considered in the valence theory; for example, the representation of the d-orbitals for carbon atom. Furthermore, the existence of covalent bonds in molecules causes polarisation of the adjacent atoms, which can be considered in calculations by including basis functions of orbitals representing high values of l. The p-type orbitals can be included to model the polarisation of 1s-type orbitals and d-type orbitals can be included to model the polarisation of p-type orbitals. Thus, the double-zeta plus polarisation (DZPP) basis functions have been introduced.

The contracted Gaussian basis sets can be constructed using the least-squares fit of N primitive Gaussians to a set of STOs optimised using the atomic SCF calculations. The construction of an STO from N primitive Gaussians is written as STO-NG; for example, for $N = 3$ (which is a common choice), we obtain the contracted Gaussians set STO-3G (Hehre et al. 1969). In a second approach, the atomic SCF calculations are considered with a relatively large basis set of Gaussian primitives, which generated a set of SCF $c_{\nu\mu}$ coefficients (see Eq. (2.3)) for the primitives of every spatial atomic orbital ϕ_{μ}. The contracted Gaussian basis sets can be obtained from the coefficients of the primitive Gaussians using Eq. (2.30) (Huzinaga 1965; Dunning 1970, 1971).

Another contraction of Gaussian basis functions is the so-called 3-21G basis set (Binkley et al. 1980; Hehre et al. 1986), where one contracted Gaussian is made up of three primitives used to represent every inner-shell atomic orbital. While every valence-shell orbital is represented by two functions where one is a contracted Gaussian of two primitives and the other is a single primitive. The 6-31G* basis set stands for split-valence 6-31G basis set plus the polarisation functions in the form of six d-type functions for every atom other than H atom (Ditchfield et al. 1971; Hehre et al. 1972; Hariharan and Pople 1973; Feller 1996; Schuchardt et al. 2007; Pritchard et al. 2019). The 6-31G** is a basis set that includes 6-31G* plus a set of three p-type polarisation functions for every H atom.

2.1.3 *Numerical Atomic Orbitals*

The numerical atomic orbitals (NAOs) are another basis sets, where the radial function is given analytically as (Lehtola 2019a, b):

$$R_{nl}(r) = \frac{u_{nl}(r)}{r} \tag{2.32}$$

where $u_{nl}(r)$ is a piecewise polynomial function, described in Lehtola (2019a, b, 2020). The method NAO is a numerical approach (that is, there is no analytical expression as for the STO and GTO approaches). The numerical scheme is applied in a radial grid space for each atom. This approach is considered significantly more efficient compared with STOs and GTOs (Lehtola 2019b).

In Fig. 2.7 are presented the Hartree–Fock energies for the neon atom using different minimal basis set compared with fully numerical reference value of -128.54709811 Hartree units (h.u.) (Lehtola 2020).

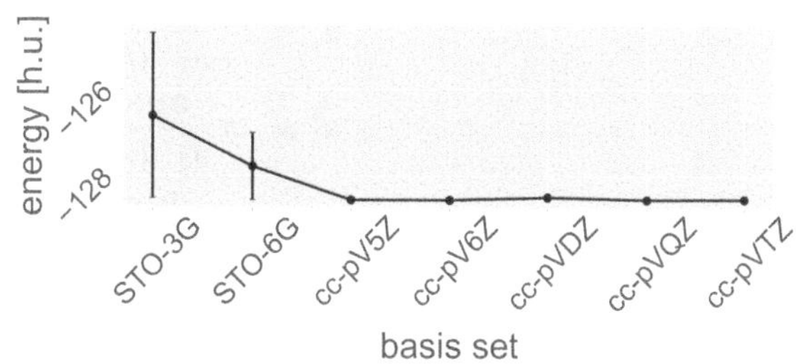

Fig. 2.7 The Hartree–Fock energies for the neon atom using different minimal basis set compared with fully numerical reference value of -128.54709811 Hartree units (h.u.) (Lehtola 2020)

2.1.4 Orbitals of Periodic Systems

2.1.4.1 The Tight-Binding Approximation

Quantum mechanics simulations of the bulk phases (such as solids and metals) are considered different to these applied in studying individual molecules or isolated macromolecular systems. The systems are characterised by the size and the shape of the so-called unit simulation cell, and the atoms are distributed within that unit cell. The bulk phase is obtained by applying the so-called periodic boundary conditions (Allen and Tildesley 1989). There exist fourteen different unit cells, known as the Bravais lattices, which are characterised by six parameters, namely the sizes of three lattice vectors ($\mathbf{a}_1, \mathbf{a}_2, \mathbf{a}_3$) and three angles ($\alpha, \beta, \gamma$).

The lattice structures are also characterised by the so-called reciprocal lattice (Leach 2001). In X-ray crystallographic structures, the reciprocal lattice are described by three vectors ($\mathbf{a}'_1, \mathbf{a}'_2, \mathbf{a}'_3$) such that $\mathbf{a}'_1$ is perpendicular to $\mathbf{a}_2$ and $\mathbf{a}_3$, and similarly the others. The magnitude is such that $\mathbf{a}'_i \cdot \mathbf{a}_i = 1$ for $i = 1, 2, 3$. Thus, the following relations are obtained

$$\begin{aligned} \mathbf{a}'_1 &= \frac{\mathbf{a}_2 \times \mathbf{a}_3}{\mathbf{a}_1 \cdot (\mathbf{a}_2\mathbf{a}_3)} \\ \mathbf{a}'_2 &= \frac{\mathbf{a}_3 \times \mathbf{a}_1}{\mathbf{a}_2 \cdot (\mathbf{a}_3\mathbf{a}_1)} \\ \mathbf{a}'_3 &= \frac{\mathbf{a}_1 \times \mathbf{a}_2}{\mathbf{a}_3 \cdot (\mathbf{a}_1\mathbf{a}_2)} \end{aligned} \tag{2.33}$$

Equation 2.33 points out that the units of reciprocal lattice vectors are 1/length, and hence the reciprocal lattice is also termed reciprocal space. Sometimes, for the computational purposes, it is more convenient to determine the reciprocal space vectors as the following (Leach 2001):

$$\begin{aligned} \mathbf{a}^*_1 &= 2\pi \frac{\mathbf{a}_2 \times \mathbf{a}_3}{\mathbf{a}_1 \cdot (\mathbf{a}_2\mathbf{a}_3)} \\ \mathbf{a}^*_2 &= 2\pi \frac{\mathbf{a}_3 \times \mathbf{a}_1}{\mathbf{a}_2 \cdot (\mathbf{a}_3\mathbf{a}_1)} \\ \mathbf{a}^*_3 &= 2\pi \frac{\mathbf{a}_1 \times \mathbf{a}_2}{\mathbf{a}_3 \cdot (\mathbf{a}_1\mathbf{a}_2)} \end{aligned} \tag{2.34}$$

Applying periodic boundary conditions for lattice would mean that the electron density (or the wave-function) has the same value at the equivalent positions on the lattice. Using the so-called Bloch's theorem, the wave-function in periodic Hartree–Fock approach for the spatial orbital i are given as linear combinations of the so-called Bloch functions (Leach 2001):

$$\psi_i^{\mathbf{k}}(\mathbf{r}) = \sum_{\mu} c_{\mu i} \Phi_{\mu}^{\mathbf{k}}(\mathbf{r}) \tag{2.35}$$

where $\Phi_{\mu}^{\mathbf{k}}(\mathbf{r})$ are the Bloch function given in terms of the atomic orbitals ϕ as

$$\Phi_{\mu}^{\mathbf{k}}(\mathbf{r}) = \sum_{\mathbf{g}} \phi_{\mu}^{\mathbf{g}}(\mathbf{r}) e^{i\mathbf{k}\cdot\mathbf{g}} \tag{2.36}$$

The subscript **k** is an index characteristic for every atom of the lattice, **g** is the lattice vector, and $\phi_{\mu}^{\mathbf{g}}(\mathbf{r})$ is the atomic orbital of the atom μ, which can be expressed as a linear combination of Gaussian functions as described above for the molecular theory. The coefficients of the linear combinations, Eq. (2.35), are determined using self-consistent field matrix equation for every **k**:

$$\mathbf{F}_{\mathbf{k}}\mathbf{C}_{\mathbf{k}} = \mathbf{S}_{\mathbf{k}}\mathbf{C}_{\mathbf{k}}\mathbf{E}_{\mathbf{k}} \tag{2.37}$$

where $\mathbf{C}_{\mathbf{k}}$ is the matrix of coefficients $c_{\mu i}$, $\mathbf{S}_{\mathbf{k}}$ is the overlap matrix of the Bloch functions for the wave-vector **k**, $\mathbf{F}_{\mathbf{k}}$ is the Fock matrix of the sum of one- and two-electron integrals, and $\mathbf{E}_{\mathbf{k}}$ is the energy matrix. The values of the wave-vectors **k** are chosen from the first Brillouin zone. When the one- and two-electron terms are expanded they involve infinite sums running over the electrons and nuclei of the lattice. Similar to the molecular HF approach, the one-electron terms represent the sum of a kinetic energy term and a potential Coulomb interaction energy term between the electrons and nuclei. Furthermore, the two-electron terms consist of Coulomb and exchange two-electron integrals (Pisani and Dovesi 1980; Doversi et al. 1983, 2000).

2.1.4.2 The Nearly Free-Electron Approximation

The tight-binding method works satisfactory for certain types of solids; however, for other periodic systems, it is found that consideration of the valence electrons as free particles is more helpful. The motion of the free valence electrons is influenced by the presence of the lattice. The wave-function form satisfying the Bloch's theorem are as the following (Leach 2001):

$$\psi^{\mathbf{k}}(\mathbf{r}) = u^{\mathbf{k}}(\mathbf{r}) \exp(i\mathbf{k}\cdot\mathbf{r}) \tag{2.38}$$

where the possible values of **k** are

$$\mathbf{k} = \left(\frac{m_1}{N_1}\mathbf{a}_1^*, \frac{m_2}{N_2}\mathbf{a}_2^*, \frac{m_3}{N_3}\mathbf{a}_3^*\right) \tag{2.39}$$

Here, m_1, m_2, m_3 are integers and $N_1 N_2 N_3 = N$ gives the number of unit cells in a crystal. The energy as a function of the wave-vector is given as

$$E_{\mathbf{k}} = \frac{\hbar^2}{2m_e}\left(k_1^2 + k_2^2 + k_3^2\right) \tag{2.40}$$

For large N ($\approx N_A$ where N_A is the Avogadro's number), typical for macroscopic systems, **k** becomes continues. In Eq. (2.38), $u^{\mathbf{k}}(\mathbf{r})$ is a periodic function of the lattice. These functions are constructed using the Fourier series expansion in the reciprocal space of the plane wave-functions (Leach 2001)

$$u^{\mathbf{k}}(\mathbf{r}) = \sum_{\mathbf{G}} u_{\mathbf{G}}^{\mathbf{k}} \exp\left(i\mathbf{G}\cdot\mathbf{r}\right) \tag{2.41}$$

where the sum runs over all the reciprocal lattice vectors **G**. For example, if $\mathbf{G} = m\mathbf{a}_1^*$, then $\exp\left(i\mathbf{G}\cdot\mathbf{r}\right)$ corresponds to the plane wave travelling perpendicular to the real-space vectors $\mathbf{a}_2$ and $\mathbf{a}_3$ and along $\mathbf{a}_1^*$ with m wavelengths fitting into one cell.

Furthermore, the external potential due to the nuclei is a periodic function in the lattice expressed as a Fourier expansion of the plane-wave functions of the reciprocal lattice as:

$$U^{\mathbf{k}}(\mathbf{r}) = \sum_{\mathbf{G}} U_{\mathbf{G}}^{\mathbf{k}} \exp\left(i\mathbf{G}\cdot\mathbf{r}\right) \tag{2.42}$$

In Eq. (2.41) and Eq. (2.42), $u_{\mathbf{G}}^{\mathbf{k}}$ and $U_{\mathbf{G}}^{\mathbf{k}}$ are Fourier coefficients.

The Schrödinger equation can take the following form (Ashcroft and Mermin 1976):

$$\left(\frac{\hbar^2}{2m_e} \mid \mathbf{k}+\mathbf{G} \mid^2 - E\right) u_{\mathbf{G}}^{\mathbf{k}} + \sum_{\mathbf{G}'} U_{\mathbf{G}+\mathbf{G}'}^{\mathbf{k}} u_{\mathbf{G}'}^{\mathbf{k}} = 0 \tag{2.43}$$

Setting all the coefficients $U_{\mathbf{G}}^{\mathbf{k}} = 0$, we obtain the equation of the free-electrons as:

$$\left(\frac{\hbar^2}{2m_e} \mid \mathbf{k}+\mathbf{G} \mid^2 - E\right) u_{\mathbf{G}}^{\mathbf{k}} = 0 \tag{2.44}$$

which has the following energy for solution:

$$E = \frac{\hbar^2}{2m_e} \mid \mathbf{k}+\mathbf{G} \mid^2 \tag{2.45}$$

and the wave-function

$$\psi(\mathbf{r}) \propto \exp\left(i\left(\mathbf{k}+\mathbf{G}\right)\cdot\mathbf{r}\right) \tag{2.46}$$

In Eq. (2.43), the summation over the reciprocal lattice vectors **G** giving as many forms of this equation for a **k** as the number of reciprocal lattice vectors in the system. Therefore, for each **G**, we obtain a solution labelled, for example, by n, called also the band index. Furthermore, there are as many Schrödinger equations as there are vectors **k**, and hence the energy levels vary with both **k** and n. However, for band structure calculations **k** is restricted to the first Brillouin zone avoiding the double states counting.

2.1.4.3 The Plane Waves Basis Set Functions

In the DFT approach for studying the periodic systems, such as solid states, the plane waves are used as the basis set. An advantage of the plane waves is that they are natural periodic functions. Here, every orbital is expressed as a linear combination of plane waves characterised by the reciprocal lattice vectors as

$$\psi_i^{\mathbf{k}}(\mathbf{r}) = \sum_{\mathbf{G}} c_{i,\mathbf{k}+\mathbf{G}} \exp\left(i\left(\mathbf{k}+\mathbf{G}\right)\cdot\mathbf{r}\right) \tag{2.47}$$

The Kohn-Sham equations of the DFT approach reduce to the following:

$$\begin{aligned}\sum_{\mathbf{G}'}\Bigg[\frac{\hbar^2}{2m_e}\mid \mathbf{k}+\mathbf{G}\mid \delta_{\mathbf{G}\mathbf{G}'} + V_{\text{ion}}(\mathbf{G}-\mathbf{G}') + V_{\text{elec}}(\mathbf{G}-\mathbf{G}')\\ + V_{\text{XC}}(\mathbf{G}-\mathbf{G}')\Bigg] c_{i,\mathbf{k}+\mathbf{G}'} = \varepsilon_i c_{i,\mathbf{k}+\mathbf{G}}\end{aligned} \tag{2.48}$$

Here, V_{ion}, V_{elec}, and V_{XC} are, respectively, the electron-nuclei, electron-electron, and the exchange-correlation functionals. $\delta_{\mathbf{G}\mathbf{G}'}$ denotes the delta function, which is one if $\mathbf{G} = \mathbf{G}'$; otherwise, it is zero.

In practice, the Fourier series sum over $\mathbf{G}'$ runs an infinite number of the reciprocal lattice vectors for macroscopic lattice systems. Besides, there is an infinite number of **k** vectors within the first Brillouin zone. Since the main focus is in the valence electrons of the atoms, which are responsible for chemical covalent bonds and other physical properties, the core electrons are only slightly influenced by the atoms of the medium. Thus, only the valence electrons are considered practically in calculations, whereas the core electrons are integrated with nucleus. Nevertheless, that could cause problems if the valence electrons near the nucleus are represented by the plane wave-functions due to the fast oscillations (Leach 2001). That is related to the fact that wave-functions of these valence electrons and of the core electrons must be orthogonal. Therefore, large values of the kinetic energies are obtained and a number of large plane waves must be used. However, in practice, the so-called pseudo-potentials are used replacing the true potential interaction of the electrons with nuclei in the core regions, which are much weaker (Heiner 1970).

2.2 Quantum Mechanics Software

The basis set functions of different families are described in Pritchard et al. (2019). Python software exists, as a database called Basis Set Exchange (BSE) (Pritchard et al. 2019). The interactive graphical user interfaces (GUI) designs allow users to explore and experiment with electronic quantum calculations visually and intuitively.

A Python-based simulation of chemistry framework exists under the name PySCF (Sun et al. 2018a). Other quantum mechanics calculations and simulations packages include DALTON (Aidas et al. 2014) and Car–Parrinello ab-initio molecular dynamics (CPMD) (CPMD 2001). Here, the Interactive Python Graphical User Interface Basis Set Exchange and Quantum Mechanics is introduced, which

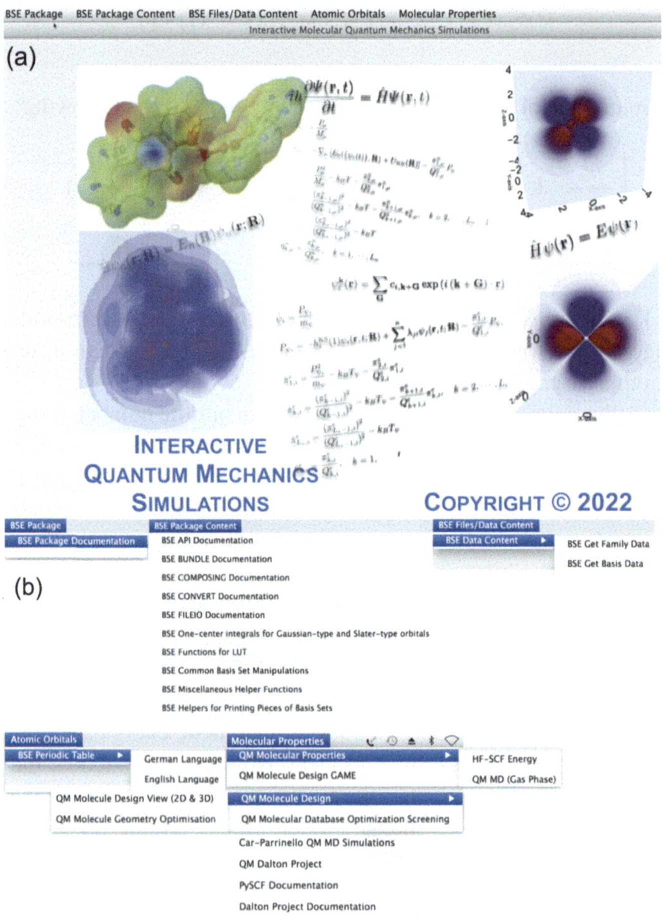

Fig. 2.8 **a** The interactive GUI design to the quantum mechanics simulations. **b** The menus and sub-menus

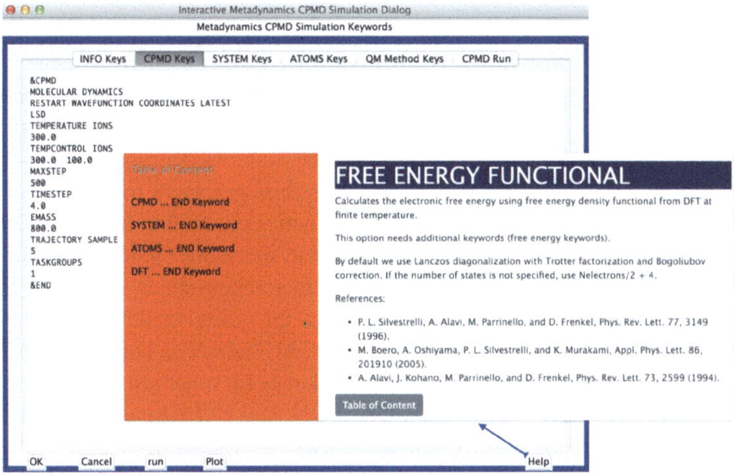

Fig. 2.9 The interactive GUI design to the quantum mechanics simulations using CPMD package

implements all the theoretical aspects of quantum calculations described in Chap. 1 and this chapter. It is a piece of software to filter the information from the basis set exchange database and calculate molecular properties using quantum mechanics electronic calculations.

Figure 2.8a shows the front page of the interactive GUI design to the basis set exchange data library. The design includes the interactive information query about the BSE package, content, data, and a new feature added to visualise spatial orbitals for every chemical element from the periodic table. Each menu has a sub-menu related to particular information in the BSE database, as shown in Fig. 2.8b.

This software presents a graphical interface for querying data from the BSE database in Python (Pritchard et al. 2019). All the existing basis sets of different families can be prepared in formats, such as NWCHEM, GAUSSIAN94, DALTON, QCHEM, JSON, and others, for use in other software packages for quantum mechanics simulations. The current version presents an interface to the PySCF, DALTON, and CPMD packages. For example, Fig. 2.9 presents a GUI design to the quantum mechanics simulations using CPMD package, including the documentation as a HyperText Markup Language (HTML) format textbook.

Furthermore, the software can be used to draw various types of atomic orbitals (such as s-, p-, d-, and f-type of orbitals) using the functions of *graph_objects* module of *plotly* package; besides, one-centre integrals can be calculated. Moreover, the software can be used in the classrooms for interactive teaching to demonstrate the visualisation and shape of the atomic orbitals for different periodic table elements. The most significant impacts include the design of the quantum mechanics molecular gaming, visualisation of atomic and molecular orbitals, geometry optimisation, graphical interface to basis set exchange database, quantum mechanics properties, and new molecule design and property calculations.

2.3 Summary

We discussed the basis set functions for standard molecular quantum mechanics electronic calculations. Besides, the basis set functions for quantum periodic systems (such as solids and metals) were introduced.

Furthermore, we introduced an interactive Python graphical user interface software for basis set exchange and quantum mechanics computer simulations, which implements all the theoretical aspects of quantum mechanics calculations in computer simulations of single (bio)molecular system and periodic systems.

This software presents a graphical interface for querying data from the BSE database in various formats, such as NWCHEM, GAUSSIAN94, DALTON, QCHEM, JSON, and others, for use in other software packages for quantum mechanics simulations. It is an interface to the PySCF, DALTON, and CPMD packages, including the documentation as an HTML format textbook.

Moreover, the software can be used to draw various types of atomic orbitals (such as s-, p-, d-, and f-type of orbitals). The software will be the instructor's toolbox in the classrooms for interactive teaching to demonstrate the visualisation and shape of the atomic orbitals for different periodic table elements. The most significant impacts include the design of the quantum mechanics molecular gaming, visualisation of atomic and molecular orbitals, geometry optimisation, graphical interface to basis set exchange database, quantum mechanics properties, and new molecule design and property calculations.

References

K. Aidas, C. Angeli, K.L. Bak, V. Bakken, R. Bast, L. Boman, O. Christiansen, R. Cimiraglia, S. Coriani, P. Dahle, E.K. Dalskov, U. Ekström, T. Enevoldsen, J.J. Eriksen, P. Ettenhuber, B. Fernández, L. Ferrighi, H. Fliegl, L. Frediani, K. Hald, A. Halkier, C. Hättig, H. Heiberg, T. Helgaker, A.C. Hennum, H. Hettema, E. Hjertenæs, S. Høst, I.-M. Høyvik, M.F. Iozzi, B. Jansik, H.J.Aa. Jensen, D. Jonsson, P. Jørgensen, J. Kauczor, S. Kirpekar, T. Kjærgaard, W. Klopper, S. Knecht, R. Kobayashi, H. Koch, J. Kongsted, A. Krapp, K. Kristensen, A. Ligabue, O.B. Lutnæs, J.I. Melo, K.V. Mikkelsen, R.H. Myhre, C. Neiss, C.B. Nielsen, P. Norman, J. Olsen, J.M.H. Olsen, A. Osted, M.J. Packer, F. Pawlowski, T.B. Pedersen, P.F. Provasi, S. Reine, Z. Rinkevicius, T.A. Ruden, K. Ruud, V. Rybkin, P. Salek, C.C.M. Samson, A. Sánchez de Merás, T. Saue, S.P.A. Sauer, B. Schimmelpfennig, K. Sneskov, A.H. Steindal, K.O. Sylvester-Hvid, P.R. Taylor, A.M. Teale, E.I. Tellgren, D.P. Tew, A.J. Thorvaldsen, L. Thøgersen, O. Vahtras, M.A. Watson, D.J.D. Wilson, M. Ziolkowski, H. Ågren, The Dalton quantum chemistry program system. WIREs Comput. Mol. Sci. **4**, 269–284 (2014)

M.P. Allen, D.J. Tildesley, *Computer Simulation of Liquids* (Oxford University Press, 1989)

N.W. Ashcroft, N.D. Mermin, *Solid State Physics* (Holt, Rinehart & Winston, New York, 1976)

P. Atkins, R. Friedman, *Molecular Quantum Mechanics*, 4th edn. (Oxford University Press, 2005)

J.S. Binkley, J.A. Pople, W.J. Hehre, Self-consistent molecular orbital methods. 21. Small split-valence basis sets for first-row elements. J. Am. Chem. Soc. **102**, 939–947 (1980). https://doi.org/10.1021/ja00523a008

S.F. Boys, Electronic wave functions - I. A general method of calculation for the stationary states of any molecular system. Proc. R. Soc. A **200**, 542–554 (1950)

CPMD. *Car-Parrinello Molecular Dynamics Simulations*. Copyright IBM Corp 1990–2008, Copyright MPI für Festkörperforschung Stuttgart (Wiley, 1997–2001)

R. Ditchfield, W.J. Hehre, J.A. Pople, Self-consistent molecular-orbital methods. IX. An extended gaussian-type basis for molecular-orbital studies of organic molecules. J. Chem. Phys. **54**, 724–728 (1971). https://doi.org/10.1063/1.1674902

R. Doversi, C. Pisani, C. Roetti, V.R. Saunders, The treatment of coulomb interactions in Hartree-Fock calculations of periodic-systems. Phys. Rev. B **28**, 5781–5792 (1983)

R. Doversi, R. Orlando, C. Roetti, V.R. Saunders, The periodic Hartree-Fock method and its implementation in the CRYSTAL code. Phys. Status Solidi B **217**, 63–88 (2000)

T.H. Dunning, Gaussian basis functions for use in molecular calculations. III. Contraction of (10s6p) atomic basis sets for the first-row atoms. J. Chem. Phys. **55**(2), 716–723 (1971)

T.H. Dunning, Gaussian basis functions for use in molecular calculations. I. Contraction of (9s5p) atomic basis sets for the first-row atoms. J. Chem. Phys. **53**(7), 2823–2833 (1970)

C. Eckart, The theory and calculation of screening constants. Phys. Rev. **36**, 878–892 (1930)

D. Feller, The role of databases in support of computational chemistry calculations. J. Comput. Chem. **17**, 1571–1586 (1996). https://doi.org/10.1002/(SICI)1096-987X(199610)17:13<1571::AID-JCC9>3.0.CO;2-P

P.M.W. Gill, B.G. Johnson, J.A. Pople, Two-electron repulsion integrals over Gaussian s-functions. Int. J. Quantum Chem. **40**(6), 745–752 (1991)

P.C. Hariharan, J.A. Pople, The influence of polarization functions on molecular orbital hydrogenation energies. Theor. Chim. Acta **28**, 213–222 (1973). https://doi.org/10.1007/bf00533485

W.J. Hehre, R. Ditchfield, J.A. Pople, Self-consistent molecular orbital methods. XII. Further extensions of gaussian-type basis sets for use in molecular orbital studies of organic molecules. J. Chem. Phys. **56**, 2257–2261 (1972). https://doi.org/10.1063/1.1677527

W.J. Hehre, L. Radom, P.v.R. Schleyer, J.A. Pople, *Ab Initio Molecular Orbital Theory* (Wiley, New York, 1986)

W.J. Hehre, R.F. Stewart, J.A. Pople, Self-consistent molecular-orbital methods. I. Use of Gaussian expansions of slater-type atomic orbitals. J. Chem. Phys. **51**, 2657–2664 (1969). https://doi.org/10.1063/1.1672392

V. Heiner, The Pseudo-potential concept. Solid State Phys. **24**, 1–36 (1970)

W. Heitler, F. London, Wechselwirkung neutraler Atome und homopolare Bindung nach der Quantenmechanik. Z. Phys. **44**, 455–472 (1927)

S. Huzinaga, Gaussian-type functions for poly-atomic systems. I. J. Chem. Phys. **42**, 1293–1302 (1965)

A.F. Jalbout, F. Nazari, L. Tucker, Gaussian-based calculations in molecular science. J. Mol. Struct. (Theochem) **671**, 1–21 (2004)

A.R. Leach, *Molecular Modelling, Principles and Applications*, 2nd edn. (Prentice Hall, Pearson Education Limited, Edingburgh Gate, 2001)

S. Lehtola, Fully numerical Hartree-Fock and density functional calculations. I. Atoms. Int. J. Quantum Chem. **119**, e25945 (2019a)

S. Lehtola, A review on non-relativistic, fully numerical electronic structure calculations on atoms and diatomic molecules. Int. J. Quantum Chem. **119**, e25968-31 (2019b)

S. Lehtola, Fully numerical calculations on atoms with fractional occupations and range-separated exchange functionals. Phys. Rev. A **101**, 012516–15 (2020)

J.E. Lennard-Jones, The electronic structure of some diatomic molecules. Trans. Faraday Soc. **25**, 668–686 (1929)

L.E. McMurchie, E.R. Davidson, One- and two-electron integrals over Cartesian Gaussian functions. J. Comput. Phys. **26**, 218–231 (1978)

S. Obara, A. Saika, Efficient recursive computation of molecular integrals over Cartesian Gaussian functions. J. Chem. Phys. **84**, 3963–3974 (1986)

C. Pisani, R. Dovesi, Exact-exchange Hartree-Fock calculations for periodic systems. I. Illustration of the method. Int. J. Quant. Chem. **XVII**, 501–516 (1980)

B.P. Pritchard, D. Altarawy, B. Didier, T.D. Gibson, T.L. Windus, A new basis set exchange: an open, up-to-date resource for the molecular sciences community. J. Chem. Inf. Model **59**(11), 4814–4820 (2019)

K.L. Schuchardt, B.T. Didier, T. Elsethagen, L. Sun, V. Gurumoorthi, J. Chase, J. Li, T.L. Windus, Basis set exchange: a community database for computational sciences. J. Chem. Inf. Model. **47**, 1045–1052 (2007). https://doi.org/10.1021/ci600510j

J.C. Slater, Atomic shielding constants. Phys. Rev. **36**, 57–64 (1930)

Q. Sun, T.C. Berkelbach, N.S. Blunt, G.H. Booth, S. Guo, Z. Li, J. Liu, J. McClain, E.R. Sayfutyarova, S. Sharma, S. Wouters, G.K.L. Chan, PySCF: the Python-based simulations of chemistry framework. WIREs Comput. Mol. Sci. **8**, e1340 (2018)

F. Weigend, R. Ahlrichs, Balanced basis sets of split valence, triple zeta valence and quadruple zeta valence quality for H to Rn: design and assessment of accuracy. Phys. Chem. Chem. Phys. **7**(18), 3297–3305 (2005)

C. Zener, Analytic atomic wave functions. Phys. Rev. **36**, 51–56 (1930)

Chapter 3
Semi-empirical Quantum Mechanics Computer Simulations

The chapter aims to introduce the semi-empirical quantum mechanics models and their use in computer simulations.

In this chapter, we focus on the semi-empirical quantum mechanics approaches and their use in computer simulations. The semi-empirical models discussed here are based on the so-called neglect of diatomic differential overlap approximation, for which the reader may refer to the following literature for more details (and the references therein) (Leach 2001; Husch et al. 2018).

3.1 Introduction

In ab-initio QM calculations, the elements of the Fock matrix, Eq. (1.60) (see Chap. 1), depend on the basis set functions, namely ϕ_μ, ϕ_ν, ϕ_α, and ϕ_β that could be centred on the same atom, on the atoms that are covalently bonded, or on atoms that are not bonded. Therefore, $F_{\nu\mu}$ are split into three groups, namely the diagonal elements $F_{\nu\nu}$, the off-diagonal elements $F_{\nu\mu}$ (that is, ϕ_μ and ϕ_ν are on the same atoms), and the off-diagonal elements $F_{\nu\mu}$ (that is, ϕ_μ and ϕ_ν are on different atoms).

Another common feature to all the semi-empirical methods is that the overlap integral matrix $S_{\nu\mu}$ (Eq. (1.59), Chap. 1) equals the identity matrix **I**, and thus

$$S_{\nu\mu} = \delta_{\nu\mu} \tag{3.1}$$

That is justifiable because some of its diagonal elements are zero since we use orthogonal basis sets on each atom; furthermore, off-diagonal elements that corresponds to the overlap of basis sets on different atoms are set to zero. Therefore, the Roothaan–Hall equation, Eq. (1.64) (Chap. 1), reduces to

H. Kamberaj, *Computer Simulations in Molecular Biology*, Scientific Computation,
https://doi.org/10.1007/978-3-031-34839-6_3

$$\mathbf{FC} = \mathbf{CE} \tag{3.2}$$

Equation 3.2 represents a standard matrix form of an eigenvalue problem. However, when calculating the Fock's matrix **F** not all overlap integral elements are set to zero.

Following discussions introduce all the semi-empirical methods used in quantum mechanics calculations of electronic structures.

3.2 Neglected Differential Overlap

In the neglected differential overlap (NDO), the differential of the overlap integrals is also set to zero; that is, from Eq. (1.59) (see Chap. 1), we have

$$\phi_\nu(\mathbf{r}; \mathbf{R})\phi_\mu(\mathbf{r}; \mathbf{R})\, d\mathbf{r} = 0 \tag{3.3}$$

assuming that ϕ_ν are real functions. Then, Eqs. (3.1) and (3.2) are obtained.

Furthermore, based on the NDO approach, the two-electron repulsion integrals $(\nu\mu|\alpha\beta)$ (see Eq. (1.73), Chap. 1) are also simplified. In particular, we can write

$$(\nu\mu|\alpha\beta) = (\nu\nu|\alpha\alpha)\delta_{\nu\mu}\delta_{\alpha\beta} \tag{3.4}$$

Moreover, the elements of the Fock's matrix, Eq. (1.71) (Chap. 1), simplify as

$$F_{\nu\mu} = \begin{cases} H_{\nu\nu}^{\text{core}} + \sum\limits_{\alpha \neq \nu = 1}^{K} P_{\alpha\alpha}(\nu\nu|\alpha\alpha) + \dfrac{1}{s} P_{\nu\nu}(\nu\nu|\nu\nu), & \mu = \nu \\ H_{\nu\mu}^{\text{core}} - \dfrac{1}{s} P_{\nu\mu}(\nu\nu|\mu\mu), & \mu \neq \nu \end{cases} \tag{3.5}$$

Note that the total wave-function and molecular properties calculated should not change upon a basis set transformation; for example, the canonical transformation to an orthogonal basis set function. For instance, consider the molecular orbitals are written as a linear combination of atomic orbitals, as in Eq. (1.30) (see Chap. 1), where ϕ_μ are atomic orbitals. Now, consider an alternative basis set functions $\{\phi'_\mu\}$, which is a linear combination of the atomic orbital basis set functions $\{\phi_\mu\}$:

$$\psi_i = \sum_{\mu=1}^{K} c'_{\mu i}\phi'_\mu \tag{3.6}$$

$$\phi'_\mu = \sum_{\nu=1}^{K} t_{\nu\mu}\phi_\nu$$

In that case, if NDO approach applies, then the same results will be obtained for the two basis set functions only if the overlap in the new basis set functions is ignored.

3.3 Complete Neglected Differential Overlap

The complete neglected differential overlap (CNDO) is suggested in Pople et al. (1965b). The two-electron integrals $(\nu\nu|\alpha\alpha)$ (where ν and α belong to two different atoms I and J) were taken equal to a parameter Γ_{IJ}, which depends on the type of atoms and their distance, but not on the type of the orbital. It represents the average electrostatic repulsion energy between an electron on atom I and another electron on atom J. Then, the Fock's matrix elements are

$$F_{\nu\nu} = H_{\nu\nu}^{\text{core}} + \left[P_{II} - \frac{1}{s}P_{\nu\nu}\right]\Gamma_{II} + \sum_{J\neq I=1}^{M} P_{JJ}\Gamma_{IJ}, \tag{3.7}$$

$$F_{\nu\mu} = H_{\nu\mu}^{\text{core}} - \frac{1}{s}P_{\nu\mu}\Gamma_{II}, \quad \mu, \nu \in I, \tag{3.8}$$

and

$$F_{\nu\mu} = H_{\nu\mu}^{\text{core}} - \frac{1}{s}P_{\nu\mu}\Gamma_{IJ}, \quad \mu, \nu \in I, J \tag{3.9}$$

where $F_{\nu\nu}$ are the diagonal elements, Γ_{II} denotes the average electron-electron repulsion energy in an atom, and P_{II} is the total electron density of atom I:

$$P_{II} = \sum_{\alpha=1;\alpha\in I}^{K} P_{\alpha\alpha} \tag{3.10}$$

Here, $H_{\nu\nu}^{\text{core}}$ and $H_{\nu\mu}^{\text{core}}$ are the sum of kinetic and electron-nucleus interaction energies for electrons moving in the field of the parent nucleus and other nuclei in the system. Thus, in CNDO, $H_{\nu\nu}^{\text{core}}$ is written as

$$H_{\nu\nu}^{\text{core}} = V_{\nu\nu} - \sum_{J\neq I=1}^{M} V_{IJ} \tag{3.11}$$

where the first term includes the integral involving the atom I where ϕ_ν belongs and the second term all other integrals (J):

$$\begin{aligned} V_{\nu\nu} &= \left(\nu| - \frac{\hbar^2}{2m_e}\nabla^2 - k_e e^2 \frac{Z_I}{|\mathbf{r} - \mathbf{R}_I|}|\nu\right) \\ V_{IJ} &= k_e e^2 \left(\nu|\frac{Z_J}{|\mathbf{r} - \mathbf{R}_J|}|\nu\right) \end{aligned} \tag{3.12}$$

Therefore, $V_{\nu\nu}$ is the energy of the orbital ϕ_ν in the field of its own nucleus I and the core electrons and $-V_{IJ}$ is the energy of electron in the field of other electrons (which is the same for every orbital ϕ_ν of I).

If ϕ_μ and ϕ_ν are orbitals of the same atom I, then

$$H_{\nu\mu}^{\text{core}} = \left(\nu \mid -\frac{\hbar^2}{2m_e}\nabla^2 - k_e e^2 \frac{Z_I}{\mid \mathbf{r} - \mathbf{R}_I \mid} \mid \mu\right) - k_e e^2 \sum_{J \neq I = 1}^{M} \left(\nu \middle| \frac{Z_J}{\mid \mathbf{r} - \mathbf{R}_J \mid} \middle| \mu\right) = 0 \tag{3.13}$$

Here, $H_{\nu\mu}^{\text{core}}$ equals zero because of the orthogonality of atomic orbitals and NDO approximation. If ϕ_μ and ϕ_ν are orbitals of different atoms I and J, then

$$H_{\nu\mu}^{\text{core}} = \left(\nu \middle| -\frac{\hbar^2}{2m_e}\nabla^2 - k_e e^2 \frac{Z_I}{\mid \mathbf{r} - \mathbf{R}_I \mid} - k_e e^2 \frac{Z_J}{\mid \mathbf{r} - \mathbf{R}_J \mid} \middle| \mu\right) + k_e e^2 \sum_{L \neq I, J = 1}^{M} \left(\nu \middle| \frac{Z_L}{\mid \mathbf{r} - \mathbf{R}_L \mid} \middle| \mu\right) \tag{3.14}$$

where first term is the so-called resonance integral (β_{IJ}), which causes bonding, and the second term is the interaction with a third atom L, which is ignored. Thus,

$$H_{\nu\mu}^{\text{core}} = \beta_{IJ} S_{\nu\mu} \tag{3.15}$$

In Eq. (3.15), $S_{\nu\mu}$ is the overlap integral. Finally, in CNDO, we have

$$F_{\nu\nu} = V_{\nu\nu} - \sum_{J \neq I = 1}^{M} V_{IJ} + \left[P_{II} - \frac{1}{s} P_{\nu\nu}\right] \Gamma_{II} + \sum_{J \neq I = 1}^{M} P_{JJ} \Gamma_{IJ}, \tag{3.16}$$

$$F_{\nu\mu} = -\frac{1}{s} P_{\nu\mu} \Gamma_{II}, \quad \mu, \nu \in I, \tag{3.17}$$

and

$$F_{\nu\mu} = \beta_{IJ} S_{\nu\mu} - \frac{1}{s} P_{\nu\mu} \Gamma_{IJ}, \quad \mu, \nu \in I, J \tag{3.18}$$

It is known (Pople et al. 1965a, b) that in first version of CNDO (named CNDO/1) attraction energies between two neutral atoms are too large and predicted equilibrium distances too short; therefore, a second version (namely CNDO/2) has been developed (Pople and Segal 1966), which counts for the effects due to electrons on one of the atoms (I) penetrating the valence shell of another atom (J), and thus they experience some nuclear attraction force. Therefore, the net charge on the atom J equals: $Q_J = eZ_J - eP_{JJ}$, where eZ_J is the nuclear charge and eP_{JJ} is the total electron charge density. Then, the total electron density is

$$P_{JJ} = Z_J - \frac{Q_J}{e} \tag{3.19}$$

Substituting Eq. (3.19) into Eq. (3.16), we obtain

$$F_{\nu\nu} = V_{\nu\nu} + \left[P_{II} - \frac{1}{s}P_{\nu\nu}\right]\Gamma_{II} + \sum_{J\neq I=1}^{M}\left[(Z_J\Gamma_{IJ} - V_{IJ}) - \frac{Q_J}{e}\Gamma_{IJ}\right] \tag{3.20}$$

Here, $-(Q_J/e)\Gamma_{IJ}$ represents the contribution from the net charge on atom J, which is zero if the nucleus charge is balanced by the electron charge. Furthermore, $(Z_J\Gamma_{IJ} - V_{IJ})$ defines the penetration integral. The resonance integral β_{IJ} depends in the atoms I and J, which is written in terms of the single-atom values:

$$\beta_{IJ} = \frac{1}{2}(\beta_I + \beta_J) \tag{3.21}$$

where $\beta_{I/J}$ are chosen from fitting the results of minimal basis set ab-initio calculations on diatomic molecules.

3.4 Intermediate Neglect of Differential Overlap

Intermediate neglect of differential overlap (INDO) (Pople et al. 1967) is developed to overcome some of the problems of the CNDO approach (for example, all the two-electron integrals $(\nu\mu|\alpha\mu)$ equal zero and $(\nu\nu|\mu\mu)$, or $(\nu\nu|\nu\nu)$ are equal to Γ_{II}), such as taking into account the relative spin of two interacting electrons, including the electrons in the same atom. Therefore, the Fock's matrix elements are written for each spin as follows for diagonal elements:

$$F^{\sigma}_{\nu\nu} = V_{\nu\nu} + \sum_{\alpha\in I}\sum_{\beta\in I}\left[P_{\alpha\beta}(\nu\nu|\alpha\beta) - P^{\sigma}_{\alpha\beta}(\nu\alpha|\nu\beta)\right] + \sum_{J\neq I=1}^{M}(P_{JJ} - P_{II})\,\Gamma_{IJ}, \tag{3.22}$$

for two orbital electrons ϕ_ν and ϕ_μ in the same atom I as:

$$F^{\sigma}_{\nu\mu} = V_{\nu\mu} + \sum_{\alpha\in I}\sum_{\beta\in I}\left[P_{\alpha\beta}(\nu\mu|\alpha\beta) - P^{\sigma}_{\alpha\beta}(\nu\alpha|\mu\beta)\right], \tag{3.23}$$

and for two electrons with orbitals ϕ_ν and ϕ_μ on two different atoms:

$$F^{\sigma}_{\nu\mu} = \beta_{IJ} S_{\nu\mu} - P^{\sigma}_{\nu\mu}\Gamma_{IJ} \tag{3.24}$$

In the case of closed-shell system, $P^{\sigma}_{\nu\mu} = P^{\delta}_{\nu\mu} = \frac{1}{s}P_{\nu\mu}$ (for $s = 2$).

3.5 Neglect of Diatomic Differential Overlap

The neglect of diatomic differential overlap (NDDO) is another level of approximation (Pople et al. 1965b), which is considered the basis of the most modern semi-empirical quantum chemistry methods (Husch and Reihe 2018; Wu et al. 2019). In NDDO approach, the two-electron, two-centre integrals, $(\nu\mu|\alpha\beta)$, where μ and ν belong to the same atom I and α and β are to the same atom J are approximated.

The diagonal elements of the Fock's matrix are

$$\begin{aligned} F_{\nu\nu} = H^{\text{core}}_{\nu\nu} &+ \sum_{\alpha\in I}\sum_{\beta\in I}\left[P_{\alpha\beta}(\nu\nu|\alpha\beta) - \frac{1}{s}P_{\alpha\beta}(\nu\alpha|\nu\beta)\right] \\ &+ \sum_{J\neq I}\sum_{\alpha\in J}\sum_{\beta\in J} P_{\alpha\beta}(\nu\nu|\alpha\beta), \end{aligned} \tag{3.25}$$

if μ and ν are in the same atom I:

$$\begin{aligned} F_{\nu\mu} = H^{\text{core}}_{\nu\mu} &+ \sum_{\alpha\in I}\sum_{\beta\in I}\left[P_{\alpha\beta}(\nu\mu|\alpha\beta) - \frac{1}{s}P_{\alpha\beta}(\nu\alpha|\mu\beta)\right] \\ &+ \sum_{J\neq I}\sum_{\alpha\in J}\sum_{\beta\in J} P_{\alpha\beta}(\nu\mu|\alpha\beta), \end{aligned} \tag{3.26}$$

and if $\nu \in I$ and $\mu \in J$, then

$$F_{\nu\mu} = H^{\text{core}}_{\nu\mu} - \frac{1}{s}\sum_{\alpha\in J}\sum_{\beta\in I} P_{\alpha\beta}(\nu\beta|\mu\alpha) \tag{3.27}$$

Regarding the computational costs, the INDO is only slightly more expensive than CNDO approach, whereas the complexity of two-electron and two-centre integrals is increased about 100 times for each pair of heavy atoms in the system for NDDO approach (Leach 2001).

3.6 Modified INDO

Modified INDO (MINDO/3) is introduced in Bingham et al. (1975a, b, c, d) for different systems. This method differs from the semi-empirical methods, mentioned above, because it is parametrised, it uses the experimental data, and it also performs geometry optimisation.

MINDO/3 employs s and p basis set functions, and thus the Fock's matrix diagonal elements are

$$F_{\nu\nu} = V_{\nu\nu} + \sum_{\alpha\in I}\left[P_{\alpha\alpha}(\nu\nu|\alpha\alpha) - \frac{1}{s}P_{\alpha\alpha}(\nu\alpha|\nu\alpha)\right] + \sum_{J\neq I}[P_{JJ} - Z_J]\,\Gamma_{IJ}, \quad (3.28)$$

if both μ and ν are in the same atom I:

$$F_{\nu\mu} = -\frac{1}{s}P_{\nu\mu}(\nu\mu|\nu\mu), \quad (3.29)$$

and if $\nu \in I$ and $\mu \in J$, then

$$F_{\nu\mu} = H_{\nu\mu}^{\text{core}} - \frac{1}{s}P_{\nu\mu}(\nu\mu|\nu\mu) = H_{\nu\mu}^{\text{core}} - \frac{1}{s}P_{\nu\mu}\Gamma_{IJ} \quad (3.30)$$

In Eq. (3.30), the two-centre repulsion integrals, Γ_{IJ}, are calculated as

$$\Gamma_{IJ} = \frac{k_e e^2}{\sqrt{R_{IJ}^2 + \frac{1}{4}\left(\frac{k_e e^2}{\langle g_I\rangle} + \frac{k_e e^2}{\langle g_J\rangle}\right)^2}} \quad (3.31)$$

where $\langle g_{I/J}\rangle$ are the average of the one-centre, two-electron integrals $g_{\nu\mu} = (\nu\nu|\mu\mu)$ on either atom I or atom J. It can be seen (from Eq. (3.31)) that for $R_{IJ} \to \infty$, we have

$$\Gamma_{IJ} = \frac{k_e e^2}{R_{IJ}} \quad (3.32)$$

which represents the Coulomb's interaction law, and for $R_{IJ} \to 0$, we obtain

$$\Gamma_{IJ} = \frac{2}{\frac{1}{\langle g_I\rangle} + \frac{1}{\langle g_J\rangle}} \quad (3.33)$$

which is the average of the one-centre integrals on the two atoms. In MINDO/3, the two-centre, one-electron integrals $H_{\nu\mu}^{\text{core}}$ are as:

$$H^{\text{core}}_{\nu\mu} = S_{\nu\mu}\beta_{IJ}(I_\nu + I_\mu) \tag{3.34}$$

where $S_{\nu\mu}$ is the overlap integral, I_μ (or I_ν) is the ionisation potential for the given orbital μ (or ν), and β_{IJ} is a parameter that depends on atom I and atom J.

Furthermore, in MINDO/3, the expression for the core-core Coulomb repulsion term between two nuclei is given as

$$E^{\text{MINDO/3}}_{IJ} = Z_I Z_J \Gamma_{IJ} \tag{3.35}$$

Note that Eq. (3.35) describes well the interactions between the two nuclei at relatively long separations; however, for short distances between the nuclei, the Coulomb's law should be used:

$$E_{IJ} = k_e e^2 \frac{Z_I Z_J}{R_{IJ}} \tag{3.36}$$

where R_{IJ} is the separation between the two nuclei. In MINDO/3, expressions given by Eqs. (3.35) and (3.36) are combined in a single expression to describe the core-core interaction a function of the electron-electron repulsion integrals as:

$$E^{\text{MINDO/3}}_{IJ} = Z_I Z_J \left[\Gamma_{IJ} + \left(\frac{k_e e^2}{R_{IJ}} - \Gamma_{IJ} \right) \exp\left(-\xi_{IJ} R_{IJ}\right) \right] \tag{3.37}$$

where ξ_{IJ} is a parameter that depends on the type of atoms I and J. Besides, for bonds OH and NH, the following expression is used:

$$E^{\text{MINDO/3}}_{\text{XH}} = Z_{\text{X}} Z_{\text{H}} \left[\Gamma_{\text{XH}} + \left(\frac{k_e e^2}{R_{\text{XH}}} - \Gamma_{\text{XH}} \right) \xi_{\text{XH}} R_{\text{XH}} \exp\left(-\xi_{\text{XH}} R_{\text{XH}}\right) \right] \tag{3.38}$$

The MINDO/3 uses a different strategy for obtaining the parameters compared with other semi-empirical methods. Besides, some parameters that were fixed under NDO, INDO, CNDO, and NNDO are allowed varying with MINDO/3 procedure. $V_{\nu\nu}$ and β_{IJ} are assumed to be parameters. Moreover, in MINDO/3 parametrisation procedure, the experimental data (such as heat of formation and molecular geometries) were used; in contrast, for other semi-empirical methods, the ab-initio data and data from atomic spectra were used.

3.7 Modified Neglect of Diatomic Differential Overlap

Modified neglect of diatomic differential overlap (MNDO) is a modern semi-empirical method proposed, which is based on the NDDO approximation (Dewar and Thiel 1976, 1977a, b). The MNDO model uses one s-type and three p-type basis functions for carbon, nitrogen, fluorine, and oxygen; also, only one s-type of basis function is used for hydrogen (Dewar and Thiel 1977a). In Stewart (2004), the MNDO model is extended for a larger number of chemical elements.

The diagonal elements of the Fock's matrix are:

$$F_{\nu\nu} = H_{\nu\nu}^{\text{core}} + \sum_{\mu \in I} \left[P_{\mu\mu}(\nu\nu|\mu\mu) - \frac{1}{s} P_{\mu\mu}(\nu\mu|\nu\mu) \right] + \sum_{J \neq I = 1}^{M} \sum_{\alpha \in J} \sum_{\beta \in J} P_{\alpha\beta}(\nu\nu|\alpha\beta) \tag{3.39}$$

with

$$H_{\nu\nu}^{\text{core}} = V_{\nu\nu} - \sum_{J \neq I = 1}^{M} V_{\nu\nu}^{(J)} \tag{3.40}$$

For μ and ν both on atom I, the Fock's matrix elements are given as:

$$F_{\nu\mu} = H_{\nu\mu}^{\text{core}} + \frac{3}{2} P_{\nu\mu}(\nu\mu|\nu\mu) - \frac{1}{s} P_{\nu\mu}(\nu\nu|\mu\mu) + \sum_{J \neq I = 1}^{M} \sum_{\alpha \in J} \sum_{\beta \in J} P_{\alpha\beta}(\nu\mu|\alpha\beta) \tag{3.41}$$

where

$$H_{\nu\mu}^{\text{core}} = - \sum_{J \neq I = 1}^{M} V_{\nu\mu}^{(J)} \tag{3.42}$$

For $\nu \in I$ and $\mu \in J$, the Fock's matrix elements are given as:

$$F_{\nu\mu} = H_{\nu\mu}^{\text{core}} - \frac{1}{s} \sum_{\alpha \in J} \sum_{\beta \in I} P_{\alpha\beta}(\nu\beta|\mu\alpha) \tag{3.43}$$

where

$$H_{\nu\mu}^{\text{core}} = \frac{1}{2} S_{\nu\mu}(\beta_\nu + \beta_\mu) \tag{3.44}$$

Here, $V_{\nu\nu}^{(J)}$ and $V_{\nu\mu}^{(J)}$ denote the two-centre, one-electron attractive interaction resonance integrals between an electron distribution $\phi_\nu\phi_\nu$ and $\phi_\nu\phi_\mu$, respectively, on atom I and the core of atom J:

$$V_{\nu\nu}^{(J)} = -k_e e^2 Z_J (\nu_I \nu_I | s_J s_J) \tag{3.45}$$
$$V_{\nu\mu}^{(J)} = -k_e e^2 Z_J (\nu_I \mu_I | s_J s_J)$$

Furthermore, the core-core repulsion terms are differently defined in MNDO model than in MINDO/3; besides, OH and NH bonds are determined separately as:

$$E_{IJ}^{\text{MNDO}} = k_e e^2 Z_I Z_J (s_I s_I | s_J s_J) f_{IJ}^{\text{MNDO}} \tag{3.46}$$

and

$$E_{\text{XH}}^{\text{MNDO}} = k_e e^2 Z_{\text{X}} Z_{\text{H}} (s_{\text{X}} s_{\text{X}} | s_{\text{H}} s_{\text{H}}) f_{\text{XH}}^{\text{MNDO}} \tag{3.47}$$

where the scaling factors are given as

$$f_{IJ}^{\text{MNDO}} = 1 + \exp(-\xi_I R_{IJ}) + \exp(-\xi_J R_{IJ}) \tag{3.48}$$

and

$$f_{\text{XH}}^{\text{MNDO}} = 1 + \xi_{\text{X}} R_{\text{XH}} \exp(-\xi_{\text{X}} R_{\text{XH}}) + \exp(-\xi_{\text{H}} R_{\text{XH}}) \tag{3.49}$$

In Eq. (3.46),

$$(s_I s_I | s_J s_J) \equiv \Gamma_{IJ} \tag{3.50}$$

where Γ_{IJ} is given by Eq. (3.33).

Note that the use of (s, p)-type of the basis set in MNDO indicates that the method can not be used to most of the transition metals because they require the use of d-type basis set; however, d-type orbitals have been included in the new versions, the so-called MNDO/d, of the MNDO (Thiel and Voityuk 1991, 1996). Besides, the MNDO method does not accurately represent the intermolecular interactions involving hydrogen bonds because it overestimates the repulsion between atoms at the distances approximately equal to the sum of their van der Waals radii (Leach 2001).

Figure 3.1a, b presents the scaling factor f_{IJ}^{MNDO} for different values of $\xi_{I/J}$, respectively, $1.0\,\text{Å}^{-1}$, $3.0\,\text{Å}^{-1}$, $10.0\,\text{Å}^{-1}$ for H_2 molecule (see Fig. 3.1a). In Fig. 3.1b, it is shown the core-core repulsion energy E_{IJ}^{MNDO} for the same values of $\xi_{I/J}$, E_{IJ}^{MNDO} for $f_{IJ}^{\text{MNDO}} = 1$, E_{IJ} in atomic units (a.u.) is Coulomb's law energy using Eq. (3.36), $(s_I s_I | s_J s_J)$ calculated from Eq. (3.33) with $\langle g_I \rangle = \langle g_J \rangle = 0.83$ a.u., and $Z_I = Z_J = 1$.

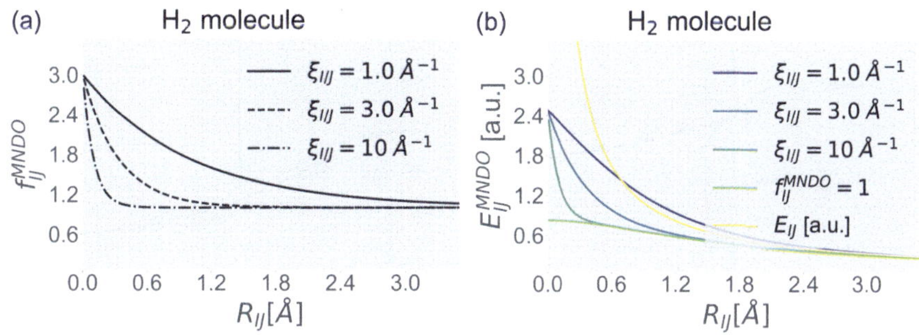

Fig. 3.1 **a** The scaling factor f_{IJ}^{MNDO} for different values of $\xi_{I/J}$, respectively, $1.0\,\text{Å}^{-1}$, $3.0\,\text{Å}^{-1}$, $10.0\,\text{Å}^{-1}$ for H_2 molecule. **b** The core-core repulsion energy E_{IJ}^{MNDO} for the same values of $\xi_{I/J}$, E_{IJ}^{MNDO} for $f_{IJ}^{\text{MNDO}} = 1$, E_{IJ} in atomic units (a.u.) is Coulomb's law energy using Eq. (3.36), $(s_I s_I | s_J s_J)$ calculated from Eq. (3.33) with $\langle g_I \rangle = \langle g_J \rangle = 0.83$ a.u., and $Z_I = Z_J = 1$

3.8 Austin Model *x*

The Austin model 1 (AM1) differs from the MNDO model on the scaling factor of the pairwise core-core repulsion interaction energies (Dewar et al. 1985). The AM1 allows to overcome some of the issues with the MNDO method in calculations of the repulsion interactions between atoms at distances approximately equal to the sum of their van der Waals radii by modifying the core-core repulsion and attractive terms using Gaussian functions. In particular, the attractive Gaussian functions were centred in the region where the repulsion interactions were too large and they were such that they reduced the repulsion terms; whereas the repulsion functions were centred at smaller internuclear separations. Thus, the expression for core-core term was modified as (Leach 2001; Husch et al. 2018):

$$\begin{aligned} E_{IJ}^{\text{AM1}} &= k_e e^2 Z_I Z_J (s_I s_I | s_J s_J) f_{IJ}^{\text{AM1}} \\ &= E_{IJ}^{\text{MNDO}} + k_e e^2 \frac{Z_I Z_J}{R_{IJ}} (s_I s_I | s_J s_J) \Delta f_{IJ}^{\text{AM1}} \end{aligned} \tag{3.51}$$

where the scaling factors f_{IJ}^{AM1} and $\Delta f_{IJ}^{\text{AM1}}$ are given as

$$\begin{aligned} f_{IJ}^{\text{AM1}} &= f_{IJ}^{\text{MNDO}} + \frac{1}{R_{IJ}} \Delta f_{IJ}^{\text{AM1}} \\ \Delta f_{IJ}^{\text{AM1}} &= \sum_{k=1}^{N_I} K_{I_k} \exp\left(-L_{I_k} (R_{IJ} - M_{I_k})^2\right) \\ &\quad + \sum_{k=1}^{N_J} K_{J_k} \exp\left(-L_{J_k} (R_{IJ} - M_{J_k})^2\right) \end{aligned} \tag{3.52}$$

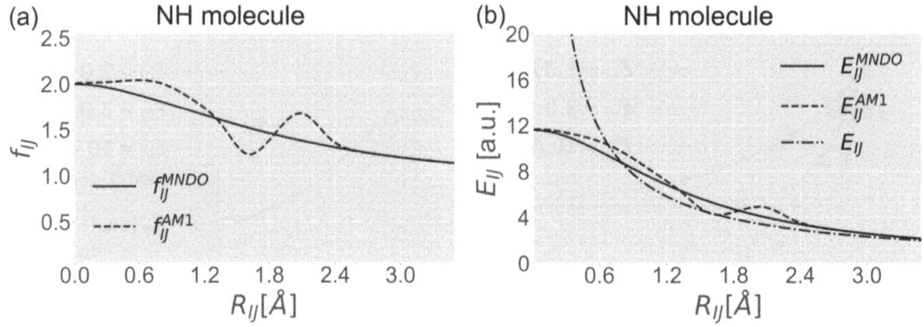

Fig. 3.2 **a** The scaling factors f_{XH}^{MNDO} and f_{XH}^{AM1} for $\xi_I = \xi_J = 1.0$ Å^{-1}, for NH molecule. **b** The core-core repulsion energies E_{IJ}^{MNDO} and E_{IJ}^{AM1} for the same values of $\xi_{I/J}$, E_{IJ} in atomic units (a.u.) is Coulomb's law energy using Eq. (3.36), $(s_I s_I|s_J s_J)$ calculated from Eq. (3.33) with $\langle g_I \rangle = \langle g_J \rangle = 0.83$ a.u., and $Z_I = 7$, $Z_J = 1$. We used three Gaussian functions: $(K_{I_1} = 0.10; K_{I_2} = -0.15; K_{I_3} = 0.15)$, $(K_{J_1} = 0.10; K_{J_2} = -0.15; K_{J_3} = 0.15)$, $(L_{I_1} = 5.0; L_{I_2} = 20.0; L_{I_3} = 20.0)$, $(L_{J_1} = 5.0; L_{J_2} = 20.0; L_{J_3} = 20.0)$, $(M_{I_1} = 0.80; M_{I_2} = 1.60; M_{I_3} = 2.10)$, and $(M_{J_1} = 0.80; M_{J_2} = 1.60; M_{J_3} = 2.10)$

where $N_{I/J}$ is the number of Gaussian functions centred on atom I or J, and L_{I_k} and L_{J_k} are parameters used to adjust the width of Gaussian functions, which can be taken having the same value, as their values are not critical (Leach 2001); the only restriction is that L_{I_k} must be positive to obtain finite values of f_{IJ}^{AM1} for large R_{IJ}. Besides, $\xi_{I/J}$, (K_{I_k}, K_{J_k}) and (M_{I_k}, M_{J_k}) parameters have to be optimised for each atom. The sign of K_{I_k} determines the monotonicity of f_{IJ}^{AM1}; that is, if $K_{I_k} > 0$, then kth Gaussian increases f_{IJ}^{AM1} at a given R_{IJ}; otherwise, if $K_{I_k} < 0$, Gaussian decreases f_{IJ}^{AM1} at a given R_{IJ}.

Figure 3.2a shows the scaling factors f_{XH}^{MNDO} and f_{XH}^{AM1} for $\xi_I = \xi_J = 1.0$ Å^{-1}, for NH molecule. Figure 3.2b presents plots of the core-core repulsion energies E_{IJ}^{MNDO} and E_{IJ}^{AM1} for the same values of $\xi_{I/J}$, E_{IJ} in atomic units (a.u.) is Coulomb's law energy using Eq. (3.36), $(s_I s_I|s_J s_J)$ calculated from Eq. (3.33) with $\langle g_I \rangle = \langle g_J \rangle = 0.83$ a.u., and $Z_I = 7$, $Z_J = 1$. We used three Gaussian functions: $(K_{I_1} = 0.10; K_{I_2} = -0.15; K_{I_3} = 0.15)$, $(K_{J_1} = 0.10; K_{J_2} = -0.15; K_{J_3} = 0.15)$, $(L_{I_1} = 5.0; L_{I_2} = 20.0; L_{I_3} = 20.0)$, $(L_{J_1} = 5.0; L_{J_2} = 20.0; L_{J_3} = 20.0)$, $(M_{I_1} = 0.80; M_{I_2} = 1.60; M_{I_3} = 2.10)$, and $(M_{J_1} = 0.80; M_{J_2} = 1.60; M_{J_3} = 2.10)$.

It can be seen that f_{IJ}^{AM1} varies quite a lot in the range $R_{IJ} \in [0, 3.5]$ Å, and hence E_{IJ}^{AM1} is quite involving function. For instance, $E_{IJ}^{\text{AM1}} > E_{IJ}$ for 0.7 Å $< R_{IJ} <$ 1.6 Å and 1.7 Å $< R_{IJ} <$ 3.0 Å, and, as expected for $R_{IJ} \to \infty$, $E_{IJ}^{\text{AM1}} \to E_{IJ}$. Therefore, as also pointed out (Husch et al. 2018), the scaling factor f_{IJ}^{AM1} function provides a degree of flexibility and it allows for fine-tuning to have a better agreement with experimental data. At the same time, the addition of Gaussian function increases the degree of arbitrariness, as observed originally in Dewar et al. (1985).

Another parametrisation of AM1 was introduced (Rocha et al. 2006), the so-called Recife Model 1 (RM1), using the same methodology as for AM1 (Dewar et al. 1985). Besides, AM1 method has been parametrised for d orbitals under the name

AM1/d (Voityuk and Rösch 2000) using the same philosophy of parametrisation as for generalisation of MNDO to MNDO/d. The scaling factor f_{IJ}^{AM1} is also slightly modified to represent the heavy elements of the periodic table as:

$$\tilde{f}_{IJ}^{\mathrm{AM1}} = 1 + 2\delta_{IJ} \exp\left(-\tilde{\xi}_{IJ} R_{IJ}\right) + \sum_{k=1}^{N_I} K_{I_k} \exp\left(-L_{I_k}(R_{IJ} - M_{I_k})^2\right) \quad (3.53)$$
$$+ \sum_{k=1}^{N_J} K_{J_k} \exp\left(-L_{J_k}(R_{IJ} - M_{J_k})^2\right)$$

where δ_{IJ} and $\tilde{\xi}_{IJ}$ are defined as in Voityuk and Rösch (2000). A re-parametrisation of AM1/d model, named AM1* has also been introduced (Winget et al. 2003; Winget and Clark 2005; Kayi and Clark 2007, 2009a, b, 2010a, b; Kayi 2010; Kayi and Clark 2011).

3.9 The Parametrised Model x

There exist three very popular semi-empirical orbital models, named PMx (for $x = 3, 6, 7$), which are based on the NDDO model (Stewart 1989, 2007, 2012). PM3 Hamiltonian is composed of the same terms as for AM1 model; however, the parameters for the PM3 model are derived using an automated parametrisation procedure (Stewart 1989). Since many parameters of AM1 model were defined by chemical intuition, there are several parameters of PM3 that are significantly different from their values in AM1. It is important to note that both AM1 and PM3 model use the same functional form and they both predict different thermodynamic and structural properties to approximately the same level of accuracy. However, the rotational barrier of the amide bond is too low and in some cases almost does not exist (Leach 2001). It is an open debate over the relative merits of AM1 and PM3 approaches to the parametrisation.

Note that PMx models consider of the parameters (which are considered above as element-dependent) independent on the atom-type and the element-dependent parameters are derived individually (Stewart 1989).

In the PM6 model, the basis set s, p, and d are considered, and there are several feature modifications relative to PM3 and to AM1 (Stewart 2007). In the PM6 model, the core-core repulsion energy is parametrised as follows:

$$E_{IJ}^{\mathrm{PM6}} = \sum_{I=1}^{N_I} \sum_{J=I+1}^{N_J} \left[Z_I Z_J \Gamma_{IJ} f_{IJ}^{\mathrm{PM6}} + 10^{-8}\mathrm{eV} \left(\frac{Z_I^{1/3} Z_J^{1/3}}{R_{IJ}} \right)^{12} \right] \quad (3.54)$$

In Eq. (3.54), there are two modification compared with core-core repulsion energy of MNDO/d model, namely the scaling factor f_{IJ}^{PM6} and the second term added for the pairwise interaction, which has a singularity for $R_{IJ} = 0$, and hence it is a more physical expression. The second term looks like the repulsion term of a Lennard-Jones potential, which was intentionally designed that way (Stewart 2007). The scaling constant 10^{-8} electron volt (eV) is empirically chosen (Stewart 2007). Thus, the second term dominate the expression of E_{IJ}^{PM6} for small R_{IJ}.

The scaling factor f_{IJ}^{PM6} was modelled similar to the scaling factor of AM1/d for heavy atoms (Voityuk and Rösch 2000), and it has the following expression (Husch et al. 2018)

$$
\begin{aligned}
f_{IJ}^{\mathrm{PM6}} = 1 + 2\delta_{IJ} \exp\left(-\tilde{\xi}_{IJ}\left(R_{IJ} + 0.0003\ \text{Å}^{-5} R_{IJ}^6\right)\right) \\
+ \sum_{k=1}^{N_I} K_{I_k} \exp\left(-L_{I_k}(R_{IJ} - M_{I_k})^2\right) \\
+ \sum_{k=1}^{N_J} K_{J_k} \exp\left(-L_{J_k}(R_{IJ} - M_{J_k})^2\right)
\end{aligned}
\tag{3.55}
$$

It can be seen that the number of Gaussian functions is limited to one per atom-type. In comparison to AM1/d approach (see Eq. (3.52)), a term $0.0003\ \text{Å}^{-5} R_{IJ}^6$ is added to the exponential scaling factor, as shown in Eq. (3.55), which makes it to agree well with experimental results of the rare-gas compounds (Stewart 2007).

Figure 3.3 presents the plots of the core-core repulsion energy E_{IJ}^{PM6} and its first $(E_{IJ}^{\mathrm{PM6}}(1))$ and second term $(E_{IJ}^{\mathrm{PM6}}(2))$ in the expression given by Eq. (3.54) for $\xi_{IJ} = 1.0\ \text{Å}^{-1}$ and $\delta_{IJ} = 1.0$ of H_2 molecule, E_{IJ} in atomic units (a.u.) is Coulomb's law energy using Eq. (3.36), $(s_I s_I | s_J s_J)$ calculated from Eq. (3.33) with $\langle g_I \rangle = \langle g_J \rangle = 0.83$ a.u., and $Z_I = 1$, $Z_J = 1$. We used one Gaussian function: $K_{I_1} = K_{J_1} = 0.10$, $L_{I_1} = L_{J_1} = 5.0$, $M_{I_1} = M_{J_1} = 0.80$.

In the PM6 model, there also are special expressions that apply for certain classes of atom-types or for calculation of scaling factor f_{IJ}^{PM6} between certain pairs of atoms, such as C−H, N−H, O−H, C−C, and Si−O (Husch et al. 2018).

The next model is the PM7 model, which is a modification of the PM6 model (Stewart 2012). The most significant modifications relates to the core-core repulsion interaction energy. It was discovered (Stewart 2008) that the two-centre electron repulsion interactions decrease to the exact value in the limit of large separations when semi-empirical orbital models applied to periodic calculations. Thus, the electron repulsion interaction Γ_{IJ} (Eq. (3.31)) is modified using the following approximation:

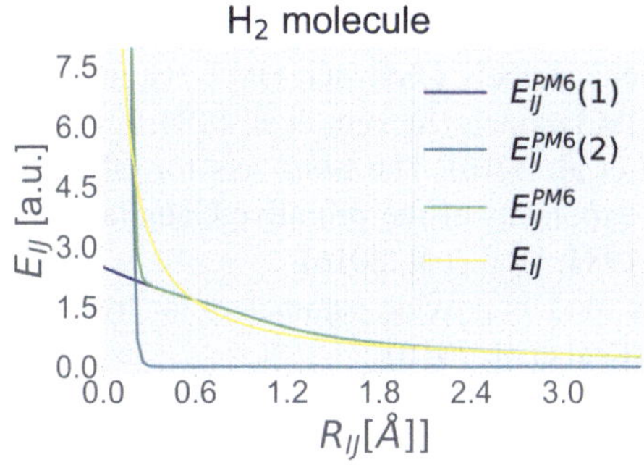

Fig. 3.3 The core-core repulsion energy E_{IJ}^{PM6} and its first ($E_{IJ}^{\text{PM6}}(1)$) and second term ($E_{IJ}^{\text{PM6}}(2)$) in the expression given by Eq. (3.54) for $\xi_{IJ} = 1.0\ \text{Å}^{-1}$ and $\delta_{IJ} = 1.0$ of H_2 molecule, E_{IJ} in atomic units (a.u.) is Coulomb's law energy using Eq. (3.36), $(s_I s_I|s_J s_J)$ calculated from Eq. (3.33) with $\langle g_I\rangle = \langle g_J\rangle = 0.83$ a.u., and $Z_I = 1$, $Z_J = 1$. We used one Gaussian function: $K_{I_1} = K_{J_1} = 0.10$, $L_{I_1} = L_{J_1} = 5.0$, $M_{I_1} = M_{J_1} = 0.80$

$$\Gamma_{IJ}^{\text{PM7}} = k_e e^2 \Bigg[\frac{1}{R_{IJ}} \exp\left(-0.22\ \text{Å}^{-2}\left(R_{IJ} - 7.00\ \text{Å}\right)^2\right) + \left(1 - \exp\left(-0.22\ \text{Å}^{-2}\left(R_{IJ} - 7.00\ \text{Å}\right)^2\right)\right) \times \frac{1}{\sqrt{R_{IJ}^2 + \frac{1}{4}\left(\frac{k_e e^2}{\langle g_I\rangle} + \frac{k_e e^2}{\langle g_J\rangle}\right)^2}} \Bigg] \tag{3.56}$$

where the value 7.00 Å is arbitrary chosen and it belongs to imaginary bond lengths because the typical bond length are smaller (Stewart 2012). The core-core repulsion energy of the PM7 model is:

$$E_{IJ}^{\text{PM7}} = \sum_{I=1}^{N_I} \sum_{J=I+1}^{N_J} \left[Z_I Z_J \Gamma_{IJ}^{\text{PM7}} f_{IJ}^{\text{PM6}} + 10^{-8}\text{eV} \left(\frac{Z_I^{1/3} Z_J^{1/3}}{R_{IJ}} \right)^{12} \right] \tag{3.57}$$

Other empirical corrections are suggested (Korth 2010; Rezác and Hobza 2012) to improve predictions of Eq. (3.57) for hydrogen bonding and dispersion interactions.

3.10 The Orthogonalisation-Corrected Model *x*

The orthogonalisation-corrected model x (OMx, for $x = 1, 2, 3$) activate one s-type basis set function for hydrogen atom and one s-type and three p-type basis set functions for carbon (C), nitrogen (N), oxygen (O), and fluorine (F) (Kolb and

Thiel 1993; Weber and Thiel 2000; Dral et al. 2016). The basis functions consist of three Gaussian functions, namely OM1-3G, OM2-3G, and OM3-3G, respectively, for OM1, OM2, and OM3 models (Stevens et al. 1984; Kolb and Thiel 1993; Weber and Thiel 2000; Dral et al. 2016). The basis sets for OMx-3G are from ECP-3G basis sets, where the exponents of the primitive Gaussian functions are scaled by $(\xi^{Z_I})^2$ (Stevens et al. 1984; Dral et al. 2016).

Note that the OMx models provide parameters for hydrogen, carbon, nitrogen, oxygen, and fluorine (Dral et al. 2016).

3.11 Summary

In this chapter, we discussed the semi-empirical QM models and their parametrisation. The computational effectivity of NDDO semi-empirical models depends on the solution of SCF equations. Nevertheless, the current semi-empirical QM models enable calculations of electronic energies several orders faster than pure QM methods, such HF and DFT models, discussed in Chap. 1. That is due to reduction on the number of the electron-electron repulsion integrals to be computed; however, this results in the essential errors in the electron-electron repulsion integrals for the Löwdin orthogonalised basis.

Therefore, it is recommended to combine the NDDO approximation with other electronic approximations, such as the one-electron matrix and the core-core repulsion energy, for a more accurate semi-empirical QM model (Husch et al. 2018).

References

R.C. Bingham, M.J.S. Dewar, D.H. Lo, Ground states of molecules. XXV. MINDO/3. An improved version of the MINDO semi-empirical SCFMO method. J. Am. Chem. Soc. **97**, 1285–1293 (1975a)

R.C. Bingham, M.J.S. Dewar, D.H. Lo, Ground states of molecules. XXVI. MINDO/3. Calculations for hydrocarbons. J. Am. Chem. Soc. **97**, 1294–1301 (1975b)

R.C. Bingham, M.J.S. Dewar, D.H. Lo, Ground states of molecules. XXVII. MINDO/3. Calculations for CHON species. J. Am. Chem. Soc. **97**, 1302–1306 (1975c)

R.C. Bingham, M.J.S. Dewar, D.H. Lo, Ground states of molecules. XXVIII. MINDO/3. Calculations for compounds containing carbon, hydrogen, fluorine and chlorine. J. Am. Chem. Soc. **97**, 1307–1310 (1975d)

M.J.S. Dewar, W. Thiel, A semiempirical model for the two-center repulsion integrals in the NDDO approximation. Theor. Chim. Acta **46**, 89–104 (1976)

M.J.S. Dewar, W. Thiel, Ground states of molecules. 38. the MNDO method. Approximations and parameters. J. Am. Chem. Soc. **99**, 4899–4907 (1977a)

M.J.S. Dewar, W. Thiel, Ground states of molecules. 39. MNDO results for molecules containing hydrogen, carbon, nitrogen, and oxygen. J. Am. Chem. Soc. **99**, 4907–4917 (1977b)

M.J.S. Dewar, E.G. Zoebisch, E.F. Healy, J.J.P. Stewart, Development and use of quantum mechanical molecular models. 76. AM1: a new general purpose quantum mechanical molecular model. J. Am. Chem. Soc. **107**, 3902–3909 (1985)

P.O. Dral, X. Wu, L. Spörkel, A. Koslowski, W. Weber, R. Steiger, M. Scholten, W. Thiel, Semiempirical quantum-chemical orthogonalization-corrected methods: theory, implementation, and parameters. J. Chem. Theory Comput. **12**, 1082–1096 (2016)
T. Husch, M. Reihe, Comprehensive analysis of the neglect of diatomic differential overlap approximation. J. Chem. Theory Comput. **14**(10), 5169–5179 (2018)
T. Husch, A.C. Vaucher, M. Reiher, Semiempirical molecular orbital models based on the neglect of diatomic differential overlap approximation. Int. J. Quantum Chem. **118**, e25799 (2018)
H. Kayi, AM1* parameters for gold. J. Mol. Model. **16**, 1029–1038 (2010)
H. Kayi, T. Clark, AM1* parameters for copper and zinc. J. Mol. Model. **13**, 965–979 (2007)
H. Kayi, T. Clark, AM1* parameters for vanadium and chromium. J. Mol. Model. **15**, 1253–1269 (2009a)
H. Kayi, T. Clark, AM1* parameters for bromine and iodine. J. Mol. Model. **15**, 295–308 (2009b)
H. Kayi, T. Clark, AM1* parameters for manganese and iron. J. Mol. Model. **16**, 1109–1126 (2010a)
H. Kayi, T. Clark, AM1* parameters for cobalt and nickel. J. Mol. Model. **16**, 29–47 (2010b)
H. Kayi, T. Clark, AM1* parameters for palladium and silver. J. Mol. Model. **11**, 2585–2600 (2011)
M. Kolb, W. Thiel, Beyond the MNDO model: methodical considerations and numerical results. J. Comput. Chem. **14**, 775–789 (1993)
M. Korth, Third-generation hydrogen-bonding corrections for semiempirical QM methods and force fields. J. Chem. Theory Comput. **6**, 3808–3816 (2010)
A.R. Leach, *Molecular Modelling, Principles and Applications*, 2nd edn. (Prentice Hall, Pearson Education Limited, Edingburgh Gate, 2001)
J.A. Pople, D.L. Beveridge, P.A. Dobosh, Approximate self-consistent molecular orbital theory. V Intermediate neglect of differential overlap. J. Chem. Phys. **47**(6), 2026–2033 (1967)
J.A. Pople, D.P. Santry, G.A. Segal, Approximate self-consistent molecular orbital theory. II Calculations with complete neglect of differential overlap. J. Chem. Phys. **43**, 136 (1965a)
J.A. Pople, D.P. Santry, G.A. Segal, Approximate self-consistent molecular orbital theory. I. Invariant procedures. J. Chem. Phys. **43**(10), 129–135 (1965b)
J.A. Pople, G.A. Segal, Approximate self-consistent molecular orbital theory. III CNDO results for ab_2 and ab_3 systems. J. Chem. Phys. **44**(9), 3289–3296 (1966)
J. Rezác, P. Hobza, Advanced corrections of hydrogen bonding and dispersion for semiempirical quantum mechanical methods. J. Chem. Theory Comput. **8**, 141–151 (2012)
G.B. Rocha, R.O. Freire, A.M. Simas, J.J.P. Stewart, RM1: a reparameterization of AM1 for H, C, N, O, P, S, F, Cl, Br, and I. J. Comput. Chem. **27**, 1101–1111 (2006)
W.J. Stevens, H. Basch, M. Krauss, Compact effective potentials and efficient shared-exponent basis sets for the first- and second-row atoms. J. Chem. Phys. **81**, 6026–6033 (1984)
J.J.P. Stewart, Optimization of parameters for semiempirical methods I. Method. J. Comput. Chem. **10**, 209–220 (1989)
J.J.P. Stewart, Optimization of parameters for semiempirical methods IV: extension of MNDO, AM1, and PM3 to more main group elements. J. Mol. Model. **10**, 155–164 (2004)
J.J.P. Stewart, Optimization of parameters for semiempirical methods V: modification of NDDO approximations and application to 70 elements. J. Mol. Model. **13**, 1173–1213 (2007)
J.J.P. Stewart, Application of the PM6 method to modeling the solid state. J. Mol. Model. **14**, 499–535 (2008)
J.J.P. Stewart, Optimization of parameters for semiempirical methods VI: more modifications to the NDDO approximations and re-optimization of parameters. J. Mol. Model. **19**, 1–32 (2012)
W. Thiel, A.A. Voityuk, Extension of the MNDO formalism to *d*-orbitals: integral approximations and preliminary numerical results. Theor. Chim. Acta **81**, 391–404 (1991)
W. Thiel, A.A. Voityuk, Erratum: Extension of the MNDO formalism to *d*-orbitals: integral approximations and preliminary numerical results. Theor. Chim. Acta **93**, 315 (1996)
A.A. Voityuk, N. Rösch, AM1/d parameters for molybdenum. J. Phys. Chem. A **104**, 4089–4094 (2000)
W. Weber, W. Thiel, Orthogonalization corrections for semiempirical methods. Theor. Chem. Acc. **103**, 495–506 (2000)

P. Winget, T. Clark, AM1* parameters for aluminum, silicon, titanium and zirconium. J. Mol. Model. **11**, 439–456 (2005)

P. Winget, A.H.C. Horn, C. Selcuki, B. Martin, T. Clark, AM1* parameters for phosphorus, sulfur and chlorine. J. Mol. Model. **9**, 408–414 (2003)

X. Wu, P.O. Dral, A. Koslowski, W. Thiel, Big data analysis of ab initio molecular integrals in the neglect of diatomic differential overlap approximation. J. Comput. Chem. **40**(4), 638–649 (2019)

Chapter 4
Machine Learning Quantum Mechanics Computer Simulations

The chapter aims to discuss machine learning quantum mechanics methods and their use in molecular dynamics simulations.

This chapter introduces the machine learning quantum mechanics methods and their use in computer simulations. In particular, the focus is on the use of machine learning approach in predicting the potential electronic energy surface and its gradient for QM molecular dynamics simulations.

4.1 Introduction

In ab-initio QM molecular dynamics simulations, QM calculations of potential electronic energy surface $E_0(\mathbf{R})$ is computationally very costly because of the expensive SCF approach (see Algorithm 1 and Algorithm 2, Chap. 1) (Andreoni and Curioni 2000; Marx and Hutter 2000, 2009; Geudtner et al. 2012). An alternative approach to semi-empirical methods, described in Chap. 3, is to use the machine-learning (ML) approach (Huang et al. 2015), as also discussed in Schütt et al. (2017) and Noé et al. (2020). There are several ML approaches used for the prediction of the exchange and correlation energies characterised by the quantum mechanical features at a mean-field level, such as HF and DFT (Welborn et al. 2018; Cheng et al. 2019a, b; Fabrizio et al. 2019; Qiao et al. 2020; Chen et al. 2020; Dick and Fernandez-Serra 2020; Brunken and Reiher 2020). These ML models predict the electron density (Cheng et al. 2019b; Fabrizio et al. 2019), the wave function (Welborn et al. 2018; Cheng et al. 2019a; Qiao et al. 2020) or ground state energy (Chen et al. 2020; Brunken and Reiher 2020; Dick and Fernandez-Serra 2020). In particular, the molecular-orbital based

H. Kamberaj, *Computer Simulations in Molecular Biology*, Scientific Computation,
https://doi.org/10.1007/978-3-031-34839-6_4

ML approach, the so-called OrbNet (Qiao et al. 2020) or MOB-ML (Welborn et al. 2018; Cheng et al. 2019b), is used to calculate molecular properties from the local molecular orbitals (Welborn et al. 2018; Cheng et al. 2019a). The feature vectors included the quantum mechanics Hartree–Fock operator matrix elements, $F_{\mu\nu}$, Coulomb $J_{\mu\nu}$, and exchange $K_{\mu\nu}$. Thus, the OrbNet has a detailed quantum mechanics information and it includes both diagonal and off-diagonal elements within the features; besides, it considers the intra- and inter-atom elements, including the intra- and inter-shell elements, and thus the long-range contributions are present. Furthermore, the OrbNet uses Graphical Neural Network (GNN) approach for the ML optimisation procedure, described in Qiao et al. (2020).

Other ML approaches that use the quantum mechanics property of molecules as feature input vectors include the so-called NeuralXC (Dick and Fernandez-Serra 2020) and DeePHF (Chen et al. 2020). The electronic structure calculations are obtained by quantum mechanics properties at low cost. In contrast to OrbNet, both NeuralXC and DeePHF are based on the atomic orbitals and utilise the electronic density. For NeuralXC the calculations are performed using DFT with cc-pVDZ (Woon and Dunning Jr 1993) atomic orbital basis sets with program PySCF (Sun et al. 2018b) (as described in Chen et al. 2020), which are considered low-level theory, and it uses shell-averages of the atomic orbitals; whereas DeePHF uses the electronic density and orbitals obtained from HF theory. Besides, in contrast to OrbNet, the DeePHF approach does not include interactions between different shells of the same atom and a weighting function is needed a priory based on the inter-atomic distances. Furthermore, the rotational invariant condition of the features is applied differently; for example, in NeuralXC, the rotational invariance is enforced by calculating the trace of the local density matrix such that information content is not preserved. In the DeePHF, the rotational invariance of the features is implemented based on the eigenvalues of the local density matrix, whereas in OrbNet, that is enforced using the symmetric adapted atomic orbitals such that it preserves the information content.

Moreover, there are differences between the three approaches regarding the ML regression methods employed. In NeuralXC (Dick and Fernandez-Serra 2020) and DeePHF (Chen et al. 2020) approaches, the dense neural network is utilised (Behler and Parrinello 2007). Interestingly, in the framework of the quantum mechanics predictions, other applications have also been developed using the atom or geometry feature representations and either kernel-based or neural-network ML approaches (Lubbers et al. 2018; Kamath et al. 2018; Gastegger et al. 2018; Unke and Meuwly 2018; Collins et al. 2018; Zhou et al. 2019; Yang et al. 2019).

4.2 Electronic Potential-Energy Surface Machine Learning Approach

In the supervised ML, an artificial neural network (ANN) with $\mathcal{K}$ hidden layers, the output Y_α is defined as

$$Y_\alpha = f_{\mathbf{P}}^{\text{linear}} \underbrace{\left(\sum_{\beta=1}^{L_O} f_{\mathbf{P}}^{\text{ann}} \underbrace{\left(\sum_{l_{\mathcal{K}}=1}^{L_{\mathcal{K}}} f_{\mathbf{P}}^{\text{ann}} \underbrace{\left(\cdots f_{\mathbf{P}}^{\text{ann}} \underbrace{\left(\sum_{l_2=1}^{L_2} f_{\mathbf{P}}^{\text{ann}} \underbrace{\left(\sum_{l_1=1}^{L_1} f_{\mathbf{P}}^{\text{ann}} \underbrace{\left(\sum_{j=1}^{L_i} X_j W_{jl_1} + b_{l_1} \right)}_{\text{input layer}} W_{l_1 l_2} + b_{l_2} \right)}_{\text{1st hidden layer}} \cdots \right)}_{\text{2nd hidden layer}} \cdots \right)}_{\cdots} W_{l_{\mathcal{K}}\beta} + b_\beta \right)}_{(\mathcal{K})\text{th hidden layer}} W_{\beta\alpha} + b_\alpha \right)}_{\text{output layer}} \tag{4.1}$$

where $f_{\mathbf{P}}^{\text{linear}}$ is a linear function (typically, $f(x) = x$) and $f_{\mathbf{P}}^{\text{ann}}$ are non-linear function with existing analytical first derivative. Equation 4.1 determined a machine learning transformation of the input feature descriptor vector according to:

$$M_{\mathbf{P}} : \mathbf{X} \rightarrow \mathbf{Y} \tag{4.2}$$

In general, the ANN is characterised by an input vector $\mathbf{X}$ of dimension L_i, $\mathcal{K}$ hidden layers of $l_{L_1}^{(1)}, l_{L_2}^{(2)}, \ldots, l_{L_{\mathcal{K}}}^{(\mathcal{K})}$ neurons each, and an output vector $\mathbf{Y}$ of dimension m. Here, $\mathbf{W}$ and $\mathbf{b}$ are considered as free parameters, which need to be optimised for a given training data used as inputs and given outputs, which are known. The so-called loss function is minimised to optimise these parameters (Huang et al. 2015; White 2022):

$$L(\mathbf{W}, \mathbf{b}) = \mathcal{F}\left(\mathbf{Y}^0, \mathbf{Y}\right) \tag{4.3}$$

where $\mathbf{Y}^0$ represents the true output vector and $\mathcal{F}$ is a convex function (such as mean square error or mean absolute error). For that, the gradients of $L(\mathbf{W}, \mathbf{b})$ for $\mathbf{W}$ and $\mathbf{b}$ are calculated:

$$\begin{aligned} \Delta \mathbf{W} &= -\left(\frac{\partial L(\mathbf{W}, \mathbf{b})}{\partial \mathbf{W}}\right)_{\mathbf{b}} \\ \Delta \mathbf{b} &= -\left(\frac{\partial L(\mathbf{W}, \mathbf{b})}{\partial \mathbf{b}}\right)_{\mathbf{W}} \end{aligned} \tag{4.4}$$

A common problem of the machine learning approaches is the over-fitting (Huang et al. 2015; Ciriano and Bender 2018), which is avoided by the following regularisation terms:

$$\begin{aligned} \Delta' \mathbf{W} &= \gamma_w (\Delta \mathbf{W} + \gamma_1 \mathbf{W}) \\ \Delta' \mathbf{b} &= \gamma_w (\Delta \mathbf{b} + \gamma_1 \mathbf{b}) \end{aligned} \tag{4.5}$$

where γ_w is called learning rate for the gradient and γ_1 is called the regulation strength.

The aim of a ML approach in calculating electronic potential-energy surface is in determining the non-linear function $M_{\mathbf{P}}$ such that (White 2022)

$$M_{\mathbf{P}} : \mathcal{T}(\mathbf{R}) \to \mathcal{E}(\mathbf{R}) \tag{4.6}$$

where $\mathcal{E}(\mathbf{R})$ represents the estimated function of either the electronic potential-energy surface function $\bar{E}_0^{\mathrm{HF/KS}}(\mathbf{R})$ or the difference between the electronic potential-energy surface functions calculated from two methods, such as a cheap (semi-empirical $E_0^{\text{semi-emp}}(\mathbf{R})$) and an expensive (Hartree–Fock or Kohn–Sham $E_0^{\mathrm{HF/KS}}(\mathbf{R})$) (Böselt et al. 2021)

$$\mathcal{E}(\mathbf{R}) = E_0^{\mathrm{HF/KS}}(\mathbf{R}) - E_0^{\text{semi-emp}}(\mathbf{R}) \tag{4.7}$$

and thus

$$\bar{E}_0^{\mathrm{HF/KS}}(\mathbf{R}) = E_0^{\text{semi-emp}}(\mathbf{R}) + \mathcal{E}(\mathbf{R}) \tag{4.8}$$

Here, $E_0^{\text{semi-emp}}(\mathbf{R})$ is exactly calculated within the accuracy of the semi-empirical approach used.

$\mathcal{T}$ is a transformation function of the nuclei configuration $\mathbf{R}$ (which creates the inputs for the ML):

$$\mathcal{T} : \mathbf{R} \to \mathbf{X} \tag{4.9}$$

P denotes the set of ML parameters to be optimised (Böselt et al. 2021). Then, the estimated electronic energy $\mathcal{E}(\mathbf{R})$ is used with MD simulations to propagate the nuclei coordinates as described above.

Furthermore, high-dimensional neural network potentials (HDNNPs) (Behler 2011, 2015) are suggested to enforce the correct parity and to reduce the training set. In HDNNP approach, multiple ANN (or NNP, standing for neural network potential) are set up, one for each atom type in the system (such as C, N, H, O, and others), and the energies $\mathcal{E}_i$ obtained from individual NNP are summed up to give the total estimated energy $\mathcal{E}(\mathbf{R}) = \sum_t \mathcal{E}_t(\{\mathbf{R}_t\})$ (where $\{\mathbf{R}_t\}$ denotes the set of atoms of the same type). That ensures that the total energy is invariant under coordinates swapping of the atoms of the same type. Moreover, the energy of the system becomes also invariant under overall translation and rotation by introducing the following symmetric functions for $\mathcal{T}$ (Behler 2011, 2015; Böselt et al. 2021):

$$X_{i,\text{radian}} = \mathcal{T}_{\text{radial}}(\mathbf{R}) \equiv \sum_{j=1}^{M_t} \exp\left(-\gamma_2(R_{ij} - \gamma_3)^2\right) F_C(R_{ij}; R_{\text{cut}}) \tag{4.10}$$

$$\begin{aligned} X_{i,\text{angle}} = \mathcal{T}_{\text{angle}}(\mathbf{R}) \equiv 2^{\gamma_4 - 1} & \sum_{j \neq k \neq i = 1}^{M_t} (1 - \cos(\theta_{ijk} - \gamma_5))^{\gamma_4} \\ & \times \exp\left(-\gamma_6 \left(\frac{R_{ij} - R_{ik}}{2} - \gamma_7\right)^2\right) \\ & \times F_C(R_{ij}; R_{\text{cut}}) F_C(R_{ik}; R_{\text{cut}}) \end{aligned}$$

where $i = 1, 2, \ldots, M_t$ number of atoms of the same type $t = 1, 2, \ldots$, $R_{ij} = \mid \mathbf{R}_i - \mathbf{R}_j \mid$, and θ_{ijk} is the angle between atoms $i - j - k$. $\gamma_1, \ldots, \gamma_7$ are a new set of adjustable parameters. In Eq. 4.10, F_C is a cutoff function, which equals one if there are no periodic boundary conditions; otherwise (Böselt et al. 2021):

$$F_C(R_{ij}; R_{\text{cut}}) = \begin{cases} 0.5\left[1 + \cos\left(\dfrac{\pi R_{ij}}{R_{\text{cut}}}\right)\right], & R_{ij} \leq R_{\text{cut}} \\ 0, & R_{ij} > R_{\text{cut}} \end{cases} \tag{4.11}$$

Note that other descriptors can also be used, such as the so-called Coulomb matrix, **C**, to encode the molecular features, containing both the geometrical information of the three-dimensional structure and the type of atom (Rupp et al. 2012). For any two atoms i and j in a given input molecule, the matrix element C_{ij} is as follows:

$$C_{ij} = \begin{cases} \dfrac{Z_i^{2.4}}{2}, & i = j \\ \dfrac{Z_i Z_j}{r_{ij}}, & i \neq j \end{cases} \tag{4.12}$$

where Z_i is the atomic number of the ith atom and r_{ij} is the distance between the atoms i and j. The fingerprint represented by the Coulomb matrix, **C**, has some advantages, such as it takes into account the three-dimensional molecular structure, and it is invariant under rotation and translation of the structure. To calculate **C** for a given molecular structure, we need the nuclear charges for each atom and the Cartesian coordinates of the atomic positions taken from the equilibrium geometry. However, note that **C** is not invariant under the permutations of the atom order in a molecule. Therefore, the spectrum of eigenvalues of matrix **C** can be used as a fingerprint of the molecule, since they are invariant under both rotation/translation and permutations of the rows and columns.

A third feature descriptor that can be employed is the sum over bonds, which is a numerical descriptor representing the vector of bond types present in a molecule, similar to Yang et al. (2019). If N_b is the number of unique bonds in the dataset of the compounds studied, then a vector with dimensions N_b is constructed for each molecule with entry either zeros or the integers giving the frequency of appearance for each bond type in molecular structure. This fingerprint descriptor vector has a unique length within the dataset. Then, the vector descriptor of the sum over bonds is concatenated at the end of the Coulomb matrix descriptor.

Since for MD simulations the gradient of energy needs to be calculated, it may be more efficient to learn about the gradients (Chmiela et al. 2017, 2018). In that case, the loss function, Eq. 4.3, is written as:

$$L(\mathbf{W}, \mathbf{b}) = \mathcal{F}(E_0(\mathbf{R}), \mathcal{E}(\mathbf{R})) + \mathcal{F}(\nabla_{\mathbf{R}} E_0(\mathbf{R}), \nabla_{\mathbf{R}} \mathcal{E}(\mathbf{R})) \tag{4.13}$$

4.3 Electron Density Machine Learning Approach

Interestingly, these feature descriptors have nothing to do with quantum mechanics information of molecular systems. Although, other feature descriptors of molecules, which are based on the local density matrix, have been suggested for ML prediction of ground state energy of electrons (Unke and Meuwly 2018; Fabrizio et al. 2019; Chen et al. 2020). The information about the local environment of an atom μ of the molecule up to a cutoff distance R_{cut} can be described by density function:

$$\rho_\mu(\mathbf{r}) = \sum_{\nu, |\mathbf{r}_\nu| \leq R_{\text{cut}}} Z_\nu \delta(|\ \mathbf{r} - \mathbf{r}_\nu\ |) = |\ \psi_\mu(\mathbf{r})\ |^2 \tag{4.14}$$

where $\mathbf{r}$ is the relative vector to atom μ, ψ_μ is the atomic orbital, and δ is the delta function.

For that the atomic basis set functions are introduced:

$$\eta^\mu_{nlm}(\mathbf{r}) = R^\mu_n(r; R_{\text{cut}}) Y^\mu_{lm}(\theta, \phi) \tag{4.15}$$

where $R_n^\mu(r; R_{\text{cut}})$ is the radial function of atom μ indexed by $n \in [0, \mathcal{N} - 1]$ and $Y_{lm}^\mu(\theta, \phi)$ are the spherical harmonics of atom μ of degree $l \in [0, \mathcal{L} - 1]$ and order $m \in [-l, l]$. (r, θ, ϕ) are the spherical coordinates of atom μ. Projection of atomic orbital ψ_i onto the basis set function centred at atom μ is given as

$$C_{inlm}^\mu = \int \eta_{nlm}^\mu(\mathbf{r})\psi_i(\mathbf{r})\,d\mathbf{r} = \int \eta_{nlm}^\mu(\mathbf{r}) \sum_{\nu=1}^{K} c_{i\nu}\phi_\nu(\mathbf{r})\,d\mathbf{r} \qquad (4.16)$$
$$= \sum_{\nu=1}^{K} c_{i\nu} \int \eta_{nlm}^\mu(\mathbf{r})\phi_\nu(\mathbf{r})\,d\mathbf{r}$$

where ϕ_ν are the orbital basis set functions. The density matrix elements in the basis set $\{\eta_{nlm}^\mu\}$ is defined as

$$D_{n_1l_1m_1;n_2l_2m_2}^{\alpha\beta} = \sum_i C_{in_1l_1m_1}^\alpha C_{in_2l_2m_2}^\beta \qquad (4.17)$$
$$= \sum_i \left[\sum_{j=1}^{K} c_{ij} \int \eta_{n_1l_1m_1}^\alpha(\mathbf{r}_1)\phi_j(\mathbf{r}_1)\,d\mathbf{r}_1\right]\left[\sum_{k=1}^{K} c_{ik} \int \eta_{n_2l_2m_2}^\beta(\mathbf{r}_2)\phi_k(\mathbf{r}_2)\,d\mathbf{r}_2\right]$$

Then, the local density matrix can be defined as

$$\left[D_{nl}^\alpha\right]_{m_1m_2} = D_{nlm_1;nlm_2}^{\alpha\alpha} = \sum_i C_{inlm_1}^\alpha C_{inlm_2}^\alpha \qquad (4.18)$$

The eigenvalues of local density matrix are determined to enforce the rotation and translation invariance of the basis set $\{\eta_{nlm}^\mu\}$:

$$\mathbf{X}_{nl}^\alpha = (\lambda_1, \lambda_2, \ldots, \lambda_{2l+1}) \qquad (4.19)$$

where λ_i are the eigenvalues of matrix $\left[D_{nl}^\alpha\right]_{m_1m_2}$. Thus, for every n and l we obtain the feature descriptor vector for atom α, $\mathbf{X}^\alpha$. Then, after determining the set of parameters $\mathbf{P} = (\mathbf{W}, \mathbf{b})$ using a ML approach, the ground state energy is defines as

$$E_0 = \sum_{t=1}^{N_t} E^\alpha = \sum_{t=1}^{N_t} \sum_{\alpha=1}^{L_o} E_\alpha^t \qquad (4.20)$$

where t runs over all atom types, N_t, and α runs over all number of atoms of the same type t (which equals the number of outputs of NNP, determined using Eq. 4.1).

Further extensions have also been suggested (Chen et al. 2020) to improve the approach using the kernel functions $f_k(\hat{h}_i^{\text{HF}}(1))$, where $\hat{h}_i^{\text{HF}}(1)$ is the Hartree–Fock one-electron Hamiltonian operator. Then,

$$\left[D_{nlk}^{\alpha}\right]_{m_1 m_2} = \sum_i C_{inlm_1}^{\alpha} C_{inlm_2}^{\alpha} f_k(\varepsilon_i) \tag{4.21}$$

where $\varepsilon_i = \langle \psi_i | \hat{h}_i^{\mathrm{HF}}(1) | \psi_i \rangle$. For example, $f_0(\varepsilon_i) = 1$, $f_1(\varepsilon_i) = \varepsilon_i$, and $f_2(\varepsilon_i) = e^{\varepsilon_i}$ (Chen et al. 2020).

Moreover, non-local information can be taken into account by determining non-local density matrix

$$\left[D_{nlk}^{\alpha\beta}\right]_{m_1 m_2} = w(r_{\alpha\beta}) \sum_i C_{inlm_1}^{\alpha} C_{inlm_2}^{\beta} f_k(\varepsilon_i) \tag{4.22}$$

where $w(r_{\alpha\beta})$ are weight functions of the distance between two atoms, which can be suggested to be of the form given by Eq. 4.12. Ensuring that the eigenvalues are real, the following non-local density matrix can be introduced

$$D_{nlk}^{\alpha;\text{non-local}} = \frac{1}{2} \sum_{\beta} \left(D_{nlk}^{\alpha\beta} + D_{nlk}^{\beta\alpha} \right) \tag{4.23}$$

4.4 Molecular Orbital Machine Learning Approach

Molecular orbital machine learning (MOB-ML) uses molecular orbitals to predict correlation energies (Welborn et al. 2018; Cheng et al. 2019a, b; Husch et al. 2021). Therefore, MOB-ML estimates the correlation energy at the cost of calculation of the molecular orbitals. Based on quantum calculations, the correlation energy is sum of the contributions from pairs of occupied orbitals ψ_i and ψ_j:

$$E_0^{\mathrm{corr}} = \sum_{i=1}^{n} \sum_{j=1}^{n} \varepsilon_{ij} \tag{4.24}$$

The MOB-ML aims to determine the non-linear function $M_{\mathbf{P}}$ such that:

$$M_{\mathbf{P}} : \mathcal{T}(\psi_i^v, \psi_j^v, \psi_i^o, \psi_j^o) \rightarrow \tilde{\varepsilon}_{ij} \tag{4.25}$$

where $\tilde{\varepsilon}_{ij}$ are ML predicted values of ε_{ij} and $\mathcal{T}$ is a transformation of the pairs of molecular orbitals (ψ_i^v, ψ_j^v) and (ψ_i^o, ψ_j^o) to create the feature descriptor matrix (Cheng et al. 2019b; Husch et al. 2021):

$$\mathbf{f}_{ij} = \mathcal{T}(\psi_i^v, \psi_j^v, \psi_i^o, \psi_j^o) \tag{4.26}$$

Here, $\mathbf{f}_{ij}$ are the elements of the Fock (**F**), Coulomb (**J**), and exchange (**K**) matrices between ψ_i^v and ψ_j^v (the set of the so-called virtual orbitals) and ψ_i^o and ψ_j^o (the set of the so-called occupied orbitals):

$$
\begin{aligned}
\mathbf{f}_{ij} = \big(&F_{ii}, F_{ij}, F_{jj}, \mathbf{F}_i^o, \mathbf{F}_j^o, \mathbf{F}_{ij}^{vv}, \\
&J_{ii}, J_{ij}, J_{jj}, \mathbf{J}_i^o, \mathbf{J}_j^o, \mathbf{J}_i^v, \mathbf{J}_j^v, \mathbf{J}_{ij}^{vv}, \\
&K_{ij}, \mathbf{K}_i^o, \mathbf{K}_j^o, \mathbf{K}_i^v, \mathbf{K}_j^v, \mathbf{K}_{ij}^{vv}\big)
\end{aligned}
\tag{4.27}
$$

Here, the terms of the matrix elements denoted with superscripts ^{o} represent the rows of the corresponding occupied-occupied block, these with superscripts ^{v} represent the rows of the corresponding occupied-virtual block, and the superscripts ^{vv} represent the corresponding virtual-virtual block. Furthermore, the virtual-virtual block is represented by only the upper triangle part and the diagonal elements of matrix **K** are omitted (Cheng et al. 2019b). (Note that diagonal elements of matrices **K** and **J** are equal.) Moreover, two independent machine learning optimisation are performed, respectively, for $\mathbf{f}_{ii}$ and $\mathbf{f}_{ij}$:

$$
\tilde{\varepsilon}_{ij} = \begin{cases} M_{\mathbf{P}}^{(1)}(\mathbf{f}_{ii}), \ i = j \\ M_{\mathbf{P}}^{(2)}(\mathbf{f}_{ij}), \ i \neq j \end{cases}
\tag{4.28}
$$

where only the upper triangle part of $\mathbf{f}_{ij}$ is considered. There exist several technical improvement of the feature selection to increase the training efficiency (Cheng et al. 2019b), such as local molecular orbital selection, occupied local molecular orbital symmetrisation, local molecular orbital sorting, orbital localisation, and feature selection.

4.5 Summary

We introduced the ML quantum mechanics methods used in computer simulations. In particular, the focus is on the use of ML approach in predicting the potential electronic energy surface and its gradient for QM molecular dynamics simulations.

Furthermore, ML approaches used for the prediction of the exchange and correlation energies characterised by the quantum mechanical features at a mean-field level (such as HF and DFT) were described for predicting the electron density.

Moreover, the ML approaches used to predict the wave function or ground state energy were discussed. In particular, the molecular-orbital based ML approach (OrbNet and MOB-ML) calculates molecular properties from the local molecular orbitals.

References

W. Andreoni, A. Curioni, New advances in chemistry and material science with CPMD and parallel computing. Parallel Comput. **26**, 819–842 (2000)

J. Behler, Atom-centered symmetry functions for constructing high-dimensional neural network potentials. J. Chem. Phys. **134**, 074106 (2011)

J. Behler, Constructing high-dimensional neural network potentials: a tutorial review. Int. J. Quantum Chem. **115**, 1032–1050 (2015)

J. Behler, M. Parrinello, Generalized neural-network representation of high-dimensional potential-energy surfaces. Phys. Rev. Lett. **98**, 146401 (2007)

L. Böselt, M. Thürlemann, S. Riniker, Machine learning in QM/MM molecular dynamics simulations of condensed-phase systems. J. Chem. Theory Comput. **17**, 2641–2658 (2021)

C. Brunken, M. Reiher, Self-parametrizing system-focused atomistic models. J. Chem. Theory Comput. **16**, 1646–1665 (2020)

Y. Chen, L. Zhang, H. Wang, E. Weinan, Ground state energy functional with Hartree-Fock efficiency and chemical accuracy. J. Phys. Chem. A **124**(35), 7155–7165 (2020)

L. Cheng, N.B. Kovachki, M. Welborn, T.F. Miller III., Regression clustering for improved accuracy and training costs with molecular-orbital-based machine learning. J. Chem. Theory Comput. **15**, 6668–6677 (2019a)

L. Cheng, M. Welborn, A.S. Christensen, T.F. Miller III., A universal density matrix functional from molecular orbital-based machine learning: transferability across organic molecules. J. Chem. Phys. **150**, 131103 (2019b)

S. Chmiela, A. Tkatchenko, H.E. Sauceda, I. Poltavsky, K.T. Schütt, K.R. Müller, Machine learning of accurate energy conserving molecular force fields. Sci. Adv. **3**, e1603015 (2017)

S. Chmiela, H.E. Sauceda, K.R. Müller, A. Tkatchenko, Towards exact molecular dynamics simulations with machine-learned force fields. Nat. Commun. **9**, 3887 (2018)

I.C.-Ciriano, A. Bender, Deep confidence: a computationally efficient framework for calculating reliable prediction errors for deep neural networks. J. Chem. Inf. Model. **59**(3), 1269–1281 (2018)

C.R. Collins, G.J. Gordon, O.A. von Lilienfeld, D.J. Yaron, Constant size descriptors for accurate machine learning models of molecular properties. J. Chem. Phys. **148**, 241718–11 (2018)

S. Dick, M. Fernandez-Serra, Machine learning accurate exchange and correlation functionals of the electronic density. Nat. Commun. **11**, 3509 (2020)

A. Fabrizio, A. Grisafi, B. Meyer, M. Ceriotti, C. Corminboeuf, Electron density learning of non-covalent systems. Chem. Sci. **10**, 9424–9432 (2019)

M. Gastegger, L. Schwiedrzik, M. Bittermann, F. Berzsenyi, P. Marquetand, wACSF-Weighted atom-centered symmetry functions as descriptors in machine learning potentials. J. Chem. Phys. **148**, 241709–11 (2018)

G. Geudtner, P. Calaminici, J. Carmona-Espindola, J.M. del Campo, V.D. Dominguez-Soria, R. Flores-Morena, G.U. Gamboa, A. Goursot, A.M. Köster, J.U. Reveles, T. Mineva, J.M. Vasquez-Perez, A. Vela, B. Zuniga-Gutierrez, D.R. Salahub, demon2k. WIREs Comput. Mol. Sci. **2**, 548–555 (2012)

G. Huang, G. Bin Huang, S. Song, K. You, Trends in extreme learning machines: a review. Neural Netw. **61**, 32–48 (2015)

T. Husch, J. Sun, L. Cheng, S.J.R. Lee, T.F. Miller III., Improved accuracy and transferability of molecular-orbital-based machine learning: organics, transition-metal complexes, non-covalent interactions, and transition states. J. Chem. Phys. **154**, 064108 (2021)

A. Kamath, R.A. Vargas-Hernández, R.V. Krems, T.Jr. Carrington, S. Manzhos, Neural networks vs Gaussian process regression for representing potential energy surface: a comparative study of fit quality and vibrational spectrum accuracy. J. Chem. Phys. **148**, 241702–241707 (2018)

N. Lubbers, J.S. Smith, K. Barros, Hirarchical modeling of molecular energies using a deep neural network. J. Chem. Phys. **148**, 241715–8 (2018)

D. Marx, J. Hutter, *Ab-Initio Molecular Dynamics* (Cambridge University Press, 2009)

D. Marx, J. Hutter, Ab-initio molecular dynamics: thenry and implementation, in *Modern Methods and Algorithms of Quantum Chemistry*, ed. by J. Grotendorst (Forschungcentrum Jülich, 2000), pp. 301–449

F. Noé, A. Tkatchenko, K.R. Müller, C. Clementi, Machine learning for molecular simulation. Annu. Rev. Phys. Chem. **71**, 361–390 (2020)

Z. Qiao, M. Welborn, A. Anandkuar, F.R. Manby, T.F. Miller III., OrbNet: Deep learning for quantum chemistry using symmetry-adapted atomic-orbital. J. Chem. Phys. **153**, 124111 (2020)

M. Rupp, A. Tkatchenko, K.R. Müller, O.A. von Lilienfeld, Fast and accurate modeling of molecular atomization energies with machine learning. Phys. Rev. Lett. **108**, 058301 (2012)
K.T. Schütt, F. Arbabzadah, S. Chmiela, K.R. Müller, A. Tkatchenko, Quantum-chemical insights from deep tensor neural networks. Nat. Commun. **8**, 13890 (2017)
Q. Sun, T.C. Berkelbach, N.S. Blunt, G.H. Booth, S. Guo, Z. Li, J. Liu, J.D. McClain, E.R. Sayfutyarova, S. Sharma, PySCF: the python-based simulations of chemistry framework. Wiley Interdiscip. Rev.: Comput. Mol. Sci. **8**(1), e1340 (2018)
O.T. Unke, M. Meuwly, A reactive, scalable, and transferable model for molecular energies from a neural network approach based on local information. J. Chem. Phys. **148**, 241708–15 (2018)
M. Welborn, L. Cheng, T.F. Miller III., Transferability in machine learning for electronic structure via the molecular orbital basis. J. Chem. Theory Comput. **14**, 4772–4779 (2018)
A.D. White, Deep learning for molecules and materials. Living J. Comput. Mol. Sci. **3**(1), 1499 (2022)
E.D. Woon, T.H. Dunning Jr, Gaussian basis sets for use in correlated molecular calculations. III. The atoms aluminum through argon. J. Chem. Phys. **98**(2), 1358–1371 (1993)
K. Yang, K. Swanson, W. Jin, C. Coley, Ph. Eiden, H. Gao, A.G.-Perez, T. Hopper, B. Kelley, M. Mathea, A. Palmer, V. Settels, T. Jaakkola, K. Jensen, R. Barzilay, Analyzing learned molecular representations for property prediction. J. Chem. Inf. Model. **59**(8), 3370–3388 (2019)
Zh. Zhou, S. Kearnes, L. Li, R.N. Zare, P. Riley, Optimization of molecules via deep-reinforcement learning. Sci. Rep. **9**, 10752 (2019)

Chapter 5
Classical Molecular Dynamics Simulations of Biomolecules

The chapter aims to introduce the classical molecular dynamics simulations of biomolecules in main statistical ensembles.

Molecular dynamics simulations at atomic level have widely been used in studying macromolecular systems, such as protein, DNA and their complexes, mainly because the laws of classical statistical mechanics can largely govern the processes involved at the experimental conditions. Macromolecules, such as proteins, are characterised by dynamics with time scales ranging from nanoseconds to milliseconds. In this chapter, we discuss the molecular dynamics method as one of the most common computer simulation approach used to study molecular systems. In particular, we will present the equations of motion in the most relevant statistical ensembles used in the molecular dynamics simulations of molecular systems. For further reading, other literature can also be recommended (Kamberaj 2020).

5.1 Protein Data Bank

The Protein Data Bank (PDB) is a web-based archive-information service about the three-dimensional structures of proteins, nucleic acids, and their complexes.[1] RCSB PDB is a member of the wwPDB used to annotate PDB data structures. It builds upon experimental data, and it is used as a resource for research and educational purposes in different areas of science, such as molecular biology, structural biology, biophysics and computational biology.

[1] www.rcsb.org.

H. Kamberaj, *Computer Simulations in Molecular Biology*, Scientific Computation,
https://doi.org/10.1007/978-3-031-34839-6_5

In Table 5.1, we show the information about the structures of biomolecular systems obtained using experimental methods, such as X-ray crystallography, nuclear magnetic resonance (NMR), electron microscopy (EM), multiple methods, neutron scattering, and other methods.

Figure 5.1 presents the percentage of the biological molecular structures obtained using different methods. The results indicates that the most used methods for obtaining three-dimensional structures of biomolecules are X-ray, NMR and EM methods.

Table 5.1 The dataset of biomolecular systems according to Protein Data Bank

Molecular type	X-ray	NMR	EM	Multiple methods	Neutron	Other	Total
Protein (only)	138684	11701	5124	168	69	32	155778
Protein/Oligosaccharide	8214	31	771	5	0	0	9021
Protein/NA	7283	270	1750	3	0	0	9306
Nucleic acid (only)	2172	1359	58	8	2	1	3600
Other	149	31	3	0	0	0	183
Oligosaccharide (only)	11	6	0	1	0	4	22

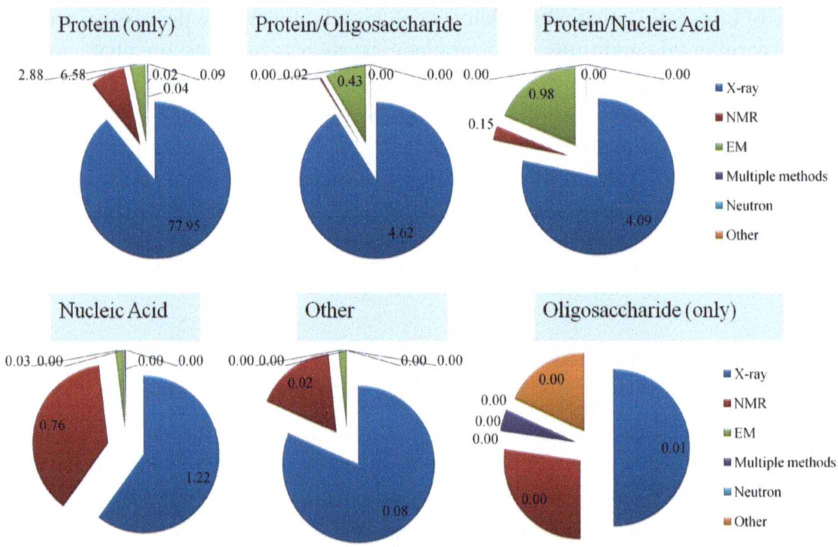

Fig. 5.1 The percentage of biomolecular structures using different experimental methods

5.2 Molecular Dynamics Simulations of Biomolecules

The dynamics of molecular dynamics (MD) technique are typically driven by Newton's second law (for $i = 1, 2, \ldots, N$) (Karplus and McCammon 2002):

$$m_i \ddot{\mathbf{r}}_i = \sum_{j=1 \neq i}^{N} \mathbf{F}_{ji} = -\nabla_{\mathbf{r}_i} U(\mathbf{r}), \tag{5.1}$$

where $\mathbf{r} = (\mathbf{r}_1, \ldots, \mathbf{r}_N)$ is the configuration vector of a system of N atoms of masses m_i (for $i = 1, 2, \ldots, N$), and $U(\mathbf{r})$ is the potential energy function. Equation 5.1 represents a second order differential equation with time t. Numerically, that equation is solved by introducing the ith atom momentum $\mathbf{p}_i = m_i \dot{\mathbf{r}}_i$, and obtaining two first order differential equations as:

$$\begin{aligned} \dot{\mathbf{r}}_i &= \frac{\mathbf{p}_i}{m_i}, \\ \dot{\mathbf{p}}_i &= -\nabla_{\mathbf{r}_i} U(\mathbf{r}) = \mathbf{F}_i. \end{aligned} \tag{5.2}$$

In Eq. (5.2), $\mathbf{F}_i$ denotes the net force acting on the ith atom, given as:

$$\mathbf{F}_i = \sum_{j=1 \neq i}^{N} \mathbf{F}_{ji}. \tag{5.3}$$

Here, $\mathbf{F}_{ji}$ denotes the force between two atoms, namely i and j. Solving Eq. (5.2) concerning the momentum $\mathbf{p}_i$ and position $\mathbf{r}_i$ for each atom, we obtain a trajectory in the phase space of a system of $6N$ degrees of freedom. The numerical integrator schemes are represented as the discrete flow maps, $\boldsymbol{\Phi}_{\Delta t}$, which propagate the phase space vector $\mathbf{X} \equiv (\mathbf{r}, \mathbf{p})$ every time step Δt for a time T:

$$\mathbf{X}(t + \Delta t) = \boldsymbol{\Phi}_{\Delta t} \mathbf{X}(t), \tag{5.4}$$

where $t = 0, \Delta t, \ldots, T - \Delta t$. The following velocity-Verlet algorithm is used to integrate numerically the equations of motion (Kamberaj 2020):

$$\begin{aligned} \mathbf{p}_i \left(t + \frac{\Delta t}{2}\right) &= \mathbf{p}_i (t) + \frac{\Delta t}{2} \mathbf{F}_i (t), \\ \mathbf{r}_i (t + \Delta t) &= \mathbf{r}_i (t) + \Delta t \mathbf{p}_i \left(t + \frac{\Delta t}{2}\right), \\ \mathbf{p}_i (t + \Delta t) &= \mathbf{p}_i \left(t + \frac{\Delta t}{2}\right) + \frac{\Delta t}{2} \mathbf{F}_i (t + \Delta t). \end{aligned} \tag{5.5}$$

Equations (5.5) enable propagation of the momenta and positions every time step during a MD simulation (see also Ref. Allen and Tildesley 1989) given the initial positions and momenta. Usually, the positions are taken either from experimental data (such as X-ray and NMR) or theoretically, and momenta are generated from Maxwell-Boltzmann distribution.

In MD simulation run of biological molecules, the temperature of the system, which is defined as (Kamberaj 2020):

$$T = \langle 2\frac{K}{3Nk_B}\rangle, \tag{5.6}$$

where K is the kinetic energy and k_B denotes Boltzmann's constant, increases naturally when the biomolecule undergoes exothermic conformation changes. That is known as the *temperature-related phenomenon* (Allen and Tildesley 1989).

5.2.1 *Advanced Molecular Dynamics Techniques*

Because of the temperature-related phenomenon and that the experimental results are often obtained at constant temperature and pressure, advanced molecular dynamics techniques have been developed, as described in details in Ref. Kamberaj (2020).

MD techniques are introduced to control the temperature by allowing the system to exchange energy with an external bath, called thermostat (Kamberaj 2020). The most used techniques for controlling the temperature during the MD simulations are Nosé-Hoover thermostats (Nosé and Klein 1983; Nosé 1984a; Hoover 1985; Nosé 1991), Berendsen thermostat (Berendsen et al. 1984), and Langevin dynamics (Pastor 1994). Implementations of all these thermostats are discussed in Ref. Kamberaj (2020).

A popular way of controlling the temperature in MD simulations of biomolecular systems is using the chain of thermostats approach (Martyna et al. 1996). The equations of motion for a massive chain of thermostats are given as (Kamberaj 2020):

$$\begin{aligned}
\dot{q}_i &= \frac{p_i}{m_i}, \\
\dot{p}_i &= F_i - \frac{\xi_i^{(1)}}{Q_i^{(1)}} p_i, \\
\dot{\xi}_i^{(1)} &= \frac{p_i^2}{m} - k_B T - \frac{\xi_i^{(2)}}{Q_i^{(2)}} \xi_i^{(1)}, \\
\dot{\xi}_i^{(k)} &= \frac{(\xi^{(k-1)})^2}{Q_i^{(k-1)}} - k_B T - \frac{\xi_i^{(k+1)}}{Q_i^{(k+1)}} \xi_i^{(k)}, \quad k = 2, 3, \ldots, M-1, \\
\dot{\xi}_i^{(M)} &= \frac{(\xi_i^{(M-1)})^2}{Q_i^{(M-1)}} - k_B T,
\end{aligned} \tag{5.7}$$

$$\dot{s}_i^{(k)} = \frac{\xi_i^{(k)}}{Q_i^{(k)}}, \quad k = 1, 2, \ldots, M,$$

for $i = 1, 2, \ldots, g$, where g is the number of degrees of freedom. In Eq. (5.7), $(s_i^{(k)}, \xi_i^{(k)})$, $k = 1, 2, \ldots, M$, are the thermostat variables and their conjugated momenta associated with every degree of freedom i, and M is the length of the chain of thermostats. $Q_i^{(k)}$, for $k = 1, 2, \ldots, M$ and $i = 1, 2, \ldots, g$, denote the mass associated to the thermostat degrees of freedom. They are optimised to be:

$$Q_i^{(k)} = g_i k_B T \left(\tau_i^{(k)}\right)^2, \tag{5.8}$$

where $\tau_i^{(k)}$ is the relaxation time of the heat exchange between the system and thermostat, and g_i is the number of degrees of freedom associated with the group i (which is equal to one if a chain of thermostats is coupled to only one degree of freedom). The choice of that mass (and hence, $\tau_i^{(k)}$) influences the dynamics and the conservation laws on the extended system. Large values of $Q_i^{(k)}$ (or $\tau_i^{(k)}$) indicates a week coupling between the system and thermostat, and hence longer it will take to the temperature of system to equilibrate to the target temperature, T. In contrast, small values of $Q_i^{(k)}$ (or $\tau_i^{(k)}$) indicates a strong coupling and fast equilibration to the target temperature; however, small values of the mass may cause instability on the dynamics.

For example, in general, the MD simulation packages use the units of Å for length and ps for time, and hence

$$k_B = 0.00083144722\,\text{kg} \cdot \text{Å}^2 \cdot \text{ps}^{-2} \cdot \text{mol}^{-1} \cdot \text{K}^{-1}. \tag{5.9}$$

If we require a relaxation time of $\tau_i^{(k)} = 0.2\,\text{ps}$, for every degree of freedom and thermostat, which is equivalent to 100 time steps in typical MD simulation run using bond constraints (such as SHAKE (Ryckaert et al. 1977)) that use a time step of 0.002 ps, then at target temperature $T = 300\,\text{K}$ the thermostat mass coupled to every degree of freedom becomes:

$$\begin{aligned} Q_i^{(k)} &= \left(0.00083144722\,\frac{\text{kg} \cdot \text{Å}^2}{\text{ps}^2 \cdot \text{mol} \cdot \text{K}}\right)(300\,\text{K})\,(0.2\,\text{ps})^2 \\ &\approx 10\,\frac{\text{kcal}}{\text{mol}} \cdot \text{AKMA}^2, \end{aligned} \tag{5.10}$$

where AKMA is a CHARMM package internal program unit (1 AKMA = 0.0488882129 ps). It is important to note that adjustment of $Q_i^{(k)}$ depends on the number of degrees of freedom associated with that chain of thermostats; however, $\tau_i^{(k)}$ is independent on g_i, and thus it is practically more convenient to adjust $\tau_i^{(k)}$ than $Q_i^{(k)}$.

Using Nosé-Hoover chain of thermostats approach, the entire isolated system of variables extend to $(\mathbf{r}, \mathbf{p}, \mathbf{s}, \boldsymbol{\xi})$, where the new entire system variables associated with the chain of thermostats are given as:

$$\mathbf{s} = \left\{ s_i^{(k)} \right\}_{i=1,\ldots,g}^{k=1,\ldots,M}, \qquad (5.11)$$
$$\boldsymbol{\xi} = \left\{ \xi_i^{(k)} \right\}_{i=1,\ldots,g}^{k=1,\ldots,M}.$$

The conserved energy of the entire isolated system is (Kamberaj 2020):

$$E_{\text{NHC}} = \sum_{i=1}^{g} \left(\frac{p_i^2}{2m_i} + \sum_{k=1}^{M} \left(\frac{\xi_i^{(k)}}{2Q_i^{(k)}} + k_B T s_i^{(k)} \right) \right) + U(q_1, \ldots, q_g), \qquad (5.12)$$

where

$$E_{\text{system}} = \sum_{i=1}^{g} \frac{p_i^2}{2m_i} + U(q_1, \ldots, q_g), \qquad (5.13)$$

is the energy of the physical system, which exchange energy with the bath to adjust the temperature, and hence its energy fluctuates. The amount of these fluctuations is controlled by the temperature and the specific heat (that is, the degree of the energy change with temperature), and it is given as (Kamberaj 2020):

$$\langle (E_{\text{system}})^2 \rangle = -\left(\frac{\partial \langle E_{\text{system}} \rangle}{\partial \beta} \right)_{N,V} = k_B T^2 C_V, \qquad (5.14)$$

where $\beta = 1/k_B T$ is the inverse temperature.

Besides controlling the temperature, often it is required to control the pressure during the MD simulations of biomolecules. For that, two additional variables are introduced, the volume V and its conjugated momentum ξ_ϵ. In a MD simulation run, these two variables allow the system to sample an isobaric statistical ensemble. For constant pressure dynamics Andersen type of piston is used to control pressure (Andersen 1980), which along with Nosé (Nosé 1984) thermostat to control the temperature, allow the system to sample isothermal-isobaric statistical ensemble. There are several modifications on Nosé-Andersen equations of motion as introduced in Refs. Hoover (1985), Melchionna et al. (1993), Martyna et al. (1994). Here, we write the equations of motion that can be used to sample isothermal-isobaric statistical ensemble in a MD simulation run, as proposed in Ref. Kamberaj (2020):

$$\dot{\mathbf{r}}_i = \frac{\mathbf{p}_i}{m_i} + \frac{\xi_\epsilon}{W} \mathbf{r}_i, \qquad (5.15)$$
$$\dot{\mathbf{p}}_i = \mathbf{F}_i - \left(1 + \frac{d}{g} \right) \frac{\xi_\epsilon}{W} \mathbf{p}_i - \frac{\xi_i^{(1)}}{Q_i^{(1)}} \mathbf{p}_i,$$

$$\begin{aligned}
\dot{\xi}_i^{(1)} &= \frac{\mathbf{p}_i^2}{m_i} - dk_BT - \frac{\xi_i^{(2)}}{Q_i^{(2)}}\xi_i^{(1)}, \\
\dot{\xi}_i^{(k)} &= \frac{(\xi_i^{(k-1)})^2}{Q_i^{(k-1)}} - k_BT - \frac{\xi_i^{(k+1)}}{Q_i^{(k+1)}}\xi_i^{(k)}, \quad k = 2, \ldots, M-1, \\
\dot{\xi}_i^{(M)} &= \frac{(\xi_i^{(M-1)})^2}{Q_i^{(M-1)}} - k_BT, \\
\dot{s}_i^{(k)} &= \frac{\xi_i^{(k)}}{Q_i^{(k)}}, \quad k = 1, \ldots, M, \\
\dot{\xi}_\epsilon &= Vd(\mathcal{P} - p_0) + \frac{d}{g}\sum_{i=1}^{N}\frac{\mathbf{p}_i^2}{m_i} - \frac{\xi_b^{(1)}}{Q_b^{(1)}}\xi_\epsilon, \\
\frac{1}{Vd}\dot{V} &= \frac{\xi_\epsilon}{W}, \\
\dot{\xi}_b^{(1)} &= \frac{\xi_\epsilon^2}{W} - k_BT - \frac{\xi_b^{(2)}}{Q_b^{(2)}}\xi_b^{(1)}, \\
\dot{\xi}_b^{(k)} &= \frac{(\xi_b^{(k-1)})^2}{Q_b^{(k-1)}} - k_BT - \frac{\xi_b^{(k+1)}}{Q_b^{(k+1)}}\xi_b^{(k)}, \quad k = 2, \ldots, M_b - 1, \\
\dot{\xi}_b^{(M)} &= \frac{(\xi_b^{(M-1)})^2}{Q_b^{(M-1)}} - k_BT, \\
\dot{s}_b^{(k)} &= \frac{\xi_b^{(k)}}{Q_b^{(k)}}, \quad k = 1, \ldots, M_b.
\end{aligned}$$

In Eq. (5.15), $(s_i^{(k)}, \xi_i^{(k)})$ (for $k = 1, \ldots, M$) denote the chain of thermostat variables of length M coupled to the i atom in the system, and $(s_b^{(k)}, \xi_b^{(k)})$ (for $k = 1, \ldots, M_b$) are the variables of the chain of thermostats of length M_b associated with barostat degree of freedom. d denotes the dimension of the problem (for three-dimensional problems, $d = 3$), p_0 is the target pressure, and $\mathcal{P}$ is the instantaneous pressure of the physical system, given as:

$$\mathcal{P} = \frac{1}{Vd}\left(\sum_{i=1}^{N}\frac{\mathbf{p}_i^2}{m_i} + \sum_{i=1}^{N}\mathbf{F}_i \cdot \mathbf{r}_i - (Vd)\frac{\partial U}{\partial V}\right), \tag{5.16}$$

where $\mathbf{F}_i$ is the net Newtonian force acting on the ith atom. Note that when long-range interactions are applied, or short-range corrections are implemented, then the potential interaction is a function of volume V, and the last term in Eq. (5.16) is different from zero.

The thermostat masses and barostat mass W are defined as:

$$
\begin{aligned}
Q_i^{(k)} &= dk_BT\left(\tau_i^{(k)}\right)^2, \quad i=1,\ldots,N, \quad k=1,\ldots,M, \\
Q_b^{(k)} &= k_BT\left(\tau_b^{(k)}\right)^2, \quad k=1,\ldots,M_b, \\
W &= (d+1)\,k_BT\,(\tau_b)^2,
\end{aligned}
\tag{5.17}
$$

where $\tau_i^{(k)}$ is the relaxation time of coupling between the chain of thermostats and ith atom's degrees of freedom, $\tau_b^{(k)}$ is the relaxation time of the coupling between the chain of thermostats and barostat, and τ_b is the relaxation time of changes in the volume of the simulation box.

For instance, if we assume that the relaxation times are as follows:

$$
\begin{aligned}
&\left(0.2 \le \tau_i^{(k)} \le 2\right)\ \text{ps}, \quad i=1,\ldots,N, \quad k=1,\ldots,M, \\
&\left(0.2 \le \tau_b^{(k)} \le 2\right)\ \text{ps}, \quad k=1,\ldots,M_b, \\
&(0.2 \le \tau_b \le 2)\ \text{ps},
\end{aligned}
\tag{5.18}
$$

which is equivalent to 100 to 1000 time steps in a MD simulation run using bond constraints (such as SHAKE (Ryckaert et al. 1977)) with a time step of 0.002 ps, then the masses are in the following ranges:

$$
\begin{aligned}
&\left(30 \le Q_i^{(k)} \le 3000\right)\frac{\text{kcal}}{\text{mol}}\cdot\text{AKMA}^2, \quad i=1,\ldots,N, \quad k=1,\ldots,M, \\
&\left(10 \le Q_b^{(k)} \le 1000\right)\frac{\text{kcal}}{\text{mol}}\cdot\text{AKMA}^2, \quad k=1,\ldots,M_b, \\
&(40 \le W \le 4000)\,\frac{\text{kcal}}{\text{mol}}\cdot\text{AKMA}^2.
\end{aligned}
\tag{5.19}
$$

The extended system of variables is an isolated system characterised by the following extended energy, which is a conserved quantity:

$$
\begin{aligned}
E_{\text{NPTC}} = &\sum_{i=1}^{N}\left(\frac{\mathbf{p}_i^2}{2m_i} + \sum_{k=1}^{M}\frac{(\xi_i^{(k)})^2}{2Q_i^{(k)}} + dk_BTs_i^{(1)} + k_BT\sum_{k=2}^{M}s_i^{(k)}\right) \\
&+ \frac{(\xi_\epsilon)^2}{2W} + \sum_{k=1}^{M_b}\frac{(\xi_b^{(k)})^2}{2Q_b^{(k)}} + k_BT\sum_{k=1}^{M_b}s_b^{(k)} + p_0V + U(\mathbf{r}).
\end{aligned}
\tag{5.20}
$$

Furthermore, the physical system variables, $(\mathbf{r}, \mathbf{p})$, are sampling an isothermal-isobaric ensemble with a partition function (Kamberaj 2020):

$$
Z_{\text{NPT}} \propto \int d\mathbf{p}\int d\mathbf{r}\,\exp\left(-\beta E_{\text{system}}\right), \tag{5.21}
$$

where E_{system} is the energy of the system:

$$E_{\text{system}} = \sum_{i=1}^{N} \frac{\mathbf{p}_i^2}{2m_i} + U(\mathbf{r}). \tag{5.22}$$

The fluctuations of the volume for the isothermal-isobaric ensemble are defined as (Kamberaj 2020):

$$\langle (\delta V)^2 \rangle = -\left(\frac{\partial \langle V \rangle}{\partial (\beta p)} \right)_{T,E} = -k_B T \left(\frac{\partial \langle V \rangle}{\partial p} \right)_{T,E} \geq 0. \tag{5.23}$$

where p is the pressure. Equation (5.23) implies that

$$\left(\frac{\partial \langle V \rangle}{\partial p} \right)_{T,E} \leq 0, \tag{5.24}$$

which is in agreement with the thermodynamic equilibrium stability relationship.

Furthermore, the energy fluctuations are given as (Kamberaj 2020)

$$\begin{aligned} \langle (\delta E_{\text{system}})^2 \rangle &= k_B T^2 C_V \\ &- k_B T \left(T \left(\frac{\partial p}{\partial T} \right)_{N,V} - p \right) \left(\left(\frac{\partial (pT)}{\partial T} \right)_{N,V} - p \right) \left(\frac{\partial V}{\partial p} \right)_{N,T}. \end{aligned} \tag{5.25}$$

Equation (5.25) indicates that the energy fluctuations in isothermal-isobaric ensemble depend on two terms. The first term, which gives the fluctuations in canonical ensemble, and the second term, which includes the changes on the pressure, p, and pT with the temperature T at constant N and V, and the changes of the volume V with pressure p at constant N and T.

5.3 The Grand Canonical Ensemble

The grand canonical ensemble generated using MD method has been introduced in Ref. Cagin and Pettitt (1991a), Ji and Pettitt (1994), Cagin and Pettitt (1991b), Weerasinghe and Pettitt (1994), Lo and Palmer (1995), Ji et al. (1992), Palmer and Lo (1994), Lynch and Pettitt (1997).

The grand canonical ensemble MD simulation aims to use classical equations of motion to study systems in which the temperature, volume, and chemical potential are constant, but the number of particles fluctuates.

That is accomplished by introducing additional variables into the Hamiltonian function that scale particle velocity and couple a fractional particle to the system via a continuous function of a coupling parameter, λ. In this case, a continuous particle number variable ν is defined as

$$\nu = N + \lambda \tag{5.26}$$

At the moment of the simulation time, N is the number of "whole" particles and λ, which varies between zero and one, represents the extent to which a single particle, the so-called "fractional" particle, is coupled to the rest of the system.

5.3.1 Nosé-Hoover Thermostats for Grand Canonical Ensemble

An extension variable, s, is used to scale the particles momenta as described in the previous sections. The Hamiltonian function of the grand canonical ensemble may be written as

$$\begin{aligned} H_{NH}^{GCE} = \sum_{i=1}^{N} \frac{(\mathbf{p}_i')^2}{2\,s^2 m_i} + \frac{(\mathbf{p}_f')^2}{2\,s^2 m_f} + \frac{\pi_s^2}{2Q} + \frac{p_\lambda^2}{2\,W} \\ +U(\mathbf{r}_1, \ldots, \mathbf{r}_N) + V(\mathbf{r}_1, \ldots, \mathbf{r}_N : \mathbf{r}_f, \lambda) \\ +(g_N + 1)k_B T \ln(s) + \theta(\lambda) \end{aligned} \tag{5.27}$$

where T is the equilibrium temperature, k_B is the Boltzmann constant, and g_N represents the degrees of freedom for the physical system, similar to the above discussion. The functions $\theta(\lambda)$ and $(g_N + 1)k_B T \ln(s)$ are the potential energy terms for the extension variables. In the Hamiltonian function, Eq. (5.27), the third and fourth terms represent the kinetic energies of the new added variables and allow for the derivation of their equations of motion. The fictitious mass parameters for the extension variables, Q and W, have units of (energy×time^2). They represent the extent of coupling between the system and a heat bath at the constant temperature. The first and fifth terms in Eq. (5.27) are the Hamiltonian for an N particles system with momenta, $\mathbf{p}_i'$ (with respect to the scaled time t'), and positions, $\mathbf{r}_i$ for $i = 1, 2, \ldots, N$. The second term represents the kinetic energy of the fractional particle whose momentum is given by $\mathbf{p}_f'$ (with respect to the scaled time) and whose position is given by $\mathbf{r}_f$. The sixth term is the fractional particle potential energy.

The fractional particle interactions may be scaled as

$$V(\mathbf{r}_1, \ldots, \mathbf{r}_N : \mathbf{r}_f, \lambda) = f(\lambda) \sum_{i=1}^{N} u(\mathbf{r}_i, \mathbf{r}_f) \tag{5.28}$$

where the function $f(\lambda)$ must obey the boundary conditions

$$f(0) = 0, \quad f(1) = 1 \tag{5.29}$$

The potential energy function of the extension variable characterising the particle number is given by

$$\theta(\lambda) = -\left[N + h(\lambda)\right]\mu - U_{bias}(\lambda) + h(\lambda)\mu_{id} \tag{5.30}$$

where μ is the chemical potential, $U_{bias}(\lambda)$ is the bias potential, and μ_{id} is the ideal chemical potential for an $N+1$ particles system. This form is chosen to provide a sampling from a grand canonical ensemble distribution, as will be show in the following discussion. The functions $h(\lambda)$ and $f(\lambda)$ obey to the same boundary conditions and they are not necessarily identical. The ideal chemical potential may be defined either with or without a rotational motion contribution. Note that the rotational motion will add to the value of the total chemical potential constant term, which, in practice, corresponds to a change in a thermodynamic reference state, and hence it will not affect the simulation trajectories or measured thermodynamic properties of the system.

Equations of motion can be derived from Eq. (5.27) using the Hamiltonian equations:

$$\begin{aligned}
\frac{d\mathbf{r}_i}{dt'} &= \frac{\mathbf{p}_i'}{s^2 m_i}\,, \\
\frac{d\mathbf{p}_i'}{dt'} &= -\nabla_{\mathbf{r}_i} U(\mathbf{r}_1,\ldots,\mathbf{r}_N) - f(\lambda)\nabla_{\mathbf{r}_i} u(\mathbf{r}_i,\mathbf{r}_f)\,, \\
\frac{d\mathbf{r}_f}{dt'} &= \frac{\mathbf{p}_f'}{s^2 m_f}\,, \\
\frac{d\mathbf{p}_f'}{dt'} &= -f(\lambda)\sum_{k=1}^{N}\nabla_{\mathbf{r}_f} u(\mathbf{r}_k,\mathbf{r}_f)\,, \\
\frac{ds}{dt'} &= \frac{\pi_s}{Q}\,, \\
\frac{d\lambda}{dt'} &= \frac{p_\lambda}{W}\,, \\
\frac{d\pi_s}{dt'} &= \sum_{k=1}^{N}\frac{(\mathbf{p}_k')^2}{s^2 m_k} + \frac{(\mathbf{p}_f')^2}{s^2 m_f} - (g_N + 1)k_B T\,, \\
\frac{dp_\lambda}{dt'} &= -\frac{df(\lambda)}{d\lambda}\sum_{k=1}^{N} u(\mathbf{r}_k,\mathbf{r}_f) - \frac{d\theta(\lambda)}{d\lambda}
\end{aligned} \tag{5.31}$$

These equations are transformed from virtual space to real space by making this variable change (Zare and Szebehely 1975) $dt' = sdt$, and in addition a reformulation in terms of real system variables using the following transformations:

$$\begin{aligned}
\mathbf{p}_i &= \mathbf{p}_i'/s\,,\ i = 1,\ldots,N\,, \\
\mathbf{p}_f &= \mathbf{p}_f'/s\,, \\
\hat{\pi}_s &= \pi_s/s\,, \\
\pi_\lambda &= p_\lambda/s
\end{aligned} \tag{5.32}$$

Then, a new system of non-Hamiltonian equations for the dynamics in the real variables is obtained:

$$
\begin{aligned}
\frac{d\mathbf{r}_i}{dt} &= \frac{\mathbf{p}_i}{m_i} , \\
\frac{d\mathbf{p}_i}{dt} &= -\nabla_{\mathbf{r}_i} U(\mathbf{r}_1, \ldots, \mathbf{r}_N) - f(\lambda)\nabla_{\mathbf{r}_i} u(\mathbf{r}_i, \mathbf{r}_f) - \frac{s\hat{\pi}_s}{Q}\mathbf{p}_i , \\
\frac{d\mathbf{r}_f}{dt} &= \frac{\mathbf{p}_f}{m_f} , \\
\frac{d\mathbf{p}_f}{dt} &= -f(\lambda)\sum_{k=1}^{N} \nabla_{\mathbf{r}_f} u(\mathbf{r}_k, \mathbf{r}_f) - \frac{s\hat{\pi}_s}{Q}\mathbf{p}_f , \\
\frac{ds}{dt} &= s^2\frac{\hat{\pi}_s}{Q} , \\
\frac{d\lambda}{dt} &= s^2\frac{\pi_\lambda}{W} , \\
\frac{d\hat{\pi}_s}{dt} &= \frac{1}{s}\left(\sum_{k=1}^{N} \frac{(\mathbf{p}_k)^2}{m_k} + \frac{(\mathbf{p}_f)^2}{m_f} - g_N k_B T\right) - \frac{s\hat{\pi}_s^2}{Q} , \\
\frac{d\pi_\lambda}{dt} &= -\frac{df(\lambda)}{d\lambda}\sum_{k=1}^{N} u(\mathbf{r}_k, \mathbf{r}_f) - \frac{d\theta(\lambda)}{d\lambda} - \frac{s\hat{\pi}_s}{Q}\pi_\lambda
\end{aligned}
\tag{5.33}
$$

Making another change of variables similar to what was proposed in Hoover (1985):

$$
s\hat{\pi}_s \equiv \pi_\eta , \quad \ln s \equiv \eta \tag{5.34}
$$

Then, the second-order differential equations of motion in real space are given by

$$
\begin{aligned}
\dot{\mathbf{r}}_i &= \frac{\mathbf{p}_i}{m_i} , \\
\dot{\mathbf{p}}_i &= -\nabla_{\mathbf{r}_i} U(\mathbf{r}_1, \ldots, \mathbf{r}_N) - f(\lambda)\nabla_{\mathbf{r}_i} u(\mathbf{r}_i, \mathbf{r}_f) - \mathbf{p}_i\pi_\eta/Q , \\
\dot{\mathbf{r}}_f &= \frac{\mathbf{p}_f}{m_f} , \\
\dot{\mathbf{p}}_f &= -f(\lambda)\sum_{k=1}^{N} \nabla_{\mathbf{r}_f} u(\mathbf{r}_k, \mathbf{r}_f) - \mathbf{p}_f\pi_\eta/Q , \\
\dot{\eta} &= \frac{\pi_\eta}{Q} , \\
\dot{\lambda} &= s^2\frac{\pi_\lambda}{W} , \\
\dot{\pi}_\eta &= \sum_{k=1}^{N} \frac{\mathbf{p}_k^2}{m_k} + \frac{\mathbf{p}_f^2}{m_f} - g_N k_B T ,
\end{aligned}
\tag{5.35}
$$

$$\dot{\pi}_\lambda = -\frac{df(\lambda)}{d\lambda}\sum_{k=1}^{N} u(\mathbf{r}_k, \mathbf{r}_f) - \frac{d\theta(\lambda)}{d\lambda} - \pi_\lambda \pi_\eta / Q$$

The grand canonical partition function, $Z_{\mu VT}$, may be written in terms of canonical partition functions,

$$Z_{\mu VT} = \sum_{N=0}^{\infty} Z_{NVT} \exp\left(\frac{1}{k_B T}\mu N\right) \tag{5.36}$$

The MD simulation of the extended system of the variables sample a micro-canonical ensemble with partition function is

$$\begin{aligned} \Xi &= \sum_{N=0}^{\infty} \frac{1}{N! h^{g_N}} \\ &\times \int d\pi_s \int ds \int dp_\lambda \int_0^1 d\lambda \int d\mathbf{p}' \int d\mathbf{r} \int d\mathbf{p}'_f \int d\mathbf{r}_f \\ &\times \delta\Bigg[\sum_{i=1}^{N} \frac{(\mathbf{p}'_i)^2}{2\, s^2 m_i} + \frac{(\mathbf{p}'_f)^2}{2\, s^2 m_f} + \frac{\pi_s^2}{2Q} + \frac{p_\lambda^2}{2\, W} \\ &+ U(\mathbf{r}) + V(\mathbf{r} : \mathbf{r}_f, \lambda) + (g_N + 1) k_B T \ln(s) + \theta(\lambda) - E\Bigg] \end{aligned} \tag{5.37}$$

A relationship between this approximate partition function and the exact grand canonical partition function can be obtained by first integrating over the thermostat extension variables s and π_s, following the procedures in Nosé (1984b), Nosé (1984a), Lynch and Pettitt (1997), as described in the previous sections. This involves transforming the momenta into real space using the relationships given by Eq. (5.34), and then applying an equivalence relation for the Dirac delta function as explained in the previous sections. The resulting partition function can then be rewritten as

$$\begin{aligned} \Xi &= \sum_{N=0}^{\infty} \frac{1}{N! h^{g_N}} \frac{\sqrt{2\pi Q k_B T}}{(g_N + 1) k_B T} \exp(E/k_B T) \\ &\times \int dp_\lambda \int_0^1 d\lambda \int d\mathbf{p} \int d\mathbf{r} \int d\mathbf{p}_f \int d\mathbf{r}_f \\ &\times \exp\left(-H_0^\lambda / k_B T\right) \exp\left(-H_\lambda / k_B T\right) \end{aligned} \tag{5.38}$$

where

$$H_0^\lambda = \sum_{i=1}^{N} \frac{\mathbf{p}_i^2}{2m_i} + \frac{\mathbf{p}_f^2}{2m_f} + \frac{\pi_s^2}{2Q} + U(\mathbf{r}) + V(\mathbf{r} : \mathbf{r}_f, \lambda) \tag{5.39}$$

and

$$H_\lambda = \frac{p_\lambda^2}{2\,W} + \theta(\lambda) \tag{5.40}$$

After, integrating according to p_λ, we get:

$$\begin{aligned} \Xi = \sum_{N=0}^{\infty} \frac{1}{N!h^{g_N}} \frac{\sqrt{2\pi Q k_B T}\sqrt{2\pi W k_B T}}{(g_N+1)k_B T} \exp\left(E/k_B T\right) \\ \times \int_0^1 d\lambda \int d\mathbf{p} \int d\mathbf{r} \int d\mathbf{p}_f \int d\mathbf{r}_f \\ \times \exp\left(-H_0^\lambda/k_B T\right) \exp\left(-H_\lambda/k_B T\right) \exp\left(-\theta(\lambda)/k_B T\right) \end{aligned} \tag{5.41}$$

The reduced partition function, $\Xi_{N+\lambda}$, is defined by its relationship to the full partition function,

$$\Xi = \sum_{N=0}^{\infty} \int_0^1 \Xi_{N+\lambda} d\lambda \tag{5.42}$$

The ration of the reduced partition function for the zero and one configurations is:

$$\frac{\Xi_{N+1}}{\Xi_N} = \frac{\exp\left[-\theta(1)/k_B T\right]}{\exp\left[-\theta(0)/k_B T\right]} \frac{Z_{N+1}}{Z_{N+1}^{id} Z_N^{ex}} \tag{5.43}$$

where Z_{N+1}^{id} and Z_N^{ex} are the kinetic and potential parts of the partition function. The $N+1$ term in the denominator is a direct result of not scaling the kinetic energy of the fractional particle. Substitution of $\theta(1)$ and $\theta(0)$ into Eq. (5.43) gives

$$\frac{\Xi_{N+1}}{\Xi_N} = \left[\exp\left(\left(U_{bias}(1) - U_{bias}(0)\right)/k_B T\right) \exp\left(\mu/k_B T\right) \frac{Z_{N+1}}{Z_N}\right] \tag{5.44}$$

which for $U_{bias}(0) = U_{bias}(1)$ becomes

$$\frac{\Xi_{N+1}}{\Xi_N} = \frac{Z_{(N+1)VT}}{Z_{NVT}} \exp(\mu/k_B T) \tag{5.45}$$

which is the grand canonical ratio. The bias potential, when calculating the ensemble averages, does not need to be subtracted out, using for example weighted histogram analysis method, as long as this last restriction is satisfied.

5.3.2 Nosé-Hoover Chain of Thermostats for Grand Canonical Ensemble

The Hamiltonian function of the grand canonical ensemble using the chain of thermostats approach is written as

$$\begin{aligned} H_{NHC}^{GCE} = \sum_{i=1}^{g_N} \Bigg[\frac{(p_i')^2}{2s_{1,p_i}^2 m_i} \\ + \sum_{i=1}^{g_N} \sum_{j=1}^{M_p-1} \frac{\pi_{j,p_i}^2}{2Q_{j,p_i} s_{j+1,p_i}^2} + \frac{\pi_{M_p,p_i}^2}{2Q_{M_p,p_i}} \Bigg] \\ + \sum_{i=1}^{d} \Bigg[\frac{(p_{f_i}')^2}{2s_{1,f_i}^2 m_{f_i}} + \sum_{j=1}^{M_f-1} \frac{\pi_{j,f_i}^2}{2Q_{j,f_i} s_{j+1,f_i}^2} + \frac{\pi_{M_f,f_i}^2}{2Q_{M_f,f_i}} \Bigg] \\ + \Bigg[\frac{p_\lambda^2}{2s_{1,\lambda}^2 W} + \sum_{j=1}^{M_\lambda-1} \frac{\pi_{j,\lambda}^2}{2Q_{j,\lambda} s_{j+1,\lambda}^2} + \frac{\pi_{M_\lambda,\lambda}^2}{2Q_{M_\lambda,\lambda}} \Bigg] \\ + U(\mathbf{r}_1, \ldots, \mathbf{r}_N) + V(\mathbf{r}_1, \ldots, \mathbf{r}_N : \mathbf{r}_f, \lambda) + \theta(\lambda) \\ + \sum_{i=1}^{g_N} \sum_{j=1}^{M_p} k_B T \ln(s_{j,p_i}) \\ + \sum_{i=1}^{d} \sum_{j=1}^{M_f} k_B T \ln(s_{j,f_i}) \\ + \sum_{j=1}^{M_\lambda} k_B T \ln(s_{j,\lambda}) \end{aligned} \tag{5.46}$$

where g_N is the number of degrees of freedom of the unperturbed system, d is the dimension of the problem and M is the length of the chain of thermostats. Note that we have used a massive chain of thermostats coupled to each degree of freedom of the system, and we have used the same length for the chain of thermostats for both the real and factious particle. Now, we can use the so-known Hamilton's equations of motions. Note that in the following equations, we are going to drop the subscript i for the simplicity of the notation, which runs overall degrees of freedom of the real particles, fictitious particle and λ, respectively. Thus, we obtained the following equations for each degrees of freedom of the entire system:

$$\frac{dq}{d\tau_{1,p}} = \frac{\partial H_{NHC}^{GCE}}{\partial p'} = \frac{p'}{s_{1,p}^2 m} \tag{5.47}$$

$$\frac{dp'}{d\tau_{1,p}} = -\frac{\partial H_{NHC}^{GCE}}{\partial q} = -\nabla_q U(q) - f(\lambda)\nabla_q u(q, q_f)$$

$$\frac{dq_f}{d\tau_{1,f}} = \frac{\partial H_{NHC}^{GCE}}{\partial p'_f} = \frac{p'_f}{s_{1,f}^2 m_f}$$

$$\frac{dp'_f}{d\tau_{1,f}} = -\frac{\partial H_{NHC}^{GCE}}{\partial q_f} = -f(\lambda)\nabla_{q_f} u(q, q_f)$$

$$\frac{d\lambda}{d\tau_{1,\lambda}} = \frac{\partial H_{NHC}^{GCE}}{\partial p_\lambda} = \frac{p_\lambda}{s_{1,\lambda}^2 W}$$

$$\frac{dp_\lambda}{d\tau_\lambda} = -\frac{\partial H_{NHC}^{GCE}}{\partial \lambda} = -\frac{df(\lambda)}{d\lambda} u(q, q_f) - \frac{d\theta(\lambda)}{d\lambda}$$

$$\frac{d\pi_{1,p}}{d\tau_{2,p}} = -\frac{\partial H_{NHC}^{GCE}}{\partial s_{1,p}} = \frac{(p')^2}{s_{1,p}^3 m} - \frac{k_B T}{s_{1,p}}$$

$$\frac{ds_{1,p}}{d\tau_{2,p}} = \frac{\partial H_{NHC}^{GCE}}{\partial \pi_{1,p}} = \frac{\pi_{1,p}}{s_{2,p}^2 Q_{s_{1,p}}}$$

$$\frac{d\pi_{k,p}}{d\tau_{k+1,p}} = -\frac{\partial H_{NHC}^{GCE}}{\partial s_{k,p}} = \frac{\pi_{k-1,p}^2}{s_{k,p}^3 Q_{k-1,p}} - \frac{k_B T}{s_{k,p}}$$

$$\frac{ds_{k,p}}{d\tau_{k+1,p}} = \frac{\partial H_{NHC}^{GCE}}{\partial \pi_{s_{k,p}}} = \frac{\pi_{k,p}}{s_{k+1,p}^2 Q_{k,p}}$$

$$k = 2, \ldots, M_p - 1$$

$$\frac{d\pi_{M_p,p}}{d\tau_{M_p,p}} = -\frac{\partial H_{NHC}^{GCE}}{\partial s_{M_p,p}} = \frac{\pi_{M_p-1,p}^2}{s_{M_p,p}^3 Q_{M_p-1,p}} - \frac{k_B T}{s_{M_p,p}}$$

$$\frac{ds_{M_p,p}}{d\tau_{M_p,p}} = \frac{\partial H_{NHC}^{GCE}}{\partial \pi_{M_p,p}} = \frac{\pi_{M_p,p}}{Q_{M_p,p}}$$

$$\frac{d\pi_{1,f}}{d\tau_{2,f}} = -\frac{\partial H_{NHC}^{GCE}}{\partial s_{1,f}} = \frac{(p'_f)^2}{s_{1,f}^3 m_f} - \frac{k_B T}{s_{1,f}}$$

$$\frac{ds_{1,f}}{d\tau_{2,f}} = \frac{\partial H_{NHC}^{GCE}}{\partial \pi_{s_{1,f}}} = \frac{\pi_{1,f}}{s_{2,f}^2 Q_{1,f}}$$

$$\frac{d\pi_{k,f}}{d\tau_{k+1,f}} = -\frac{\partial H_{NHC}^{GCE}}{\partial s_{k,f}} = \frac{\pi_{k-1,f}^2}{s_{k,f}^3 Q_{k-1,f}} - \frac{k_B T}{s_{k,f}}$$

$$\frac{ds_{k,f}}{d\tau_{k+1,f}} = \frac{\partial H_{NHC}^{GCE}}{\partial \pi_{k,f}} = \frac{\pi_{k,f}}{s_{k+1,f}^2 Q_{k,f}}$$

$$k = 2, \ldots, M_f - 1$$

$$\frac{d\pi_{M_f,f}}{d\tau_{M_f,f}} = -\frac{\partial H_{NHC}^{GCE}}{\partial s_{M_f,f}} = \frac{\pi_{M_f-1,f}^2}{s_{M_f,f}^3 Q_{M_f-1,f}} - \frac{k_B T}{s_{M_f,f}}$$

$$
\begin{aligned}
\frac{ds_{M_f,f}}{d\tau_{M_f,f}} &= \frac{\partial H_{NHC}^{GCE}}{\partial \pi_{M_f,f}} = \frac{\pi_{M_f,f}}{Q_{s_{M_f,f}}} \\
\frac{d\pi_{1,\lambda}}{d\tau_{2,f}} &= -\frac{\partial H_{NHC}^{GCE}}{\partial s_{1,\lambda}} = \frac{(p_\lambda)^2}{s_{1,\lambda}^3 W} - \frac{k_B T}{s_{1,\lambda}} \\
\frac{ds_{1,\lambda}}{d\tau_{2,\lambda}} &= \frac{\partial H_{NHC}^{GCE}}{\partial \pi_{1,\lambda}} = \frac{\pi_{1,\lambda}}{s_{2,\lambda}^2 Q_{1,\lambda}} \\
\frac{d\pi_{k,\lambda}}{d\tau_{k+1,\lambda}} &= -\frac{\partial H_{NHC}^{GCE}}{\partial s_{k,\lambda}} = \frac{\pi_{k-1,\lambda}^2}{s_{k,\lambda}^3 Q_{k-1,\lambda}} - \frac{k_B T}{s_{k,\lambda}} \\
\frac{ds_{k,\lambda}}{d\tau_{k+1,\lambda}} &= \frac{\partial H_{NHC}^{GCE}}{\partial \pi_{k,\lambda}} = \frac{\pi_{k,\lambda}}{s_{k+1,\lambda}^2 Q_{k,\lambda}} \\
k &= 2, \ldots, M_\lambda - 1 \\
\frac{d\pi_{M_\lambda,\lambda}}{d\tau_{M_\lambda,\lambda}} &= -\frac{\partial H_{NHC}^{GCE}}{\partial s_{M_\lambda,\lambda}} = \frac{\pi_{M_\lambda-1,\lambda}^2}{s_{M_\lambda,\lambda}^3 Q_{M_\lambda-1,\lambda}} - \frac{k_B T}{s_{M_\lambda,\lambda}} \\
\frac{ds_{M_\lambda,\lambda}}{d\tau_{M_\lambda,\lambda}} &= \frac{\partial H_{NHC}^{GCE}}{\partial \pi_{M_\lambda,\lambda}} = \frac{\pi_{M_\lambda,\lambda}}{Q_{M_\lambda,\lambda}}
\end{aligned}
$$

where

$$
\begin{aligned}
d\tau_{1,p} &= s_{1,p} dt \\
d\tau_{k,p} &= s_{k-1,p} s_{k,p} dt. \quad (k = 2, \ldots, M_p - 1) \\
d\tau_{M_p,p} &= s_{M_p,p} dt \\
d\tau_{1,f} &= s_{1,f} dt \\
d\tau_{k,f} &= s_{k-1,f} s_{k,f} dt, \quad (k = 2, \ldots, M_f - 1) \\
d\tau_{1,\lambda} &= s_{1,\lambda} dt \\
d\tau_{k,\lambda} &= s_{k-1,\lambda} s_{k,\lambda} dt, \quad (k = 2, \ldots, M_\lambda - 1) \\
d\tau_{M_\lambda,\lambda} &= s_{M_\lambda,\lambda} dt
\end{aligned} \tag{5.48}
$$

with t being the real time. Substituting these transformations into Eq. (5.47), we obtain:

$$
\begin{aligned}
\frac{dq}{dt} &= \frac{p'}{s_{1,p} m} \\
\frac{dp'}{dt} &= -s_{1,p} \nabla_q U(q) - s_{1,p} f(\lambda) \nabla_q u(q, q_f) \\
\frac{dq_f}{dt} &= \frac{p'_f}{s_{1,f} m_f} \\
\frac{dp'_f}{dt} &= -s_{1,f} f(\lambda) \nabla_{q_f} u(q, q_f)
\end{aligned} \tag{5.49}
$$

$$\frac{d\lambda}{dt} = \frac{p_\lambda}{s_{1,\lambda} W}$$

$$\frac{dp_\lambda}{dt} = -s_{1,\lambda}\frac{df(\lambda)}{d\lambda}u(q, q_f) - s_{1,\lambda}\frac{d\theta(\lambda)}{d\lambda}$$

$$\frac{d\pi_{1,p}}{dt} = s_{2,p}\frac{(p')^2}{s_{1,p}^2 m} - s_{2,p}k_B T$$

$$\frac{1}{s_{1,p}}\frac{ds_{1,p}}{dt} = \frac{\pi_{1,p}}{s_{2,p}Q_{1,p}}$$

$$\frac{d\pi_{k,p}}{dt} = s_{k+1,p}\frac{\pi_{k-1,p}^2}{s_{k,p}^2 Q_{k-1,p}} - s_{k+1,p}k_B T$$

$$\frac{ds_{k,p}}{dt} = s_{k,p}\frac{\pi_{k,p}}{s_{k+1,p}Q_{k,p}}$$

$$k = 2, \ldots, M_p - 1$$

$$\frac{d\pi_{M_p,p}}{dt} = s_{M_p,p}\frac{\pi_{M_p-1,p}^2}{s_{M_p,p}^3 Q_{M_p-1,p}} - s_{M_p,p}\frac{k_B T}{s_{M_p,p}}$$

$$\frac{ds_{M_p,p}}{dt} = s_{M_p,p}\frac{\pi_{M_p,p}}{Q_{M_p,p}}$$

$$\frac{d\pi_{1,f}}{dt} = s_{2,f}\frac{(p'_f)^2}{s_{1,f}^2 m_f} - s_{2,f}k_B T$$

$$\frac{1}{s_{1,f}}\frac{ds_{1,f}}{dt} = \frac{\pi_{1,f}}{s_{2,f}Q_{1,f}}$$

$$\frac{d\pi_{k,f}}{dt} = s_{k+1,f}\frac{\pi_{k-1,f}^2}{s_{k,f}^2 Q_{k-1,f}} - s_{k+1,f}k_B T$$

$$\frac{ds_{k,f}}{dt} = s_{k,f}\frac{\pi_{k,f}}{s_{k+1,f}Q_{k,f}}$$

$$k = 2, \ldots, M_f - 1$$

$$\frac{d\pi_{M_f,f}}{dt} = s_{M_f,f}\frac{\pi_{M_f-1,f}^2}{s_{M_f,f}^3 Q_{M_f-1,f}} - s_{M_f,f}\frac{k_B T}{s_{M_f,f}}$$

$$\frac{ds_{M_f,f}}{dt} = s_{M_f,f}\frac{\pi_{M_f,f}}{Q_{s_{M_f,f}}}$$

$$\frac{d\pi_{1,\lambda}}{dt} = s_{2,f}\frac{(p_\lambda)^2}{s_{1,\lambda}^2 W} - s_{2,\lambda}k_B T$$

$$\frac{1}{s_{1,\lambda}}\frac{ds_{1,\lambda}}{dt} = \frac{\pi_{1,\lambda}}{s_{2,\lambda}Q_{1,\lambda}}$$

$$\frac{d\pi_{k,\lambda}}{dt} = s_{k+1,\lambda}\frac{\pi_{k-1,\lambda}^2}{s_{k,\lambda}^2 Q_{k-1,\lambda}} - s_{k+1,\lambda}k_B T$$

$$\frac{ds_{k,\lambda}}{dt} = s_{k,\lambda}\frac{\pi_{k,\lambda}}{s_{k+1,\lambda}Q_{k,\lambda}}$$
$$k = 2, \ldots, M_\lambda - 1$$
$$\frac{d\pi_{M_\lambda,\lambda}}{dt} = s_{M_\lambda,\lambda}\frac{\pi^2_{M_\lambda-1,\lambda}}{s^3_{M_\lambda,\lambda}Q_{M_\lambda-1,\lambda}} - s_{M_\lambda,\lambda}\frac{k_BT}{s_{M_\lambda,\lambda}}$$
$$\frac{ds_{M_\lambda,\lambda}}{dt} = s_{M_\lambda,\lambda}\frac{\pi_{M_\lambda,\lambda}}{Q_{M_\lambda,\lambda}}$$

In the following, we propose some other transformations of the variables:

$$p = \frac{p'}{s_{1,p}}, \quad p_f = \frac{p'_f}{s_{1,f}}, \quad \pi_\lambda = \frac{p_\lambda}{s_{1,\lambda}} \tag{5.50}$$
$$\hat{\pi}_{k,p} = \frac{\pi_{k,p}}{s_{k,p}s_{k+1,p}}, \quad \hat{\pi}_{k,f} = \frac{\pi_{s_{k,f}}}{s_{k,f}s_{k+1,f}}, \quad \hat{\pi}_{k,\lambda} = \frac{\pi_{s_{k,\lambda}}}{s_{k,\lambda}s_{k+1,\lambda}}$$
$$(k = 1, \ldots, M-1)$$
$$\hat{\pi}_{M,p} = \frac{\pi_{M,p}}{s_{M,p}}, \quad \hat{\pi}_{M,f} = \frac{\pi_{M,f}}{s_{M,f}}, \quad \hat{\pi}_{M,\lambda} = \frac{\pi_{M,\lambda}}{s_{M,\lambda}}$$

Substituting the transformation given by Eq. (5.50) into Eq. (5.49), we get:

$$\frac{dq}{dt} = \frac{p}{m} \tag{5.51}$$
$$\frac{dp}{dt} = -\nabla_q U(q) - f(\lambda)\nabla_q u(q, q_f) - p\frac{1}{s_{1,p}}\frac{ds_{1,p}}{dt}$$
$$\frac{dq_f}{dt} = \frac{p_f}{m_f}$$
$$\frac{dp_f}{dt} = -f(\lambda)\nabla_{q_f} u(q, q_f) - p_f\frac{1}{s_{1,f}}\frac{ds_{1,f}}{dt}$$
$$\frac{d\lambda}{dt} = \frac{\pi_\lambda}{W}$$
$$\frac{d\pi_\lambda}{dt} = -\frac{df(\lambda)}{d\lambda}u(q, q_f) - \frac{d\theta(\lambda)}{d\lambda} - \pi_\lambda\frac{1}{s_{1,\lambda}}\frac{ds_{1,\lambda}}{dt}$$
$$\frac{d\hat{\pi}_{1,p}}{dt} = \frac{1}{s_{1,p}}\left[\frac{p^2}{m} - k_BT\right] - \hat{\pi}_{1,p}\frac{1}{s_{1,p}}\frac{ds_{1,p}}{dt} - \hat{\pi}_{1,p}\frac{1}{s_{2,p}}\frac{ds_{2,p}}{dt}$$
$$\frac{1}{s_{1,p}}\frac{ds_{1,p}}{dt} = s_{1,p}\frac{\hat{\pi}_{1,p}}{Q_{1,p}}$$
$$\frac{d\hat{\pi}_{k,p}}{dt} = s^2_{k-1,p}\frac{\hat{\pi}^2_{k-1,p}}{s_{k,p}Q_{k-1,p}} - \frac{k_BT}{s_{k,p}} - \frac{1}{s_{k,p}}\frac{ds_{k,p}}{dt} - \frac{1}{s_{k+1,p}}\frac{ds_{k+1,p}}{dt}$$
$$\frac{1}{s_{k,p}}\frac{ds_{k,p}}{dt} = s_{k,p}\frac{\hat{\pi}_{k,p}}{Q_{k,p}}$$

$$k = 2, \ldots, M_p - 1$$

$$\frac{d\hat{\pi}_{M_p,p}}{dt} = s^2_{M_p-1,p}\frac{\hat{\pi}^2_{M_p-1,p}}{s_{M_p,p}Q_{M_p-1,p}} - \frac{k_BT}{s_{M_p,p}} - \frac{1}{s_{M,p}}\frac{ds_{M,p}}{dt}$$

$$\frac{1}{s_{M,p}}\frac{ds_{M_p,p}}{dt} = s_{M_p,p}\frac{\hat{\pi}_{M_p,p}}{Q_{M_p,p}}$$

$$\frac{d\hat{\pi}_{1,f}}{dt} = \frac{1}{s_{1,f}}\left[\frac{p_f^2}{m_f} - k_BT\right] - \hat{\pi}_{1,f}\frac{1}{s_{1,f}}\frac{ds_{1,f}}{dt} - \hat{\pi}_{1,f}\frac{1}{s_{2,f}}\frac{ds_{2,f}}{dt}$$

$$\frac{1}{s_{1,f}}\frac{ds_{1,f}}{dt} = s_{1,f}\frac{\hat{\pi}_{1,f}}{Q_{1,f}}$$

$$\frac{d\hat{\pi}_{k,f}}{dt} = s^2_{k-1,f}\frac{\hat{\pi}^2_{k-1,f}}{s_{k,f}Q_{k-1,f}} - \frac{k_BT}{s_{k,f}} - \frac{1}{s_{k,f}}\frac{ds_{k,f}}{dt} - \frac{1}{s_{k+1,f}}\frac{ds_{k+1,f}}{dt}$$

$$\frac{1}{s_{k,f}}\frac{ds_{k,f}}{dt} = s_{k,f}\frac{\hat{\pi}_{k,f}}{Q_{k,f}}$$

$$k = 2, \ldots, M_f - 1$$

$$\frac{d\hat{\pi}_{M_f,f}}{dt} = s^2_{M_f-1,f}\frac{\hat{\pi}^2_{M_f-1,f}}{s_{M_f,f}Q_{M_f-1,f}} - \frac{k_BT}{s_{M_f,f}} - \frac{1}{s_{M,f}}\frac{ds_{M,f}}{dt}$$

$$\frac{1}{s_{M,f}}\frac{ds_{M_f,f}}{dt} = s_{M_f,f}\frac{\hat{\pi}_{M_f,f}}{Q_{M_f,f}}$$

$$\frac{d\hat{\pi}_{1,\lambda}}{dt} = \frac{1}{s_{1,\lambda}}\left[\frac{\pi_\lambda^2}{W} - k_BT\right] - \hat{\pi}_{1,\lambda}\frac{1}{s_{1,\lambda}}\frac{ds_{1,\lambda}}{dt} - \hat{\pi}_{1,\lambda}\frac{1}{s_{2,\lambda}}\frac{ds_{2,\lambda}}{dt}$$

$$\frac{1}{s_{1,\lambda}}\frac{ds_{1,\lambda}}{dt} = s_{1,\lambda}\frac{\hat{\pi}_{1,\lambda}}{Q_{1,\lambda}}$$

$$\frac{d\hat{\pi}_{k,\lambda}}{dt} = s^2_{k-1,\lambda}\frac{\hat{\pi}^2_{k-1,\lambda}}{s_{k,\lambda}Q_{k-1,\lambda}} - \frac{k_BT}{s_{k,\lambda}} - \frac{1}{s_{k,\lambda}}\frac{ds_{k,\lambda}}{dt} - \frac{1}{s_{k+1,\lambda}}\frac{ds_{k+1,\lambda}}{dt}$$

$$\frac{1}{s_{k,\lambda}}\frac{ds_{k,\lambda}}{dt} = s_{k,\lambda}\frac{\hat{\pi}_{k,\lambda}}{Q_{k,\lambda}}$$

$$k = 2, \ldots, M_\lambda - 1$$

$$\frac{d\hat{\pi}_{M_\lambda,\lambda}}{dt} = s^2_{M_\lambda-1,\lambda}\frac{\hat{\pi}^2_{M_\lambda-1,\lambda}}{s_{M_\lambda,\lambda}Q_{M_\lambda-1,\lambda}} - \frac{k_BT}{s_{M_\lambda,\lambda}} - \frac{1}{s_{M,\lambda}}\frac{ds_{M,\lambda}}{dt}$$

$$\frac{1}{s_{M,\lambda}}\frac{ds_{M_\lambda,\lambda}}{dt} = s_{M_\lambda,\lambda}\frac{\hat{\pi}_{M_\lambda,\lambda}}{Q_{M_\lambda,\lambda}}$$

We can introduce another change in the variables of the following form:

$$\pi_{k,x} = s_{k,x}\hat{\pi}_{k,x}, \ \eta_{k,x} = \ln s_{k,x} \tag{5.52}$$

where $x = p,\ f,\ \lambda$ and k runs over the chain length. Then, Eq. (5.51) can be written in the final form as follows:

$$
\begin{aligned}
\frac{dq}{dt} &= \frac{p}{m} \\
\frac{dp}{dt} &= -\nabla_q U(q) - f(\lambda)\nabla_q u(q, q_f) - \frac{\pi_{1,p}}{Q_{1,p}} p \\
\frac{dq_f}{dt} &= \frac{p_f}{m_f} \\
\frac{dp_f}{dt} &= -f(\lambda)\nabla_{q_f} u(q, q_f) - \frac{\pi_{1,f}}{Q_{1,f}} p_f \\
\frac{d\lambda}{dt} &= \frac{\pi_\lambda}{W} \\
\frac{d\pi_\lambda}{dt} &= -\frac{df(\lambda)}{d\lambda} u(q, q_f) - \frac{d\theta(\lambda)}{d\lambda} - \frac{\pi_{1,\lambda}}{Q_{1,\lambda}} \pi_\lambda \\
\frac{d\pi_{1,p}}{dt} &= \frac{p^2}{m} - k_B T - \frac{\pi_{2,p}}{Q_{2,p}} \pi_{1,p} \\
\frac{d\eta_{1,p}}{dt} &= \frac{\pi_{1,p}}{Q_{1,p}} \\
\frac{d\pi_{k,p}}{dt} &= \frac{\pi_{k-1,p}^2}{Q_{k-1,p}} - k_B T - \frac{\pi_{k+1,p}}{Q_{k+1,p}} \pi_{k,p} \\
\frac{d\eta_{k,p}}{dt} &= \frac{\pi_{k,p}}{Q_{k,p}} \\
k &= 2, \ldots, M_p - 1 \\
\frac{d\pi_{M_p,p}}{dt} &= \frac{\pi_{M_p-1,p}^2}{s_{M_p,p} Q_{M_p-1,p}} - k_B T \\
\frac{d\eta_{M_p,p}}{dt} &= \frac{\pi_{M_p,p}}{Q_{M_p,p}} \\
\frac{d\pi_{1,f}}{dt} &= \frac{p_f^2}{m_f} - k_B T - \frac{\pi_{2,f}}{Q_{2,f}} \pi_{1,f} \\
\frac{d\eta_{1,f}}{dt} &= \frac{\pi_{1,f}}{Q_{1,f}} \\
\frac{d\pi_{k,f}}{dt} &= \frac{\pi_{k-1,f}^2}{Q_{k-1,f}} - k_B T - \frac{\pi_{k+1,f}}{Q_{k+1,f}} \pi_{k,f} \\
\frac{d\eta_{k,f}}{dt} &= \frac{\pi_{k,f}}{Q_{k,f}} \\
k &= 2, \ldots, M_f - 1 \\
\frac{d\pi_{M_f,f}}{dt} &= \frac{\pi_{M_f-1,f}^2}{s_{M_f,f} Q_{M_f-1,f}} - k_B T \\
\frac{d\eta_{M_f,f}}{dt} &= \frac{\pi_{M_f,f}}{Q_{M_f,f}}
\end{aligned} \tag{5.53}
$$

$$\frac{d\pi_{1,\lambda}}{dt} = \frac{\pi_\lambda^2}{W} - k_B T - \frac{\pi_{2,\lambda}}{Q_{2,\lambda}}\pi_{1,\lambda}$$
$$\frac{d\eta_{1,\lambda}}{dt} = \frac{\pi_{1,\lambda}}{Q_{1,\lambda}}$$
$$\frac{d\pi_{k,\lambda}}{dt} = \frac{\pi_{k-1,\lambda}^2}{Q_{k-1,\lambda}} - k_B T - \frac{\pi_{k+1,\lambda}}{Q_{k+1,\lambda}}\pi_{k,\lambda}$$
$$\frac{d\eta_{k,\lambda}}{dt} = \frac{\pi_{k,\lambda}}{Q_{k,\lambda}}$$
$$k = 2, \ldots, M_\lambda - 1$$
$$\frac{d\pi_{M_\lambda,\lambda}}{dt} = \frac{\pi_{M_\lambda-1,\lambda}^2}{s_{M_\lambda,\lambda} Q_{M_\lambda-1,\lambda}} - k_B T$$
$$\frac{d\eta_{M_\lambda,\lambda}}{dt} = \frac{\pi_{M_\lambda,\lambda}}{Q_{M_\lambda,\lambda}}$$

5.3.3 *Grand Isothermal-Isobaric Ensemble*

Now, we are going to formulate the equations of motion in a grand isothermal-isobaric ensemble in which μ, p and T are fixed using the Nosé-Hoover formalism.

The following discussion introduces two approaches, namely the Nosé-Andersen thermostat and Nosé-Andersen chain of thermostats approach to control the temperature (T) and pressure (p). Besides, to control the chemical potential (μ) we are again going to employ the above formalism.

5.3.3.1 Nosé-Andersen Thermostats for Grand Isothermal-isobaric Ensemble

We will suppose that a thermostat with variable (s, π) is coupled to each degrees of freedom of the system, namely real particles of the system, factious particle, λ-parameter, and barostat. Consider we have a N-particles system interacting via the potential energy function $U(\mathbf{r}_1, \mathbf{r}_2, \ldots, \mathbf{r}_N)$. The Hamiltonian function for this system will be written as follows:

$$H_{NA}^{GIIE} = \sum_{i=1}^{g_N} \left[\frac{(p_i')^2}{2s_{p,i}^2 V^{2/d} m_i} + \frac{\pi_{s_{p,i}}^2}{2Q_{s_{p,i}}} + k_B T \ln s_{p,i} \right] \tag{5.54}$$
$$+ \sum_{i=1}^{d} \left[\frac{(p'_{f,i})^2}{2s_{f,i}^2 V^{2/d} m_{f,i}} + \frac{\pi_{s_{f,i}}^2}{2Q_{s_{f,i}}} + k_B T \ln s_{f,i} \right]$$
$$+ \frac{p_\lambda^2}{2\, s_\lambda^2 W_\lambda} + \frac{\pi_{s_\lambda}^2}{2Q_{s_\lambda}} + k_B T \ln s_\lambda$$

$$+U(q'_1, q'_2, \ldots, q'_{g_N})$$
$$+U'(q'_1, q'_2, \ldots, q'_{g_N} : q'_{f,1} \ldots, q'_{f,d}, \lambda)$$
$$+\theta(\lambda)$$
$$+\frac{p_V^2}{2s_b^2 W_v} + pV + \frac{\pi_{s_b}^2}{2s_b^2 Q_b} + k_B T \ln s_b$$

where the subscript i running overall degrees of freedom is omitted for simplicity of appearance. The parameters have the same meaning as previously stated and

$$q'_i = V^{1/d} q_i, \quad q'_{f,i} = V^{1/d} q_{f,i} \tag{5.55}$$

In addition, we have introduced a fictitious particle associated with the fluctuations of the volume with mass W_v. Note that we have introduced a thermostat to each degrees of freedom in the total system. Using the Hamilton's equations, we obtain:

$$\frac{dq'}{d\tau_p} = \frac{p'}{s_p^2 V^{2/d} m} \tag{5.56}$$
$$\frac{dp'}{d\tau_p} = -V^{1/d} \nabla_q U(q') - V^{1/d} f(\lambda) \nabla_q u(q', q'_f)$$
$$\frac{ds_p}{d\tau_p} = \frac{\pi_{s_p}}{Q_p}$$
$$\frac{d\pi_{s_p}}{d\tau_p} = \frac{(p')^2}{s_p^3 V^{2/d} m} - \frac{k_B T}{s_p}$$
$$\frac{dq'_f}{d\tau_f} = \frac{p'_f}{s_f^2 V^{2/d} m_f}$$
$$\frac{dp'_f}{d\tau_f} = -V^{1/d} f(\lambda) \nabla_{q_f} u(q', q'_f)$$
$$\frac{ds_f}{d\tau_f} = \frac{\pi_{s_f}}{Q_f}$$
$$\frac{d\pi_{s_f}}{d\tau_f} = \frac{(p'_f)^2}{s_f^3 V^{2/d} m_f} - \frac{k_B T}{s_f}$$
$$\frac{dq'_\lambda}{d\tau_\lambda} = \frac{p'_\lambda}{s_\lambda^2 W_\lambda}$$
$$\frac{dp'_\lambda}{d\tau_\lambda} = -\frac{df(\lambda)}{d\lambda} u(q', q'_f) - \frac{d\theta(\lambda)}{d\lambda}$$
$$\frac{ds_\lambda}{d\tau_\lambda} = \frac{\pi_{s_\lambda}}{Q_\lambda}$$
$$\frac{d\pi_{s_\lambda}}{d\tau_\lambda} = \frac{(p'_\lambda)^2}{s_\lambda^3 W_\lambda} - \frac{k_B T}{s_\lambda}$$

$$
\begin{aligned}
\frac{dV}{d\tau_b} &= \frac{p_V}{s_b^2 W_v} \\
\frac{dp_V}{d\tau_b} &= \mathcal{P} - p_0 \\
\frac{ds_b}{d\tau_b} &= \frac{\pi_{s_b}}{Q_b} \\
\frac{d\pi_{s_b}}{d\tau_b} &= \frac{(p_V)^2}{s_b^3 W_v} - \frac{k_B T}{s_b}
\end{aligned}
$$

where $\mathcal{P}$ is the instantaneous pressure, which is given as:

$$
\mathcal{P} = \frac{1}{Vd}\left(\sum_{i=1}^{g_N} \frac{p_i^2}{m_i} + \sum_{i=1}^{d} \frac{p_{f,i}^2}{m_{f,i}} \right. \\
\left. + \sum_{i=1}^{g_N} \left[- (\nabla_{q_i} U)\, q_i - (\nabla_{q_i} U')\, q_i \right] - (Vd)\frac{\partial U}{\partial V} - (Vd)\frac{\partial U'}{\partial V} \right) \tag{5.57}
$$

where explicit depends on the volume is assumed for both U and U'.

The relationships between the real time t and the scaled time for each degree of freedom are given as follows:

$$
d\tau_p = s_p dt, \quad d\tau_f = s_f dt, \quad d\tau_\lambda = s_\lambda dt, \quad d\tau_b = s_b dt \tag{5.58}
$$

where t is the real time. Substituting the transformations given by Eqs. (5.57) and (5.58) into Eq. (5.56), we obtain the equations of motion with respect to the real time t as follows:

$$
\begin{aligned}
\frac{dq}{dt} &= \frac{p'}{s_p V^{1/d} m} + \frac{1}{Vd}\frac{dV}{dt} q \\
\frac{dp'}{dt} &= -s_p V^{1/d} \nabla_q U(q') - s_p V^{1/d} f(\lambda) \nabla_q u(q', q_f') \\
\frac{ds_p}{dt} &= s_p \frac{\pi_{s_p}}{Q_p} \\
\frac{d\pi_{s_p}}{dt} &= \frac{(p')^2}{s_p^2 V^{2/d} m} - k_B T \\
\frac{dq_f}{dt} &= \frac{p_f'}{s_f V^{1/d} m_f} + \frac{1}{Vd}\frac{dV}{dt} q_f \\
\frac{dp_f'}{dt} &= -s_f V^{1/d} f(\lambda) \nabla_{q_f} u(q', q_f') \\
\frac{ds_f}{dt} &= s_f \frac{\pi_{s_f}}{Q_f}
\end{aligned} \tag{5.59}
$$

$$\frac{d\pi_{s_f}}{dt} = \frac{(p'_f)^2}{s_f^2 V^{2/d} m_f} - k_B T$$

$$\frac{dq_\lambda}{dt} = \frac{p'_\lambda}{s_\lambda W_\lambda}$$

$$\frac{dp'_\lambda}{dt} = -s_\lambda \frac{df(\lambda)}{d\lambda} u(q', q'_f) - s_\lambda \frac{d\theta(\lambda)}{d\lambda}$$

$$\frac{ds_\lambda}{dt} = s_\lambda \frac{\pi_{s_\lambda}}{Q_\lambda}$$

$$\frac{d\pi_{s_\lambda}}{dt} = \frac{(p'_\lambda)^2}{s_\lambda^2 W_\lambda} - k_B T$$

$$\frac{dV}{dt} = \frac{p_V}{s_b W_v}$$

$$\frac{dp_V}{dt} = s_b(\mathcal{P} - p_0)$$

$$\frac{ds_b}{dt} = s_b \frac{\pi_{s_b}}{Q_b}$$

$$\frac{d\pi_{s_b}}{dt} = \frac{(p_V)^2}{s_b^2 W_v} - k_B T$$

Introducing the following changes of the variables

$$p = \frac{p'}{s_p V^{1/d}}, \quad p_f = \frac{p'_f}{s_f V^{1/d}}, \quad p_\lambda = \frac{p'_\lambda}{s_\lambda}, \quad \pi_V = \frac{p_V}{s_b} \qquad (5.60)$$
$$\hat{\pi}_p = \frac{\pi_p}{s_p}, \quad \hat{\pi}_f = \frac{\pi_f}{s_f}, \quad \hat{\pi}_\lambda = \frac{\pi_\lambda}{s_\lambda}, \quad \hat{\pi}_b = \frac{\pi_b}{s_b}$$

we obtain these equations of motion with respect to the new variables and the real time:

$$\frac{dq}{dt} = \frac{p}{m} + \frac{1}{Vd}\frac{dV}{dt} q \qquad (5.61)$$

$$\frac{dp}{dt} = -\nabla_q U(q) - f(\lambda)\nabla_q u(q, q_f) - \frac{1}{Vd}\frac{dV}{dt} p - \frac{1}{s_p}\frac{ds_p}{dt} p$$

$$\frac{ds_p}{dt} = s_p^2 \frac{\hat{\pi}_{s_p}}{Q_p}$$

$$\frac{d\hat{\pi}_{s_p}}{dt} = \frac{1}{s_p}\frac{p^2}{m} - \frac{k_B T}{s_p} - \frac{1}{s_p}\frac{ds_p}{dt}\hat{\pi}_{s_p}$$

$$\frac{dq_f}{dt} = \frac{p_f}{m_f} + \frac{1}{Vd}\frac{dV}{dt} q_f$$

$$\frac{dp_f}{dt} = -f(\lambda)\nabla_{q_f} u(q, q_f) - \frac{1}{Vd}\frac{dV}{dt} p_f - \frac{1}{s_f}\frac{ds_f}{dt} p_f$$

$$\frac{ds_f}{dt} = s_f^2 \frac{\hat{\pi}_{s_f}}{Q_f}$$
$$\frac{d\hat{\pi}_{s_f}}{dt} = \frac{1}{s_f}\frac{p_f^2}{m_f} - \frac{k_B T}{s_f} - \frac{1}{s_f}\frac{ds_f}{dt}\hat{\pi}_{s_f}$$
$$\frac{dq_\lambda}{dt} = \frac{p_\lambda}{W_\lambda}$$
$$\frac{dp_\lambda}{dt} = -\frac{df(\lambda)}{d\lambda}u(q, q_f) - \frac{d\theta(\lambda)}{d\lambda} - \frac{1}{s_\lambda}\frac{ds_\lambda}{dt}p_\lambda$$
$$\frac{ds_\lambda}{dt} = s_\lambda^2 \frac{\hat{\pi}_{s_\lambda}}{Q_\lambda}$$
$$\frac{d\hat{\pi}_{s_\lambda}}{dt} = \frac{1}{s_\lambda}\frac{p_\lambda^2}{W_\lambda} - \frac{k_B T}{s_\lambda} - \frac{1}{s_\lambda}\frac{ds_\lambda}{dt}\hat{\pi}_{s_\lambda}$$
$$\frac{dV}{dt} = \frac{\pi_V}{W_v}$$
$$\frac{d\pi_V}{dt} = \mathcal{P} - p_0 - \frac{1}{s_b}\frac{ds_b}{dt}\pi_V$$
$$\frac{ds_b}{dt} = s_b^2 \frac{\hat{\pi}_{s_b}}{Q_b}$$
$$\frac{d\hat{\pi}_{s_b}}{dt} = \frac{\pi_V^2}{W_v} - \frac{k_B T}{s_b} - \frac{1}{s_b}\frac{ds_b}{dt}\hat{\pi}_{s_b}$$

Another transformation of the variables is suggested:

$$\begin{aligned}
\pi_{\eta_p} &= s_p \hat{\pi}_{s_p}, \quad \eta_p = \ln s_p \\
\pi_{\eta_f} &= s_f \hat{\pi}_{s_f}, \quad \eta_f = \ln s_f \\
\pi_{\eta_\lambda} &= s_\lambda \hat{\pi}_{s_\lambda}, \quad \eta_\lambda = \ln s_\lambda \\
\pi_{\eta_b} &= s_b \hat{\pi}_{s_b}, \quad \eta_b = \ln s_b \\
\varepsilon &= \frac{1}{d}\ln V, \quad \pi_\varepsilon = \frac{\pi_V}{d}
\end{aligned} \tag{5.62}$$

Substituting transformations of type in Eq. (5.62) into Eq. (5.61), we obtain:

$$\begin{aligned}
\frac{dq}{dt} &= \frac{p}{m} + \frac{\pi_\varepsilon}{W_v} q \\
\frac{dp}{dt} &= -\nabla_q U(q) - f(\lambda)\nabla_q u(q, q_f) - \frac{\pi_\varepsilon}{W_v} p - \frac{\pi_{\eta_p}}{Q_p} p \\
\frac{d\eta_p}{dt} &= \frac{\pi_{\eta_p}}{Q_p} \\
\frac{d\pi_{\eta_p}}{dt} &= \frac{p^2}{m} - k_B T
\end{aligned} \tag{5.63}$$

$$\frac{dq_f}{dt} = \frac{p_f}{m_f} + \frac{\pi_\varepsilon}{W_v} q_f$$
$$\frac{dp_f}{dt} = -f(\lambda)\nabla_{q_f} u(q, q_f) - \frac{\pi_\varepsilon}{W_v} p_f - \frac{\pi_{\eta_f}}{Q_f} p_f$$
$$\frac{d\eta_f}{dt} = \frac{\pi_{\eta_f}}{Q_f}$$
$$\frac{d\pi_{\eta_f}}{dt} = \frac{p_f^2}{m_f} - k_B T$$
$$\frac{dq_\lambda}{dt} = \frac{p_\lambda}{W_\lambda}$$
$$\frac{dp_\lambda}{dt} = -\frac{df(\lambda)}{d\lambda} u(q, q_f) - \frac{d\theta(\lambda)}{d\lambda} - \frac{\pi_{\eta_\lambda}}{Q_\lambda} p_\lambda$$
$$\frac{d\eta_\lambda}{dt} = \frac{\pi_{\eta_\lambda}}{Q_\lambda}$$
$$\frac{d\pi_{\eta_\lambda}}{dt} = \frac{p_\lambda^2}{W_\lambda} - k_B T$$
$$\frac{d\varepsilon}{dt} = \frac{\pi_\varepsilon}{W_v}$$
$$\frac{d\pi_\varepsilon}{dt} = Vd\,(\mathcal{P} - p_0) - \frac{\pi_{\eta_b}}{Q_b} \pi_\varepsilon$$
$$\frac{d\eta_b}{dt} = \frac{\pi_{\eta_b}}{Q_b}$$
$$\frac{d\pi_{\eta_b}}{dt} = \frac{\pi_\varepsilon^2}{W_v} - k_B T$$

It is straightforward to show that the extended system of equations of the variables samples micro-canonical ensemble with a partition function given as:

$$\begin{aligned}\Sigma_{\mu pT} &= \sum_{N=0}^{\infty} Z_{NpT} \exp(\beta\mu N) \\ &= \sum_{N=0}^{\infty} \frac{1}{N! h^{g_N}} \\ &\times \left[\prod_{i=1}^{g_N} \int d\pi_{s_{p,i}} \int ds_{p,i} \int dp_i' \int dq_i' \right] \\ &\times \left[\prod_{i=1}^{d} \int d\pi_{s_{f,i}} \int ds_{f,i} \int dp_{f,i}' \int dq_{f,i}' \right] \\ &\times \int d\pi_{s_\lambda} \int ds_\lambda \int dp_\lambda' \int d\lambda \end{aligned} \tag{5.64}$$

$$
\begin{aligned}
&\times \int d\pi_{s_b} \int ds_b \int dp_V \int dV \\
&\times \delta \left\{ \sum_{i=1}^{g_N} \left[\frac{(p_i')^2}{2s_{p,i}^2\, V^{2/d} m_i} + \frac{\pi_{s_{p,i}}^2}{2Q_{s_{p,i}}} + k_B T \ln s_{p,i} \right] \right. \\
&+ \sum_{i=1}^{d} \left[\frac{(p_{f,i}')^2}{2s_{f,i}^2\, V^{2/d} m_{f,i}} + \frac{\pi_{s_{f,i}}^2}{2Q_{s_{f,i}}} + k_B T \ln s_{f,i} \right] \\
&+ \frac{p_\lambda^2}{2\, s_\lambda^2 W_\lambda} + \frac{\pi_{s_\lambda}^2}{2Q_{s_\lambda}} + k_B T \ln s_\lambda \\
&+ U(q_1', q_2', \ldots, q_{g_N}') \\
&+ U'(q_1', q_2', \ldots, q_{g_N}' : q_{f,1}' \ldots, q_{f,d}', \lambda) + \theta(\lambda) \\
&\left. + \frac{p_V^2}{2s_b^2 W_v} + pV + \frac{\pi_{s_b}^2}{2s_b^2 Q_b} + k_B T \ln s_b - E \right\}
\end{aligned}
$$

We can integrate the extended variable dynamics and apply the variable transformations as shown above, to get:

$$
\begin{aligned}
\Sigma_{\mu p T} &= \sum_{N=0}^{\infty} \frac{1}{N! h^{g_N}} \frac{1}{(g_N + d + 2) k_B T} \\
&\times \prod_{i=1}^{g_N} \left[2\pi Q_{p,i} k_B T\right]^{1/2} \prod_{i=1}^{d} \left[2\pi Q_{f,i} k_B T\right]^{1/2} \\
&\times [2\pi Q_\lambda k_B T]^{1/2} [2\pi Q_b k_B T]^{1/2} \exp(E/k_B T) \\
&\times \left[\prod_{i=1}^{g_N} \int dp_i \int dq_i \right] \left[\prod_{i=1}^{d} \int dp_{f,i} \int dq_{f,i} \right] \\
&\times \int dp_\lambda \int d\lambda \int d\pi_\varepsilon \int dV \\
&\times \exp\left(-H_{0,\lambda}/k_B T\right) \exp(-H_\lambda/k_B T)
\end{aligned}
\tag{5.65}
$$

where

$$
\begin{aligned}
H_{0,\lambda} &= \sum_{i=1}^{g_N} \frac{p_i^2}{2m_i} + \sum_{i=1}^{d} \frac{p_{f,i}^2}{2m_{f,i}} \\
&+ U(q_1', q_2', \ldots, q_{g_N}') \\
&+ U'(q_1', q_2', \ldots, q_{g_N}' : q_{f,1}' \ldots, q_{f,d}', \lambda) \\
&+ \frac{\pi_\varepsilon^2}{2W_v} + pV
\end{aligned}
\tag{5.66}
$$

and

$$H_\lambda = \frac{p_\lambda^2}{2W_\lambda} + \theta(\lambda) \tag{5.67}$$

This indicates that a grand isothermal-isobaric ensemble is generated.

5.3.3.2 Nosé-Andersen Chain of Thermostats for Grand Isothermal-isobaric Ensemble

For the Nosé-Andersen chain of thermostats grand isothermal-isobaric ensemble the Hamiltonian function is proposed as follows:

$$\begin{aligned}
H_{NAC}^{GIIE} = \sum_{i=1}^{g_N} \Bigg[& \frac{(p_i')^2}{2s_{p_1,i}^2 V^{2/d} m_i} \\
& + \sum_{j=1}^{M-1} \frac{\pi_{p_j,i}^2}{2s_{p_{j+1},i}^2 Q_{p_j,i}} + \frac{\pi_{p_M,i}^2}{2Q_{p_M,i}} + \sum_{j=1}^{M} k_B T \ln s_{p_j,i} \Bigg] \\
& + \sum_{i=1}^{d} \Bigg[\frac{(p'_{f,i})^2}{2s_{f_1,i}^2 V^{2/d} m_{f,i}} \\
& + \sum_{j=1}^{M-1} \frac{\pi_{f_j,i}^2}{2s_{f_{j+1},i}^2 Q_{f_j,i}} + \frac{\pi_{f_M,i}^2}{2Q_{f_M,i}} + \sum_{j=1}^{M} k_B T \ln s_{f_j,i} \Bigg] \\
& + \frac{p_\lambda^2}{2s_{\lambda_1}^2 W_\lambda} \\
& + \sum_{j=1}^{M-1} \frac{\pi_{\lambda_j}^2}{2s_{\lambda_{j+1}}^2 Q_{\lambda_j}} + \frac{\pi_{\lambda_M}^2}{2Q_{\lambda_M}} + \sum_{j=1}^{M} k_B T \ln s_{\lambda_j} \\
& + \frac{p_v^2}{2s_{b_1}^2 W_v} \\
& + \sum_{j=1}^{M-1} \frac{\pi_{b_j}^2}{2s_{b_{j+1}}^2 Q_{b_j}} + \frac{\pi_{b_M}^2}{2Q_{b_M}} + \sum_{j=1}^{M} k_B T \ln s_{b_j} \\
& + U(q_1', q_2', \ldots, q_{g_N}') \\
& + U'(q_1', q_2', \ldots, q_{g_N}' : q_{f,1}' \ldots, q_{f,d}', \lambda) \\
& + \theta(\lambda) + pV
\end{aligned} \tag{5.68}$$

where for the simplicity of the appearance, it is assumed the same length of the chain for each degree of the freedom. However, in practice, different length of the chain of the thermostats can be considered. Besides, in the following, we are going to drop off the subscript i and use the Hamilton's equations of motions, and we can obtain the equations of motion as follows:

$$
\begin{aligned}
\frac{dq'}{d\tau_{p_1}} &= \frac{p'}{s_{p_1}^2 V^{2/d} m} \\
\frac{dp'}{d\tau_{p_1}} &= -V^{1/d}\nabla_q U(q) - V^{1/d} f(\lambda)\nabla_q u(q, q_f) \\
\frac{ds_{p_1}}{d\tau_{p_2}} &= \frac{\pi_{p_1}}{s_{p_2}^2 Q_{p_1}} \\
\frac{d\pi_{p_1}}{d\tau_{p_2}} &= \frac{(p')^2}{s_{p_1}^3 V^{2/d} m} - \frac{k_B T}{s_{p_1}} \\
\frac{ds_{p_k}}{d\tau_{p_{k+1}}} &= \frac{\pi_{p_k}}{s_{p_{k+1}}^2 Q_{p_k}} \\
\frac{d\pi_{p_k}}{d\tau_{p_{k+1}}} &= \frac{\pi_{p_{k-1}}^2}{s_{p_k}^3 Q_{p_{k-1}}} - \frac{k_B T}{s_{p_k}}, \quad k = 2, \ldots, M-1 \\
\frac{ds_{p_M}}{d\tau_{p_M}} &= \frac{\pi_{p_M}}{Q_{p_M}} \\
\frac{d\pi_{p_M}}{d\tau_{p_M}} &= \frac{\pi_{p_{M-1}}^2}{s_{p_M}^3 Q_{p_{M-1}}} - \frac{k_B T}{s_{p_M}} \\
\frac{dq_f'}{d\tau_{f_1}} &= \frac{p_f'}{s_{f_1}^2 V^{2/d} m_f} \\
\frac{dp_f'}{d\tau_{f_1}} &= -V^{1/d} f(\lambda)\nabla_{q_f} u(q, q_f) \\
\frac{ds_{f_1}}{d\tau_{f_2}} &= \frac{\pi_{f_1}}{s_{f_2}^2 Q_{f_1}} \\
\frac{d\pi_{f_1}}{d\tau_{f_2}} &= \frac{(p_f')^2}{s_{f_1}^3 V^{2/d} m_f} - \frac{k_B T}{s_{f_1}} \\
\frac{ds_{f_k}}{d\tau_{f_{k+1}}} &= \frac{\pi_{f_k}}{s_{f_{k+1}}^2 Q_{f_k}} \\
\frac{d\pi_{f_k}}{d\tau_{f_{k+1}}} &= \frac{\pi_{f_{k-1}}^2}{s_{f_k}^3 Q_{f_{k-1}}} - \frac{k_B T}{s_{f_k}}, \quad k = 2, \ldots, M-1 \\
\frac{ds_{f_M}}{d\tau_{f_M}} &= \frac{\pi_{f_M}}{Q_{f_M}} \\
\frac{d\pi_{f_M}}{d\tau_{f_M}} &= \frac{\pi_{f_{M-1}}^2}{s_{f_M}^3 Q_{f_{M-1}}} - \frac{k_B T}{s_{f_M}} \\
\frac{d\lambda}{d\tau_{\lambda_1}} &= \frac{p_\lambda}{s_{\lambda_1}^2 W_\lambda} \\
\frac{dp_\lambda}{d\tau_{\lambda_1}} &= -\frac{df(\lambda)}{d\lambda} u(q, q_f) - \frac{d\theta(\lambda)}{d\lambda}
\end{aligned}
\tag{5.69}
$$

$$
\begin{aligned}
\frac{ds_{\lambda_1}}{d\tau_{\lambda_2}} &= \frac{\pi_{\lambda_1}}{s_{\lambda_2}^2 Q_{\lambda_1}} \\
\frac{d\pi_{\lambda_1}}{d\tau_{\lambda_2}} &= \frac{p_\lambda^2}{s_{\lambda_1}^3 W_\lambda} - \frac{k_B T}{s_{\lambda_1}} \\
\frac{ds_{\lambda_k}}{d\tau_{\lambda_{k+1}}} &= \frac{\pi_{\lambda_k}}{s_{\lambda_{k+1}}^2 Q_{\lambda_k}} \\
\frac{d\pi_{\lambda_k}}{d\tau_{\lambda_{k+1}}} &= \frac{\pi_{\lambda_{k-1}}^2}{s_{\lambda_k}^3 Q_{\lambda_{k-1}}} - \frac{k_B T}{s_{\lambda_k}}, \quad k = 2, \ldots, M-1 \\
\frac{ds_{\lambda_M}}{d\tau_{\lambda_M}} &= \frac{\pi_{\lambda_M}}{Q_{\lambda_M}} \\
\frac{d\pi_{\lambda_M}}{d\tau_{\lambda_M}} &= \frac{\pi_{\lambda_{M-1}}^2}{s_{\lambda_M}^3 Q_{\lambda_{M-1}}} - \frac{k_B T}{s_{\lambda_M}} \\
\frac{dV}{d\tau_{b_1}} &= \frac{p_v}{s_{b_1}^2 W_v} \\
\frac{dp_v}{d\tau_{b_1}} &= \mathcal{P} - p_0 \\
\frac{ds_{b_1}}{d\tau_{b_2}} &= \frac{\pi_{b_1}}{s_{b_2}^2 Q_{b_1}} \\
\frac{d\pi_{b_1}}{d\tau_{b_2}} &= \frac{p_v^2}{s_{b_1}^3 W_v} - \frac{k_B T}{s_{b_1}} \\
\frac{ds_{b_k}}{d\tau_{b_{k+1}}} &= \frac{\pi_{b_k}}{s_{b_{k+1}}^2 Q_{b_k}} \\
\frac{d\pi_{b_k}}{d\tau_{b_{k+1}}} &= \frac{\pi_{b_{k-1}}^2}{s_{b_k}^3 Q_{b_{k-1}}} - \frac{k_B T}{s_{b_k}}, \quad k = 2, \ldots, M-1 \\
\frac{ds_{b_M}}{d\tau_{b_M}} &= \frac{\pi_{b_M}}{Q_{b_M}} \\
\frac{d\pi_{b_M}}{d\tau_{b_M}} &= \frac{\pi_{b_{M-1}}^2}{s_{b_M}^3 Q_{b_{M-1}}} - \frac{k_B T}{s_{b_M}}
\end{aligned}
$$

where the relationships between the scaled times and the real time t are given as follows:

$$
\begin{aligned}
d\tau_{p_1} &= s_{p_1} dt \\
d\tau_{p_k} &= s_{p_{k-1}} s_{p_k} dt, \quad (k = 2, \ldots, M-1) \\
d\tau_{p_M} &= s_{p_M} dt \\
d\tau_{f_1} &= s_{f_1} dt \\
d\tau_{f_k} &= s_{f_{k-1}} s_{f_k} dt, \quad (k = 2, \ldots, M-1)
\end{aligned}
\tag{5.70}
$$

$$
\begin{aligned}
d\tau_{\lambda_1} &= s_{\lambda_1} dt \\
d\tau_{\lambda_k} &= s_{\lambda_{k-1}} s_{\lambda_k} dt, \quad (k = 2, \ldots, M-1) \\
d\tau_{\lambda_M} &= s_{\lambda_M} dt \\
d\tau_{b_1} &= s_{b_1} dt \\
d\tau_{b_k} &= s_{b_{k-1}} s_{b_k} dt, \quad (k = 2, \ldots, M-1) \\
d\tau_{b_M} &= s_{b_M} dt
\end{aligned}
$$

Substituting these transformations and the relationships given by Eq. (5.70) into Eq. (5.69), we obtain:

$$
\begin{aligned}
\frac{dq}{dt} &= \frac{p'}{s_{p_1} V^{1/d} m} + \frac{1}{Vd}\frac{dV}{dt} q \\
\frac{dp'}{dt} &= -s_{p_1} V^{1/d} \nabla_q U(q) - s_{p_1} V^{1/d} f(\lambda) \nabla_q u(q, q_f) \\
\frac{ds_{p_1}}{dt} &= s_{p_1} \frac{\pi_{p_1}}{s_{p_2} Q_{p_1}} \\
\frac{d\pi_{p_1}}{dt} &= s_{p_2} \frac{(p')^2}{s_{p_1}^2 V^{2/d} m} - s_{p_2} k_B T \\
\frac{ds_{p_k}}{dt} &= s_{p_k} \frac{\pi_{p_k}}{s_{p_{k+1}} Q_{p_k}} \\
\frac{d\pi_{p_k}}{dt} &= s_{p_{k+1}} \frac{\pi_{p_{k-1}}^2}{s_{p_k}^2 Q_{p_{k-1}}} - s_{p_{k+1}} k_B T, \quad k = 2, \ldots, M-1 \\
\frac{ds_{p_M}}{dt} &= s_{p_M} \frac{\pi_{p_M}}{Q_{p_M}} \\
\frac{d\pi_{p_M}}{dt} &= \frac{\pi_{p_{M-1}}^2}{s_{p_M}^2 Q_{p_{M-1}}} - k_B T \\
\frac{dq_f}{dt} &= \frac{p'_f}{s_{f_1} V^{1/d} m_f} + \frac{1}{Vd}\frac{dV}{dt} q_f \\
\frac{dp'_f}{dt} &= -s_{f_1} V^{1/d} f(\lambda) \nabla_{q_f} u(q, q_f) \\
\frac{ds_{f_1}}{dt} &= s_{f_1} \frac{\pi_{f_1}}{s_{f_2} Q_{f_1}} \\
\frac{d\pi_{f_1}}{dt} &= s_{f_2} \frac{(p'_f)^2}{s_{f_1}^2 V^{2/d} m} - s_{f_2} k_B T \\
\frac{ds_{f_k}}{dt} &= s_{f_k} \frac{\pi_{f_k}}{s_{f_{k+1}} Q_{f_k}} \\
\frac{d\pi_{f_k}}{dt} &= s_{f_{k+1}} \frac{\pi_{f_{k-1}}^2}{s_{f_k}^2 Q_{f_{k-1}}} - s_{f_{k+1}} k_B T, \quad k = 2, \ldots, M-1
\end{aligned}
\tag{5.71}
$$

$$\frac{ds_{f_M}}{dt} = s_{f_M} \frac{\pi_{f_M}}{Q_{f_M}}$$

$$\frac{d\pi_{f_M}}{dt} = \frac{\pi_{f_{M-1}}^2}{s_{f_M}^2 Q_{f_{M-1}}} - k_B T$$

$$\frac{d\lambda}{dt} = \frac{p_\lambda}{s_{\lambda_1} W_\lambda}$$

$$\frac{dp_\lambda}{dt} = -s_{\lambda_1} \frac{df(\lambda)}{d\lambda} u(q, q_f) - s_{\lambda_1} \frac{d\theta(\lambda)}{d\lambda}$$

$$\frac{ds_{\lambda_1}}{dt} = s_{\lambda_1} \frac{\pi_{\lambda_1}}{s_{\lambda_2} Q_{\lambda_1}}$$

$$\frac{d\pi_{\lambda_1}}{dt} = s_{\lambda_2} \frac{p_\lambda^2}{s_{\lambda_1}^2 W_\lambda} - s_{\lambda_2} k_B T$$

$$\frac{ds_{\lambda_k}}{dt} = s_{\lambda_k} \frac{\pi_{\lambda_k}}{s_{\lambda_{k+1}} Q_{\lambda_k}}$$

$$\frac{d\pi_{\lambda_k}}{dt} = s_{\lambda_{k+1}} \frac{\pi_{\lambda_{k-1}}^2}{s_{\lambda_k}^2 Q_{\lambda_{k-1}}} - s_{\lambda_{k+1}} k_B T, \quad k = 2, \ldots, M-1$$

$$\frac{ds_{\lambda_M}}{dt} = s_{\lambda_M} \frac{\pi_{\lambda_M}}{Q_{\lambda_M}}$$

$$\frac{d\pi_{\lambda_M}}{dt} = \frac{\pi_{\lambda_{M-1}}^2}{s_{\lambda_M}^2 Q_{\lambda_{M-1}}} - k_B T$$

$$\frac{dV}{dt} = \frac{p_v}{s_{b_1} W_v}$$

$$\frac{dp_v}{dt} = s_{b_1} (\mathcal{P} - p_0)$$

$$\frac{ds_{b_1}}{dt} = s_{b_1} \frac{\pi_{b_1}}{s_{b_2} Q_{b_1}}$$

$$\frac{d\pi_{b_1}}{dt} = s_{b_2} \frac{p_v^2}{s_{b_1}^2 W_v} - s_{b_2} k_B T$$

$$\frac{ds_{b_k}}{dt} = s_{b_k} \frac{\pi_{b_k}}{s_{b_{k+1}} Q_{b_k}}$$

$$\frac{d\pi_{b_k}}{dt} = s_{b_{k+1}} \frac{\pi_{b_{k-1}}^2}{s_{b_k}^2 Q_{b_{k-1}}} - s_{b_{k+1}} k_B T, \quad k = 2, \ldots, M-1$$

$$\frac{ds_{b_M}}{dt} = s_{b_M} \frac{\pi_{b_M}}{Q_{b_M}}$$

$$\frac{d\pi_{b_M}}{dt} = \frac{\pi_{b_{M-1}}^2}{s_{b_M}^2 Q_{b_{M-1}}} - k_B T$$

In the following, we propose some additional transformations of the variables:

$$
\begin{aligned}
p &= \frac{p'}{s_{p_1} V^{1/d}}, \quad p_f = \frac{p'_f}{s_{f_1} V^{1/d}}, \quad \hat{\pi} = \frac{p_v}{s_{b_1}}, \quad \pi_\lambda = \frac{p_\lambda}{s_{\lambda_1}} \\
\hat{\pi}_{b_k} &= \frac{\pi_{b_k}}{s_{b_k} s_{b_{k+1}}}, \quad \hat{\pi}_{p_k} = \frac{\pi_{p_k}}{s_{p_k} s_{p_{k+1}}}, \quad \hat{\pi}_{f_k} = \frac{\pi_{s_{f_k}}}{s_{f_k} s_{f_{k+1}}}, \quad \hat{\pi}_{\lambda_k} = \frac{\pi_{s_{\lambda_k}}}{s_{\lambda_k} s_{\lambda_{k+1}}} \\
& \qquad (k = 1, \ldots, M-1) \\
\hat{\pi}_{b_M} &= \frac{\pi_{b_M}}{s_{b_M}}, \quad \hat{\pi}_{p_M} = \frac{\pi_{p_M}}{s_{p_M}}, \quad \hat{\pi}_{f_M} = \frac{\pi_{f_M}}{s_{f_M}}, \quad \hat{\pi}_{\lambda_M} = \frac{\pi_{\lambda_M}}{s_{\lambda_M}}
\end{aligned}
\tag{5.72}
$$

Substituting the transformation given by Eq. (5.72) into Eq. (5.71), we obtain:

$$
\begin{aligned}
\frac{dq}{dt} &= \frac{p}{m} + \frac{1}{Vd}\frac{dV}{dt} q \\
\frac{dp}{dt} &= -\nabla_q U(q) - f(\lambda)\nabla_q u(q, q_f) - \frac{1}{Vd}\frac{dV}{dt} p - \frac{1}{s_{p_1}}\frac{ds_{p_1}}{dt} p \\
\frac{ds_{p_1}}{dt} &= s_{p_1}^2 \frac{\hat{\pi}_{p_1}}{Q_{p_1}} \\
\frac{d\hat{\pi}_{p_1}}{dt} &= \frac{1}{s_{p_1}}\frac{p^2}{m} - \frac{k_B T}{s_{p_1}} - \frac{1}{s_{p_2}}\frac{ds_{p_2}}{dt}\hat{\pi}_{p_1} - \frac{1}{s_{p_1}}\frac{ds_{p_1}}{dt}\hat{\pi}_{p_1} \\
\frac{ds_{p_k}}{dt} &= s_{p_k}^2 \frac{\hat{\pi}_{p_k}}{Q_{p_k}} \\
\frac{d\hat{\pi}_{p_k}}{dt} &= s_{p_{k-1}}^2 \frac{\hat{\pi}_{p_{k-1}}^2}{s_{p_k} Q_{p_{k-1}}} - \frac{k_B T}{s_{p_k}} - \frac{1}{s_{p_{k+1}}}\frac{ds_{p_{k+1}}}{dt}\hat{\pi}_{p_k} - \frac{1}{s_{p_k}}\frac{ds_{p_k}}{dt}\hat{\pi}_{p_k} \\
& \qquad k = 2, \ldots, M-1 \\
\frac{ds_{p_M}}{dt} &= s_{p_M}^2 \frac{\hat{\pi}_{p_M}}{Q_{p_M}} \\
\frac{d\hat{\pi}_{p_M}}{dt} &= s_{p_{M-1}}^2 \frac{\hat{\pi}_{p_{M-1}}^2}{s_{p_M} Q_{p_{M-1}}} - \frac{k_B T}{s_{p_M}} - \frac{1}{s_{p_M}}\frac{ds_{p_M}}{dt}\hat{\pi}_{p_M} \\
\frac{dq_f}{dt} &= \frac{p_f}{m_f} + \frac{1}{Vd}\frac{dV}{dt} q_f \\
\frac{dp_f}{dt} &= -f(\lambda)\nabla_{q_f} u(q, q_f) - \frac{1}{Vd}\frac{dV}{dt} p_f - \frac{1}{s_{f_1}}\frac{ds_{f_1}}{dt} p_f \\
\frac{ds_{f_1}}{dt} &= s_{f_1}^2 \frac{\hat{\pi}_{f_1}}{Q_{f_1}} \\
\frac{d\hat{\pi}_{f_1}}{dt} &= \frac{1}{s_{f_1}}\frac{p_f^2}{m_f} - \frac{k_B T}{s_{f_1}} - \frac{1}{s_{f_2}}\frac{ds_{f_2}}{dt}\hat{\pi}_{f_1} - \frac{1}{s_{f_1}}\frac{ds_{f_1}}{dt}\hat{\pi}_{f_1} \\
\frac{ds_{f_k}}{dt} &= s_{f_k}^2 \frac{\hat{\pi}_{f_k}}{Q_{f_k}}
\end{aligned}
\tag{5.73}
$$

$$\frac{d\hat{\pi}_{f_k}}{dt} = s_{f_{k-1}}^2 \frac{\hat{\pi}_{f_{k-1}}^2}{s_{f_k} Q_{f_{k-1}}} - \frac{k_B T}{s_{f_k}} - \frac{1}{s_{f_{k+1}}} \frac{ds_{f_{k+1}}}{dt} \hat{\pi}_{f_k} - \frac{1}{s_{f_k}} \frac{ds_{f_k}}{dt} \hat{\pi}_{f_k}$$

$$k = 2, \ldots, M-1$$

$$\frac{ds_{f_M}}{dt} = s_{f_M}^2 \frac{\hat{\pi}_{f_M}}{Q_{f_M}}$$

$$\frac{d\hat{\pi}_{f_M}}{dt} = s_{f_{M-1}}^2 \frac{\hat{\pi}_{f_{M-1}}^2}{s_{f_M} Q_{f_{M-1}}} - \frac{k_B T}{s_{f_M}} - \frac{1}{s_{f_M}} \frac{ds_{f_M}}{dt} \hat{\pi}_{f_M}$$

$$\frac{d\lambda}{dt} = \frac{\pi_\lambda}{W_\lambda}$$

$$\frac{d\pi_\lambda}{dt} = -\frac{df(\lambda)}{d\lambda} u(q, q_f) - \frac{d\theta(\lambda)}{d\lambda} - \frac{1}{s_{\lambda_1}} \frac{ds_{\lambda_1}}{dt} \pi_\lambda$$

$$\frac{ds_{\lambda_1}}{dt} = s_{\lambda_1}^2 \frac{\hat{\pi}_{\lambda_1}}{Q_{\lambda_1}}$$

$$\frac{d\hat{\pi}_{\lambda_1}}{dt} = \frac{1}{s_{\lambda_1}} \frac{p_\lambda^2}{W_\lambda} - \frac{k_B T}{s_{\lambda_1}} - \frac{1}{s_{\lambda_2}} \frac{ds_{\lambda_2}}{dt} \hat{\pi}_{\lambda_1} - \frac{1}{s_{\lambda_1}} \frac{ds_{\lambda_1}}{dt} \hat{\pi}_{\lambda_1}$$

$$\frac{ds_{\lambda_k}}{dt} = s_{\lambda_k}^2 \frac{\hat{\pi}_{\lambda_k}}{Q_{\lambda_k}}$$

$$\frac{d\hat{\pi}_{\lambda_k}}{dt} = s_{\lambda_{k-1}}^2 \frac{\hat{\pi}_{\lambda_{k-1}}^2}{s_{\lambda_k} Q_{\lambda_{k-1}}} - \frac{k_B T}{s_{\lambda_k}} - \frac{1}{s_{\lambda_{k+1}}} \frac{ds_{\lambda_{k+1}}}{dt} \hat{\pi}_{\lambda_k} - \frac{1}{s_{\lambda_k}} \frac{ds_{\lambda_k}}{dt} \hat{\pi}_{\lambda_k}$$

$$k = 2, \ldots, M-1$$

$$\frac{ds_{\lambda_M}}{dt} = s_{\lambda_M}^2 \frac{\hat{\pi}_{\lambda_M}}{Q_{\lambda_M}}$$

$$\frac{d\hat{\pi}_{\lambda_M}}{dt} = s_{\lambda_{M-1}}^2 \frac{\hat{\pi}_{\lambda_{M-1}}^2}{s_{\lambda_M} Q_{\lambda_{M-1}}} - \frac{k_B T}{s_{\lambda_M}} - \frac{1}{s_{\lambda_M}} \frac{ds_{\lambda_M}}{dt} \hat{\pi}_{\lambda_M}$$

$$\frac{dV}{dt} = \frac{\hat{\pi}_v}{W_v}$$

$$\frac{d\hat{\pi}_v}{dt} = (\mathcal{P} - p_0) - \frac{1}{s_{b_1}} \frac{ds_{b_1}}{dt} \pi_v$$

$$\frac{ds_{b_1}}{dt} = s_{b_1}^2 \frac{\hat{\pi}_{b_1}}{Q_{b_1}}$$

$$\frac{d\hat{\pi}_{b_1}}{dt} = \frac{1}{s_{b_1}} \frac{\hat{\pi}_v^2}{W_v} - \frac{k_B T}{s_{b_1}} - \frac{1}{s_{b_2}} \frac{ds_{b_2}}{dt} \hat{\pi}_{b_1} - \frac{1}{s_{b_1}} \frac{ds_{b_1}}{dt} \hat{\pi}_{b_1}$$

$$\frac{ds_{b_k}}{dt} = s_{b_k}^2 \frac{\hat{\pi}_{b_k}}{Q_{b_k}}$$

$$\frac{d\hat{\pi}_{b_k}}{dt} = s_{b_{k-1}}^2 \frac{\hat{\pi}_{b_{k-1}}^2}{s_{b_k} Q_{b_{k-1}}} - \frac{k_B T}{s_{b_k}} - \frac{1}{s_{b_{k+1}}} \frac{ds_{b_{k+1}}}{dt} \hat{\pi}_{b_k} - \frac{1}{s_{b_k}} \frac{ds_{b_k}}{dt} \hat{\pi}_{b_k}$$

$$k = 2, \ldots, M-1$$

$$\frac{ds_{b_M}}{dt} = s_{b_M}^2 \frac{\hat{\pi}_{b_M}}{Q_{b_M}}$$
$$\frac{d\hat{\pi}_{b_M}}{dt} = s_{b_{M-1}}^2 \frac{\hat{\pi}_{b_{M-1}}^2}{s_{b_M} Q_{b_{M-1}}} - \frac{k_B T}{s_{b_M}} - \frac{1}{s_{b_M}} \frac{ds_{b_M}}{dt} \hat{\pi}_{b_M}$$

Another change in the variables is introduced as:

$$\pi_{x_k} = s_{x_k} \hat{\pi}_{x_k}, \ \eta_{x_k} = \ln s_{x_k} \qquad (5.74)$$
$$\varepsilon = \frac{1}{d} \ln V, \ \pi_\varepsilon = \frac{\hat{\pi}_v}{d}$$

where $x = p, f, \lambda, b$ and $k = 1, 2, \ldots, M$. Then, substituting these transformations in Eq. (5.73), we can the final form of the equations of motion for the grand isothermal-isobaric ensemble as follows:

$$\frac{dq}{dt} = \frac{p}{m} + \frac{\pi_\varepsilon}{W_v} q \qquad (5.75)$$
$$\frac{dp}{dt} = -\nabla_q U(q) - f(\lambda) \nabla_q u(q, q_f) - \frac{\pi_\varepsilon}{W_v} p - \frac{\pi_{p_1}}{Q_{p_1}} p$$
$$\frac{d\eta_{p_1}}{dt} = \frac{\pi_{p_1}}{Q_{p_1}}$$
$$\frac{d\pi_{p_1}}{dt} = \frac{p^2}{m} - k_B T - \frac{\pi_{p_2}}{Q_{p_2}} \pi_{p_1}$$
$$\frac{d\eta_{p_k}}{dt} = \frac{\pi_{p_k}}{Q_{p_k}}$$
$$\frac{d\pi_{p_k}}{dt} = \frac{\pi_{p_{k-1}}^2}{Q_{p_{k-1}}} - k_B T - \frac{\pi_{p_{k+1}}}{Q_{p_{k+1}}} \pi_{p_k}$$
$$k = 2, \ldots, M-1$$
$$\frac{d\eta_{p_M}}{dt} = \frac{\pi_{p_M}}{Q_{p_M}}$$
$$\frac{d\pi_{p_M}}{dt} = \frac{\pi_{p_{M-1}}^2}{Q_{p_{M-1}}} - k_B T$$
$$\frac{dq_f}{dt} = \frac{p_f}{m_f} + \frac{\pi_\varepsilon}{W_v} q_f$$
$$\frac{dp_f}{dt} = -f(\lambda) \nabla_{q_f} u(q, q_f) - \frac{\pi_\varepsilon}{W_v} p_f - \frac{\pi_{f_1}}{Q_{f_1}} p_f$$
$$\frac{d\eta_{f_1}}{dt} = \frac{\pi_{f_1}}{Q_{f_1}}$$
$$\frac{d\pi_{f_1}}{dt} = \frac{p_f^2}{m_f} - k_B T - \frac{\pi_{f_2}}{Q_{f_2}} \pi_{f_1}$$

$$\frac{d\eta_{f_k}}{dt} = \frac{\pi_{f_k}}{Q_{f_k}}$$

$$\frac{d\pi_{f_k}}{dt} = \frac{\pi_{f_{k-1}}^2}{Q_{f_{k-1}}} - k_B T - \frac{\pi_{f_{k+1}}}{Q_{f_{k+1}}}\pi_{f_k}$$

$$k = 2, \ldots, M-1$$

$$\frac{d\eta_{f_M}}{dt} = \frac{\pi_{f_M}}{Q_{f_M}}$$

$$\frac{d\pi_{f_M}}{dt} = \frac{\pi_{f_{M-1}}^2}{Q_{f_{M-1}}} - k_B T$$

$$\frac{d\lambda}{dt} = \frac{\pi_\lambda}{W_\lambda}$$

$$\frac{d\pi_\lambda}{dt} = -\frac{df(\lambda)}{d\lambda}u(q, q_f) - \frac{d\theta(\lambda)}{d\lambda} - \frac{\pi_{\lambda_1}}{Q_{\lambda_1}}p_\lambda$$

$$\frac{d\eta_{\lambda_1}}{dt} = \frac{\pi_{\lambda_1}}{Q_{\lambda_1}}$$

$$\frac{d\pi_{\lambda_1}}{dt} = \frac{\pi_\lambda^2}{W_\lambda} - k_B T - \frac{\pi_{\lambda_2}}{Q_{\lambda_2}}\pi_{\lambda_1}$$

$$\frac{d\eta_{\lambda_k}}{dt} = \frac{\pi_{\lambda_k}}{Q_{\lambda_k}}$$

$$\frac{d\pi_{\lambda_k}}{dt} = \frac{\pi_{\lambda_{k-1}}^2}{Q_{\lambda_{k-1}}} - k_B T - \frac{\pi_{\lambda_{k+1}}}{Q_{\lambda_{k+1}}}\pi_{\lambda_k}$$

$$k = 2, \ldots, M-1$$

$$\frac{d\eta_{\lambda_M}}{dt} = \frac{\pi_{\lambda_M}}{Q_{\lambda_M}}$$

$$\frac{d\pi_{\lambda_M}}{dt} = \frac{\pi_{\lambda_{M-1}}^2}{Q_{\lambda_{M-1}}} - k_B T$$

$$\frac{d\varepsilon}{dt} = \frac{\pi_\varepsilon}{W_\upsilon}$$

$$\frac{d\pi_\varepsilon}{dt} = Vd\left(\mathcal{P} - p_0\right) - \frac{\pi_{b_1}}{Q_{b_1}}\pi_\varepsilon$$

$$\frac{d\eta_{b_1}}{dt} = \frac{\pi_{b_1}}{Q_{b_1}}$$

$$\frac{d\pi_{b_1}}{dt} = \frac{\pi_\varepsilon^2}{W_\upsilon} - k_B T - \frac{\pi_{b_2}}{Q_{b_2}}\pi_{b_1}$$

$$\frac{d\eta_{b_k}}{dt} = \frac{\pi_{b_k}}{Q_{b_k}}$$

$$\frac{d\pi_{b_k}}{dt} = \frac{\pi_{b_{k-1}}^2}{Q_{b_{k-1}}} - k_B T - \frac{\pi_{b_{k+1}}}{Q_{b_{k+1}}}\pi_{b_k}$$

$$k = 2, \ldots, M-1$$

$$\frac{d\eta_{b_M}}{dt} = \frac{\pi_{b_M}}{Q_{b_M}}$$

$$\frac{d\pi_{b_M}}{dt} = \frac{\pi^2_{b_{M-1}}}{Q_{b_{M-1}}} - k_B T$$

To produce the grand isothermal-isobaric ensemble distribution, the following changes are proposed for each degree of freedom i (for $i = 1, 2, \ldots, g_N$):

$$\frac{dq_i}{dt} = \frac{p_i}{m_i} + \frac{\pi_\varepsilon}{W_v} q_i \tag{5.76}$$

$$\frac{dp_i}{dt} = -\nabla_{q_i} U(\mathbf{q}) - f(\lambda)\nabla_{q_i} u(\mathbf{q}, q_f) - \left(1 + \frac{d}{g_N}\right)\frac{\pi_\varepsilon}{W_v} p_i - \frac{\pi_{p_1}}{Q_{p_1}} p_i$$

$$\frac{d\eta_{p_1}}{dt} = \frac{\pi_{p_1}}{Q_{p_1}}$$

$$\frac{d\pi_{p_1}}{dt} = \frac{p_i^2}{m_i} - k_B T - \frac{\pi_{p_2}}{Q_{p_2}} \pi_{p_1}$$

$$\frac{d\eta_{p_k}}{dt} = \frac{\pi_{p_k}}{Q_{p_k}}$$

$$\frac{d\pi_{p_k}}{dt} = \frac{\pi^2_{p_{k-1}}}{Q_{p_{k-1}}} - k_B T - \frac{\pi_{p_{k+1}}}{Q_{p_{k+1}}} \pi_{p_k}$$

$$k = 2, \ldots, M-1$$

$$\frac{d\eta_{p_M}}{dt} = \frac{\pi_{p_M}}{Q_{p_M}}$$

$$\frac{d\pi_{p_M}}{dt} = \frac{\pi^2_{p_{M-1}}}{Q_{p_{M-1}}} - k_B T$$

$$\frac{dq_f}{dt} = \frac{p_f}{m_f} + \frac{\pi_\varepsilon}{W_v} q_f$$

$$\frac{dp_f}{dt} = -f(\lambda)\nabla_{q_f} u(\mathbf{q}, q_f) - \left(1 + \frac{d}{g_N}\right)\frac{\pi_\varepsilon}{W_v} p_f - \frac{\pi_{f_1}}{Q_{f_1}} p_f$$

$$\frac{d\eta_{f_1}}{dt} = \frac{\pi_{f_1}}{Q_{f_1}}$$

$$\frac{d\pi_{f_1}}{dt} = \frac{p_f^2}{m_f} - k_B T - \frac{\pi_{f_2}}{Q_{f_2}} \pi_{f_1}$$

$$\frac{d\eta_{f_k}}{dt} = \frac{\pi_{f_k}}{Q_{f_k}}$$

$$\frac{d\pi_{f_k}}{dt} = \frac{\pi^2_{f_{k-1}}}{Q_{f_{k-1}}} - k_B T - \frac{\pi_{f_{k+1}}}{Q_{f_{k+1}}} \pi_{f_k}$$

$$k = 2, \ldots, M-1$$

$$
\begin{aligned}
\frac{d\eta_{f_M}}{dt} &= \frac{\pi_{f_M}}{Q_{f_M}} \\
\frac{d\pi_{f_M}}{dt} &= \frac{\pi_{f_{M-1}}^2}{Q_{f_{M-1}}} - k_B T \\
\frac{d\lambda}{dt} &= \frac{\pi_\lambda}{W_\lambda} \\
\frac{d\pi_\lambda}{dt} &= -\frac{df(\lambda)}{d\lambda}u(\mathbf{q}, q_f) - \frac{d\theta(\lambda)}{d\lambda} - \frac{\pi_{\lambda_1}}{Q_{\lambda_1}}p_\lambda \\
\frac{d\eta_{\lambda_1}}{dt} &= \frac{\pi_{\lambda_1}}{Q_{\lambda_1}} \\
\frac{d\pi_{\lambda_1}}{dt} &= \frac{\pi_\lambda^2}{W_\lambda} - k_B T - \frac{\pi_{\lambda_2}}{Q_{\lambda_2}}\pi_{\lambda_1} \\
\frac{d\eta_{\lambda_k}}{dt} &= \frac{\pi_{\lambda_k}}{Q_{\lambda_k}} \\
\frac{d\pi_{\lambda_k}}{dt} &= \frac{\pi_{\lambda_{k-1}}^2}{Q_{\lambda_{k-1}}} - k_B T - \frac{\pi_{\lambda_{k+1}}}{Q_{\lambda_{k+1}}}\pi_{\lambda_k} \\
k &= 2, \ldots, M-1 \\
\frac{d\eta_{\lambda_M}}{dt} &= \frac{\pi_{\lambda_M}}{Q_{\lambda_M}} \\
\frac{d\pi_{\lambda_M}}{dt} &= \frac{\pi_{\lambda_{M-1}}^2}{Q_{\lambda_{M-1}}} - k_B T \\
\frac{d\varepsilon}{dt} &= \frac{\pi_\varepsilon}{W_\upsilon} \\
\frac{d\pi_\varepsilon}{dt} &= Vd\left(\mathcal{P} - p_0\right) + \frac{d}{g_N}\sum_{i=1}^{g_N}\frac{p_i^2}{m_i} - \frac{\pi_{b_1}}{Q_{b_1}}\pi_\varepsilon \\
\frac{d\eta_{b_1}}{dt} &= \frac{\pi_{b_1}}{Q_{b_1}} \\
\frac{d\pi_{b_1}}{dt} &= \frac{\pi_\varepsilon^2}{W_\upsilon} - k_B T - \frac{\pi_{b_2}}{Q_{b_2}}\pi_{b_1} \\
\frac{d\eta_{b_k}}{dt} &= \frac{\pi_{b_k}}{Q_{b_k}} \\
\frac{d\pi_{b_k}}{dt} &= \frac{\pi_{b_{k-1}}^2}{Q_{b_{k-1}}} - k_B T - \frac{\pi_{b_{k+1}}}{Q_{b_{k+1}}}\pi_{b_k} \\
k &= 2, \ldots, M-1 \\
\frac{d\eta_{b_M}}{dt} &= \frac{\pi_{b_M}}{Q_{b_M}} \\
\frac{d\pi_{b_M}}{dt} &= \frac{\pi_{b_{M-1}}^2}{Q_{b_{M-1}}} - k_B T
\end{aligned}
$$

Other statistical ensembles are also often implemented, such as grand canonical ensemble and grand isothermal-isobaric ensemble, as introduced in Ref. Kamberaj (2020). Furthermore, particular MD simulations of the systems in the human cell environment require other parameters to be controlled, such as pH.

5.4 QM/MM Computer Simulations

Combined Quantum Mechanics/Molecular Mechanics (QM/MM) simulations have also been suggested (Warshel and Levitt 1976). In QM/MM simulations, the entire system is divided into two regions. The first is the quantum mechanics region for which we know the electronic-structure information. This is also called the *active site*. The second one is the molecular mechanics region, which is much larger in size than QM region, and its degrees of freedom act as a reservoir for QM subsystem. Figure 5.2 illustrates a typical QM/MM system description.

QM region or the active site, for example in enzymatic reactions, includes the substrate and active site of an enzyme. In the enzymatic reactions, the rest of the protein and the solvent degrees of freedom form the MM region. The potential interaction energy of the QM/MM simulations are described by the following function (Pezeshki and Lin 2014; Duster et al. 2018):

$$U_{QM/MM}(\mathbf{r}) = U_{QM}(\mathbf{r}_{qm}) + U_{MM}(\mathbf{r}_{MM}) + U_{QM/MM}(\mathbf{r}_{QM/MM}) \tag{5.77}$$

In Eq. (5.77), the term $U_{MM}(\mathbf{r}_{MM})$ includes all the molecular mechanics force field interactions, such as the bond stretching, bond bending, internal rotation, and non-bonded interactions, discussed in the next chapters.

A key feature of QM/MM algorithms is coupling between QM and MM regions. Two pictures exists for that: (i) Potential interaction picture defines the QM/MM potential for a given geometry and partition of a system. (ii) Dynamics picture defines how QM and MM regions interact with one another when the atoms are in motion.

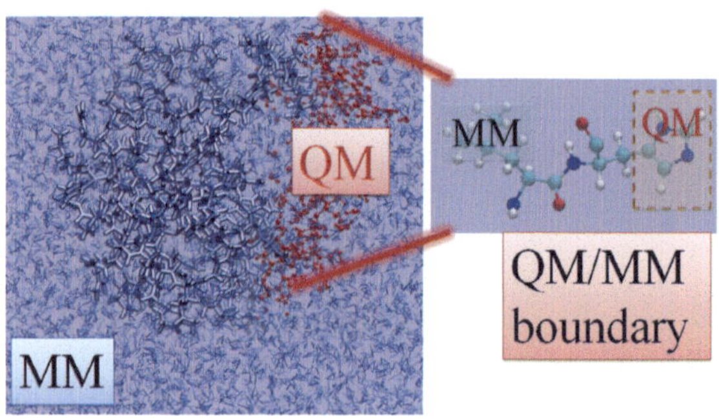

Fig. 5.2 QM/MM boundary shown as dash lines passes through a covalent bond between a QM atom and MM atom

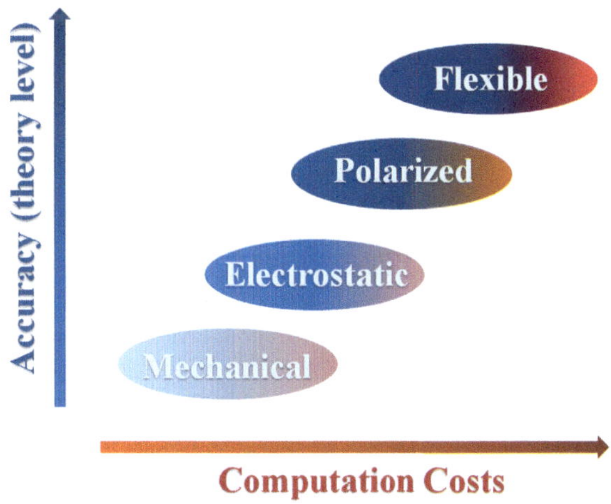

Fig. 5.3 Theory level accuracy versus computational cost of the QM/MM embedding schemes

Note that dynamic picture is coupled to potential picture since it requires an interaction potential to determine the trajectory. For that, open-boundary couplings are suggested, which characterise mutual polarisations and on-the-fly exchanges of partial charges. The van der Waals interactions are (significantly) short-ranged and included only in MM region.

Some key feature schemes of open-boundary coupling include the following, as discussed in Ref. Pezeshki and Lin (2014). The electrostatic interactions are long-range, and hence they are crucial to QM/MM methods. For treating the electrostatic interactions, the mechanical embedding approach is used where the QM calculations are as in gas phase and the QM/MM electrostatic interactions at MM level. Besides, in the electrostatic embedding, the electrostatic potential of MM region is expanded in multipoles and electrostatic interaction potential of QM region is considered with MM region.

Furthermore, the polarised embedding, which is an improvement of the electrostatic embedding, includes the classical polarisation schemes of MM region by QM region.

Moreover, the flexible embedding allows not only mutual polarisation but also partial charge transfer between QM and MM regions. Figure 5.3 presents a dependence of the theory level accuracy versus computational cost of the QM/MM embedding schemes. As indicated, the electrostatic embedding is most popular because it is a good trade off between the accuracy and the computation costs (Pezeshki and Lin 2014).

Based on the available literature Hofer and de Visser (2018) (and the references therein), there is a widespread capability of the QM/MM approach on the scope of the topics implemented. It is considered as the technique of choice to address a broad spectrum of problems in biochemistry. The QM/MM field is currently considered an active research field capable of exploring many different directions. Besides, combined with other theoretical approaches, computer infrastructure, and computationally efficient simulation software packages, it provides an excellent method in modern biochemistry. In the coming years, the QM/MM method may provide an efficient alternative technology for studying complex phenomena in biochemistry.

5.5 Software for QM/MM and Classical Computer Simulations

The QM/MM method it is still conceived under the framework of the biomolecular simulation packages (Hofer and de Visser 2018) and the references therein Warshel and Levitt (1976), Field et al. (1990), Gao (1993), Gao (1996), Warshel (2002). In this work, we have designed an interface to CHARMM software package (Brooks et al. 2009), as shown in Fig. 5.4. The QM/MM simulations are performed using either semi-empirical modelling of the QM part of the Hamiltonian or QM modelling.

In Fig. 5.5, a dialog form for the QM/MM computer simulations is presented using the semi-empirical models AM1, PM3, and MNDO, discussed in Chap. 3. Besides, different statistical ensembles are utilised and periodic boundary conditions for various simulation box shapes.

Figure 5.6 presents a dialog form for the QM/MM computer simulations in the free energy calculations at the semi-empirical approximation. In particular, the solvation QM/MM free energy perturbation and pKa calculations can be performed.

Moreover, Fig. 5.7 shows a graphical interface to the classical molecular dynamics simulations in various statistical ensembles, such NVE, NVT, NPT, and CPH (constant pH) ensembles. This simulations will be discussed in more details, including

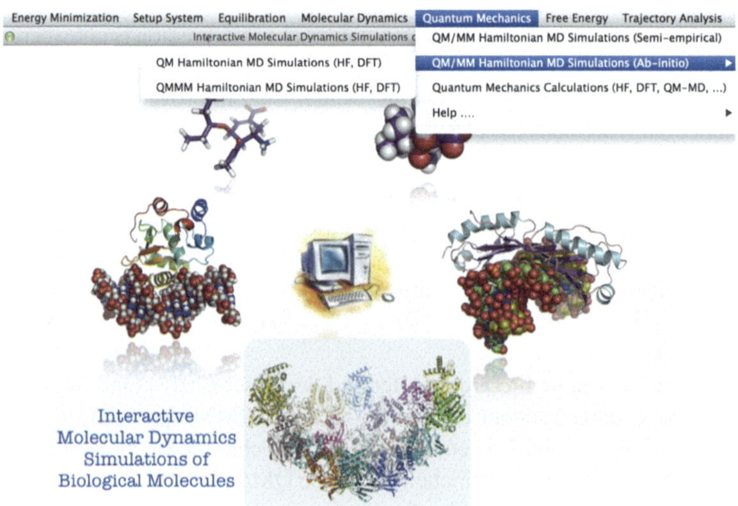

Fig. 5.4 The interface to the QM/MM computer simulations

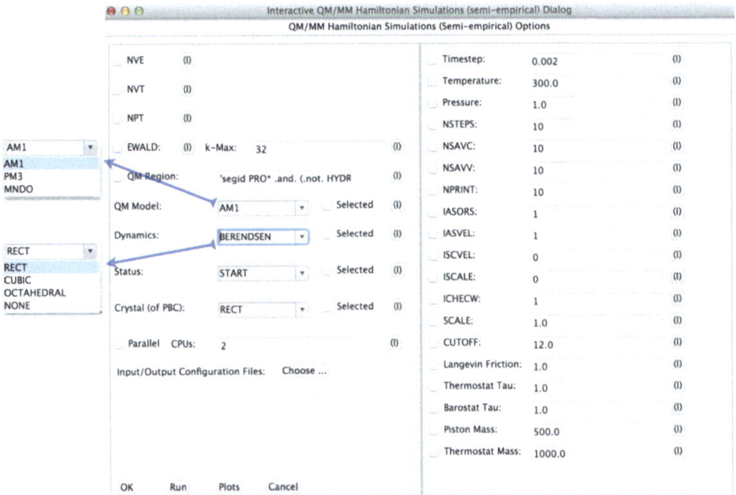

Fig. 5.5 A dialog form for the QM/MM computer simulations

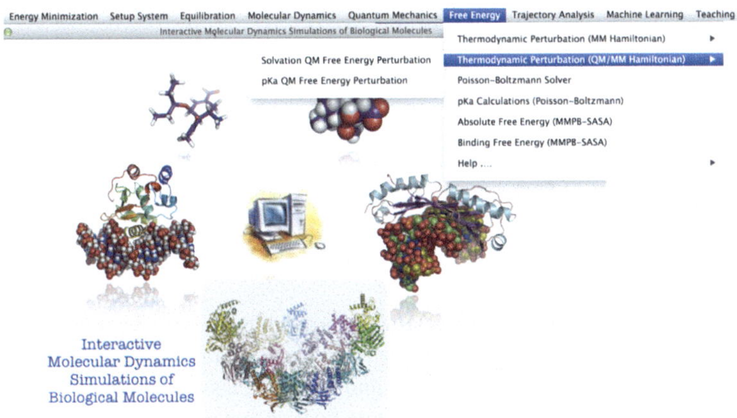

Fig. 5.6 A dialog form for the QM/MM computer simulations in the free-energy calculations at the semi-empirical approximation

case studies and mini projects, in the following chapters. In addition, a documentation is provided for controlling various macroscopic parameters (such as temperature, pressure, and non-bonded interactions), as in CHARMM software package (Brooks et al. 2009).

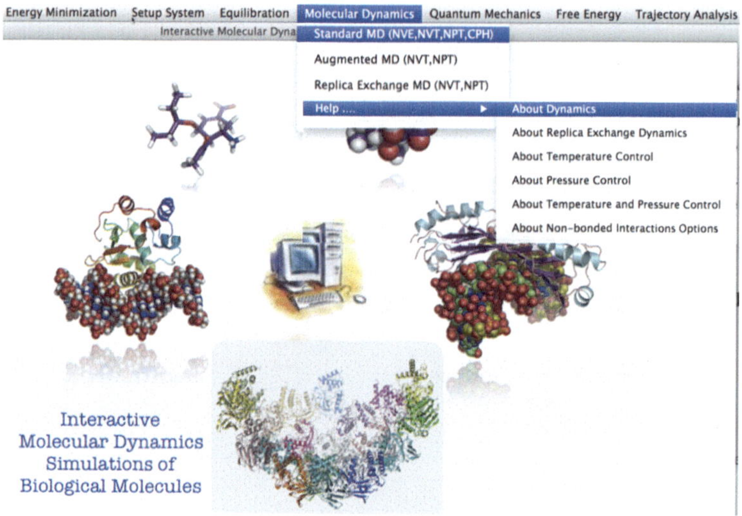

Fig. 5.7 The interface to the classical molecular dynamics simulations in various statistical ensembles

5.6 Summary

We discussed the molecular dynamics method as one of the most common computer simulation approach used to study molecular systems. In particular, we introduced the equations of motion in the most relevant statistical ensembles used in the molecular dynamics simulations of molecular systems, such as NVT, NPT, and grand canonical ensembles.

Furthermore, we introduced the QM/MM method in the context of the biomolecular computer simulations. Moreover, we introduced a graphical interface package to the CHARMM software for setting up and running classical molecular dynamics and the QM/MM simulations at various statistical ensembles and distributions.

References

M.P. Allen, D.J. Tildesley, *Computer Simulation of Liquids* (Oxford University Press, 1989)

H.C. Andersen, Molecular dynamics simulations at constant pressure and/or temperature. J. Chem. Phys. **72**, 2384–2393 (1980)

H.J.C. Berendsen, J.P.M. Postma, W.F. van Gunsteren, A. DiNola, J.R. Haak, Molecular dynamics with coupling to an external bath. J. Chem. Phys. **81**, 3684–3690 (1984)

B.R. Brooks, C.L. Brooks, A.D. MacKerell, L. Nilsson, R.J. Petrella, B. Roux, Y. Won, G. Archontis, C. Bartels, S. Boresch, A. Caflisch, L. Caves, Q. Cui, A.R. Dinner, M. Feig, S. Fischer, J. Gao, M. Hodoscek, W. Im, K. Kuczera, T. Lazaridis, J. Ma, V. Ovchinnikov, E. Paci, R.W. Pastor, C.B. Post, J.Z. Pu, M. Schaefer, B. Tidor, R.M. Venable, H.L. Woodcock, X. Wu, W. Yang, D.M. York,

M. Karplus, Charmm: the bimolecular simulation program. J. Comput. Chem. **30**(10), 1545–1614 (2009)

T. Cagin, B.M. Pettitt, Molecular dynamics with a variable number of molecules. Mol. Phys. **72**, 169 (1991)

T. Cagin, B.M. Pettitt, Grand molecular dynamics: a method for open systems. Mol. Simul. **6**, 5 (1991)

A.W. Duster, C. Hung Wang, H. Lin, Adaptive QM/MM for molecular dynamics simulations: 5. On the energy-conserved permuted adaptive-partitioning schemes. Molecules **23**, 2170 (2018)

M.J. Field, P.A. Bash, M. Karplus, A combined quantum mechanical and molecular mechanical potential for molecular dynamics simulations. J. Comput. Chem. **11**, 700 (1990)

J. Gao, Potential of mean force for the isomerization of DMF in aqueous solution: a Monte Carlo QM/MM simulation study. J. Am. Chem. Soc. **115**, 2930 (1993)

J. Gao, Hybrid quantum and molecular mechanical simulations an alternative avenue to solvent effects in organic chemistry. Acc. Chem. Res. **29**, 298 (1996)

T.S. Hofer, S.P. de Visser, Editorial: quantum mechanical/molecular mechanical approaches for the investigation of chemical systems - recent developments and advanced applications. Front. Chem. **6**, 357 (2018)

W.G. Hoover, Canonical dynamics: equilibrium phase-space distributions. Phys. Rev. A **31**, 1695 (1985)

J. Ji, B.M. Pettitt, Phase transitions of water at constant excess chemical potential. An application of grand molecular dynamics. Mol. Phys. **82**, 67–83 (1994)

J. Ji, T. Cagin, B.M. Pettitt, Dynamic simulations of water at constant chemical potential. J. Chem. Phys. **96**, 1333 (1992)

H. Kamberaj, *Molecular Dynamics Simulations In Statistical Physics: Theory And Applications* Computational Series. (Springer Nature, Switzerland, 2020)

M. Karplus, J.A. McCammon, Molecular dynamics simulations of biomolecules. Nat. Struct. Biol. **9**(9), 646–652 (2002). With corrigenda in Nat. Struct. Biol. **9**(10), 788 (2002)

C. Lo, B.J. Palmer, Alternative Hamiltonian for molecular dynamics simulations in the grand canonical ensemble. J. Chem. Phys. **102**, 925 (1995)

G.C. Lynch, B.M. Pettitt, Grand canonical ensemble molecular dynamics simulations: reformulation of extended system dynamics approaches. J. Chem. Phys. **107**, 8594 (1997)

G.J. Martyna, D.J. Tobias, M.L. Klein, Constant pressure molecular dynamics algorithms. J. Chem. Phys. **101**(5), 4177–4189 (1994)

G.J. Martyna, M.E. Tuckerman, D.J. Tobias, M.L. Klein, Explicit reversible integrators for extended systems dynamics. Mol. Phys. **87**(5), 1117–1157 (1996)

S. Melchionna, G. Ciccotti, B.L. Holian, Hoover NPT dynamics for systems varying in shape and size. Mol. Phys. **78**, 533–544 (1993)

S. Nosé, A molecular dynamics method for simulation in the canonical ensemble. Mol. Phys. **52**, 255–268 (1984)

S. Nosé, A molecular dynamics method for simulation in the canonical ensemble. Mol. Phys. **52**, 255 (1984)

S. Nosé, A unified formulation of the constant temperature molecular dynamics methods. J. Chem. Phys. **81**, 511–519 (1984)

S. Nosé, Constant temperature molecular dynamics methods. Prog. Theor. Phys. Supp. **103**, 1–46 (1991)

S. Nosé, M.L. Klein, Constant pressure molecular dynamics for molecular systems. Mol. Phys. **50**, 1055–1076 (1983)

B.J. Palmer, C. Lo, Molecular dynamics implementation of the Gibbs-Ensemble calculation. J. Chem. Phys. **101**, 10899–10907 (1994)

R.W. Pastor, *Techniques and Applications of Langevin Dynamics Simulations* In The Molecular Dynamics of Liquid Crystals. (Kluwer Academic, Dordrecht, The Netherlands, 1994)

S. Pezeshki, H. Lin, Molecular dynamics simulations of ion solvation by flexible-boundary QM/MM: on-the-fly partial charge transfer between QM and MM subsystems. J. Comp. Chem. **35**, 1778–1788 (2014)

J.P. Ryckaert, G. Ciccotti, H.J.C. Berendsen, Numerical integration of the Cartesian equations of motion of a system with constraints: molecular dynamics of n-alkanes. J. Comput. Phys. **23**, 327–341 (1977)

A. Warshel, Molecular dynamics simulations of biological reactions. Acc. Chem. Res. **35**, 385 (2002)

A. Warshel, M. Levitt, Theoretical studies of enzymic reactions: dielectric, electrostatic and steric stabilization of the carbonium ion in the reaction of lysozyme. J. Mol. Biol. **103**, 227–249 (1976)

S. Weerasinghe, B.M. Pettitt, Ideal chemical potential contribution in molecular dynamics simulations of the grand canonical ensemble. Mol. Phys. **82**, 897 (1994)

K. Zare, V. Szebehely, Time transformations in the extended phase-space. Celest. Mech. **11**, 469 (1975)

Chapter 6
Computational Molecular Modelling

The chapter aims to introduce the computational molecular modelling strategies. The focus is on the CHARMM molecular mechanics force field.

In this chapter, we discuss the computational molecular modelling strategies. In particular, the CHARMM molecular mechanics force field of biomolecules is introduced and its parametrisation. For more information about other molecular mechanics force fields and molecular modelling strategies, one can consider the following literature Leach (2001). Besides, we will discuss some new initiatives for the force field development, such as ForceBalance, machine learning, and open force field (OpenFF) approaches.

6.1 Introduction

The computational modelling is becoming an important part of biology, biochemistry, and biophysics (van Gunsteren et al. 2006). This is also due to the increase on the available computer power, computational infrastructure, and the atomistic software packages. Therefore, the comparisons between the experimental data on biological molecules and the computational modelled data is made possible for various complex phenomena, such as protein folding/unfolding, binding in protein complexes, and cell membranes processes. However, the current development of computational modelling in biological systems is affected by four problems (van Gunsteren et al. 2006), as depicted in Fig. 6.1.

The ensemble and search problems are discussed in the previous chapters and in other available literature Kamberaj (2020). The search problem relates to the convergence of the measurement due to insufficient sampling of configuration space,

H. Kamberaj, *Computer Simulations in Molecular Biology*, Scientific Computation,
https://doi.org/10.1007/978-3-031-34839-6_6

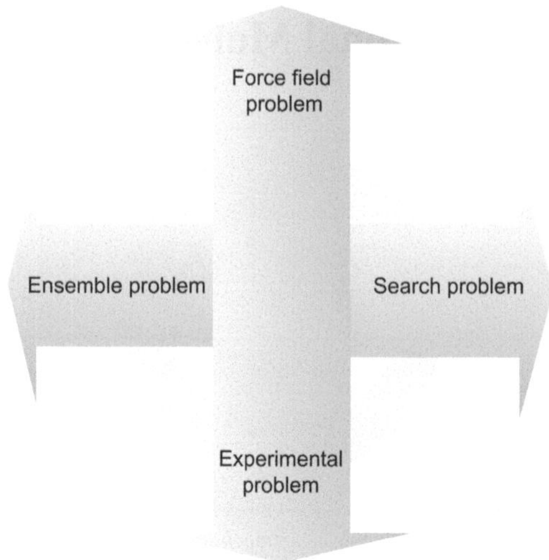

Fig. 6.1 The basic problems of the biological molecule modelling. Adapted from Ref. van Gunsteren et al. (2006)

and alleviation or aggravation of the sampling. The ensemble problem shows, for example, in calculation of the entropy and performing the averages (van Gunsteren et al. 2006).

In this chapter, the focus is on the force field problem. The experimental problem is related to other four problems discussed often in the literature Allen and Tildesley (1989), van Gunsteren et al. (2006). For instance, (i) it is impossible to perform experimental measurements in some environments; (ii) it is dangerous to perform an experiment for some phenomena; (iii) the experiment may be too expensive; or (iv) the experiment can not observe some phenomena occurring at very short time scales or very small space scales. In all these cases when the experimental measurements fail or are limited, the computer simulations data can be accessible.

6.2 CHARMM Force Field of Biomolecules

The potential energy function (or the *molecular mechanics force field*) $U(\mathbf{r})$ of the system depends on the coordinates $\mathbf{r}$ of atoms (Leach 2001). The force $\mathbf{F}_i$ acting on each atom i in the system is the negative gradient of the potential, $\mathbf{F}_i = -\nabla_{\mathbf{r}_i} U(\mathbf{r})$. There exists several force fields used in molecular dynamics simulations of biological systems, such as CHARMM, AMBER, GROMACS, OPLS, and MMx (Leach 2001). In general, the potential is a function of small number of components representing the intra- and intermolecular interactions within the system.

Potential energy function in the CHARMM molecular mechanics force field is given as MacKerell et al. (1998):

$$
\begin{aligned}
U(\mathbf{r}) = & \underbrace{\sum_{\text{bonds}} \frac{k_{i,l}}{2}\left(l_i - l_{i,0}\right)^2 + \sum_{\text{angles}} \frac{k_{i,\theta}}{2}\left(\theta_i - \theta_{i,0}\right)^2}_{\text{bond and angle stretching terms}} \\
& + \underbrace{\sum_{\text{Urey-Bradley}} \frac{k_{UB}}{2}(S - S_0)^2}_{\text{Urey-Bradley term}} \\
& + \underbrace{\sum_{\text{torsions}} \frac{V_n}{2}\left(1 + \cos(n\omega - \gamma)\right) + \sum_{\text{improper}} \frac{k_{i,\eta}}{2}\left(\eta_i - \eta_{i,0}\right)^2}_{\text{torsion and improper terms}} \\
& + \underbrace{\sum_{\text{residues}} U_{\text{CMAP}}(\Phi, \Psi)}_{\text{backbone torsional correction terms}} \\
& + \underbrace{\sum_{i=1}^{N} \sum_{j=i+1}^{N} \left[4\varepsilon_{ij} \left(\left(\frac{\sigma_{ij}}{r_{ij}}\right)^{12} - \left(\frac{\sigma_{ij}}{r_{ij}}\right)^{6} \right) + \frac{q_i q_j}{4\pi\epsilon_0 \epsilon r_{ij}} \right]}_{\text{non-bonded terms}} .
\end{aligned}
\tag{6.1}
$$

In Eq. (6.1), the first term represents the sum of interactions between pairs of covalent bonded atoms. It has a harmonic potential form and it represents the penalty in total potential energy when the bond length l_i deviated from its equilibrium value $l_{i,0}$ by $\mid l_i - l_{i,0} \mid$, where i denotes a pair A–B of covalent bonds in the system and the sum runs over all pairs, as depicted in Fig. 6.2. The bond stretching interactions are characterised by the force constants $k_{i,l}$ and the equilibrium length $l_{i,0}$.

The second term is the sum over all angle bending interactions resulting from three covalently bonded atoms A–B–C, where A and C are bonded to B, as shown in Fig. 6.3. The contribution of each angle-bending term is characterised by the force constants $k_{i,\theta}$ and the reference value $\theta_{i,0}$ of the angle $\widehat{\text{ABC}}$, as depicted in Fig. 6.3.

The Urey-Bradley term is used in CHARMM force fields to fix the geometrical distances between pairs of atoms that are not covalently bonded. It is also modelled

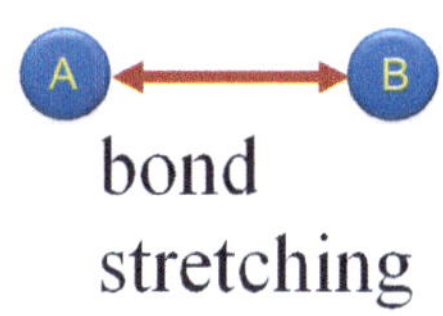

Fig. 6.2 Bond-stretching interaction term

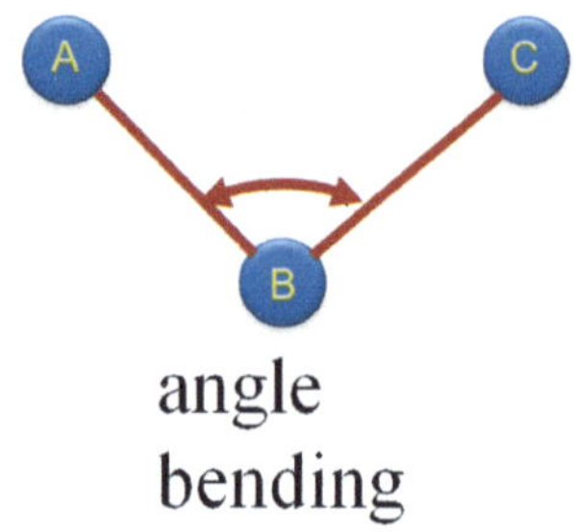

Fig. 6.3 Angle-bending interaction term

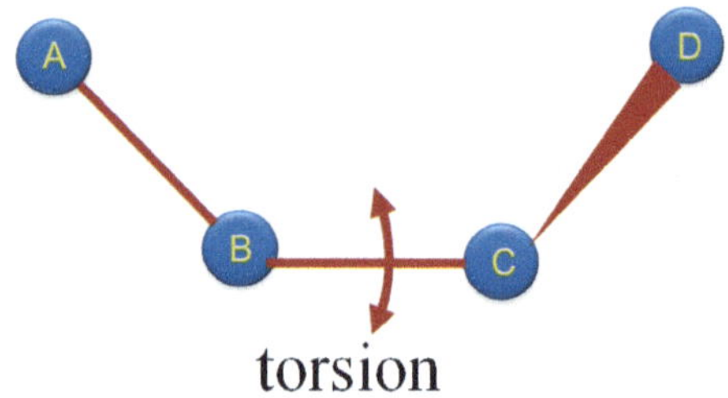

Fig. 6.4 Torsion interaction term

with a harmonic function and is characterised by two parameters, the force constant k_{UB} and the equilibrium geometrical distance S_0.

The fourth term represents the torsional potential interaction. This term parametrises the potential energy variation due to rotation around a covalent bond. Figure 6.4 shows the rotation around the covalent bond B–C of the dihedral angle formed by four atoms A, B, C, and D; ω denotes the torsional angle depicted in Fig. 6.4. The contribution from each torsion angle depends on the barrier height V_n, the multiplicity n, and γ.

Here, n is the number of minimum points in the function as the bond B–C rotates by 360° and γ determines the angle where the torsional angle passes through its global minimum value. Figure 6.5a presents a plot of U for the torsion energy (in kcal/mol) versus ω (in degrees) of H–NH1–C–CT2 (where $V_n = 5.0$ kcal/mol, $n = 2$, and $\gamma = 180°$) and Fig. 6.5b presents the same plot of HC–NH2–CT2–CD (where $V_n = 0.22$ kcal/mol, $n = 3$, and $\gamma = 0°$). The CHARMM36 force field data were used.

The fifth term represents the improper torsion angle potential energy, used to maintain a certain geometry of the molecule, such as planar geometry. It depends on the force constant $k_{i,\eta}$ (often expressed in kcal/(mol·radians^2) and the reference torsion angle $\eta_{i,0}$ (in degrees), which represents an equilibrium value.

The sixth term is the backbone torsional angles correction cross-term, which improves conformation sampling of the backbone angles (Φ, Ψ), shown in Fig. 6.6. This correction cross-term is calculated in a grid-based map and is only recently introduced to the CHARMM force field.

The last two terms are the non-bonded contributions, calculated between all pairs of atoms i and j that are in different molecules or between atoms that are separated

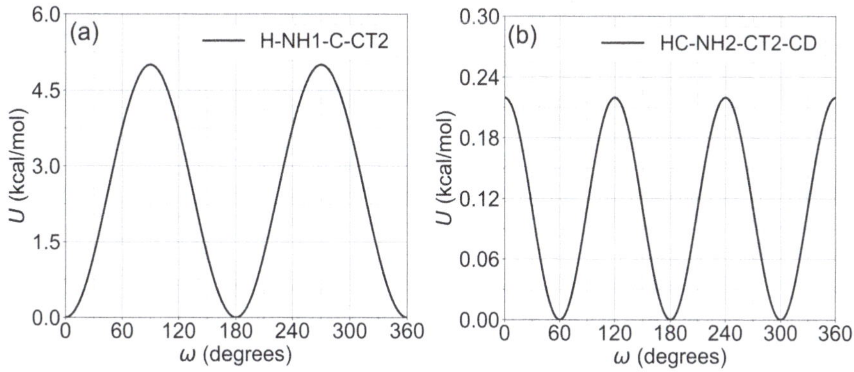

Fig. 6.5 **a** The torsion energy U (in kcal/mol) versus ω (in degrees) of H−NH1−C−CT2 (where $V_n = 5.0$ kcal/mol, $n = 2$, and $\gamma = 180°$). (**b** U of HC−NH2−CT2−CD (where $V_n = 0.22$ kcal/mol, $n = 3$, and $\gamma = 0°$). The CHARMM36 force field data were used

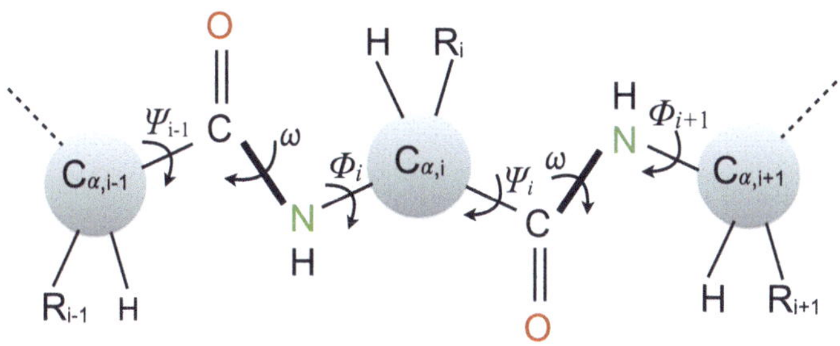

Fig. 6.6 The backbone angles (Φ, Ψ) between the residues $i-1$, i, and $i+1$

by at least three bonds in the same molecule. The non-bonded interactions are usually modelled using a Lennard-Jones (LJ) potential for van der Waals interactions and a Coulomb potential term for electrostatic interactions. The van der Waals interaction is calculated as an LJ potential with appropriate ε_{ij} and σ_{ij} parameters which are the well-depth and collision diameter, respectively. The LJ potential parameters for a pair of atoms i and j are determined using the following combination rules (Leach 2001):

$$\varepsilon_{ij} = \left(\varepsilon_i \varepsilon_j\right)^{1/2} \qquad (6.2)$$
$$\sigma_{ij} = \frac{\sigma_i + \sigma_j}{2}$$

The electrostatic contribution is calculated using the Coulomb's law from partial atomic charges (q_i and q_j) associated with each atom. Figure 6.7 represents the electrostatic (Fig. 6.7a) and van der Waals interactions (Fig. 6.7b).

In Eq. (6.1), ϵ_0 is the permittivity of free space:

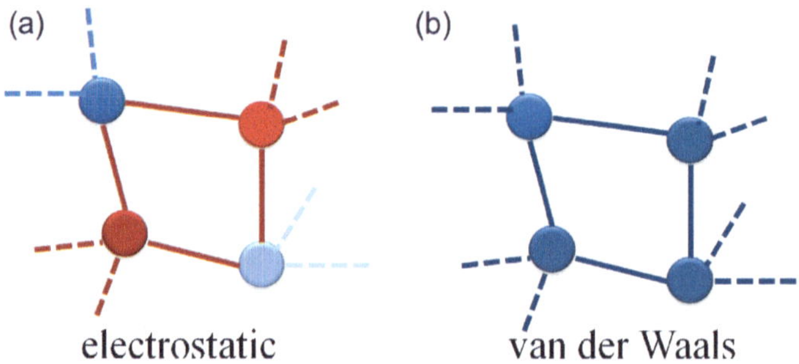

Fig. 6.7 Non-bonded interaction potential energy terms

$$\epsilon_0 = 8.8542 \times 10^{-12}\ \mathrm{C}^2/\mathrm{N}\cdot\mathrm{m}^2 \tag{6.3}$$

and the constant $1/4\pi\epsilon_0$ has a value of

$$\frac{1}{4\pi\epsilon_0} = 8.9875 \times 10^9\ \mathrm{N}\cdot\mathrm{m}^2/\mathrm{C}^2 \tag{6.4}$$

Often, this constant is expressed in the following units:

$$\frac{1}{4\pi\epsilon_0} \approx 331\ \mathrm{kcal}\cdot\mathrm{m/mol} \tag{6.5}$$

and q_i are given in the units of the absolute value of electron's charge $\mid e \mid = 1.6 \times 10^{-19}$ C. ϵ denotes the dielectric constant, which equals one for the explicit solvent simulations.

Note that molecular mechanics (MM) methods ignore the electronic motion and write the energy function of the system as a function of only nuclear positions. Therefore, MM is used to perform calculations on a system containing a significant number of atoms. In some cases, MM can provide results that are as accurate as high-level QM calculations, but for much faster computation time. However, MM can not provide properties that depend upon the electronic distribution in a molecule.

The first most important assumption in MM is the Born-Oppenheimer approximation, which allows for writing the potential energy as a function of nuclear coordinates only. MM uses simple models of interactions within a system, which include contributions from stretching of bonds, angle bending, and rotations about single bonds. An essential feature of MM is to use a set of parameters, developed for a relatively small number of cases, applied to a broader range of problems. Besides, the parameters generated from data on small molecules may also be used to study much larger molecular systems, such as *macromolecules* (proteins, nucleic acids, polymers, etc.).

Furthermore, improper potential energy terms give the improper torsion angle energy or out-of-plane bending term, often, used to maintain a planar geometry. Some force fields include cross-terms, which reflects the coupling between the internal degrees of freedom, often used to reproduce structure properties.

Moreover, the electrostatic contribution may include terms from the permanent partial charges assigned to each atom and induced dipoles, such as polarisation force fields.

It is important to note that a molecular mechanics force field is defined not only by the functional form, but also the parameters in Eq. 6.1. An essential feature is that two force fields may have the same functional form, but at the same time very different set of parameters. Moreover, different force fields may give approximately the same accuracy in calculations. Furthermore, different force fields cannot be mixed; that is, it is not strictly allowed to divide the energy into individual terms, and take some of the parameters from one force field and mix them with parameters from another force field. However, some of the terms in a molecular mechanics force field are sufficiently independent of the others (e.g., bond stretching and angle bending terms) to make this approximation acceptable in some cases.

The non-bonded interactions determine the thermodynamic equilibrium and processes, such as folding/unfolding, membrane, and micelle formation, ligand-, DNA- and protein-protein binding, solvation (in membrane and water). Therefore, there exist three main problems in the parametrisation of a mechanical force field (van Gunsteren et al. 2006): minimal free energy differences and many interactions, entropic effects, and variety of atoms and molecules.

It is also worth noting that molecular mechanics force fields are empirical such that there does not exist a correct form for the potential energy function given by Eq. 6.1. In principle, if one potential energy functional form performs better than another force field, then it is likely that force field will be favoured. There have been many efforts to compare the accuracy of different force fields. The potential energy functional form used in a force field is a compromise between accuracy and computational efficiency. Thus, we expect with increasing the computer performance, more complex functional forms will be possible to incorporate into molecular mechanics force fields. Besides this, the new potential functional forms should allow fast calculations of the first and second order derivatives of energy function with respect to atomic coordinates to use methods such as energy minimisation and molecular dynamics.

6.3 Molecular Mechanics Force Field Parameters Calibration

After choosing the specific functional form of the molecular mechanics force field, the task of determining appropriate force field set of parameters remain to be performed. The steps involved in this task include the type of data, systems, thermodynamic phase, and properties to be used in the calibration of a specific set of parameters. The choices made from different force fields (for example, CHARMM Foloppe and MacKerell 2000; MacKerell and Banavali 2000; MacKerell et al. 2004, GROMACS Hünenberger and van Gunsteren 1997, AMBER Cornell et al. 1995, MM2 Allinger 1977, MM3 Allinger et al. 1990a, b, MM4 Allinger et al. 1996b, a and OPLS Jorgensen and Tirado-Rives 1988) can be different. Often, the use of small molecules for calibration is more efficient since the parameters set can be transferred to other similar molecular systems. For large macromolecules (such as proteins, DNA, etc.) group of atoms may show different behaviour depending on the environmental conditions in folded molecules (e.g., pH, ionic concentration and temperature). Some properties may also depend more strongly only on some force field set of parameters and less strongly (or weakly) on others. In such cases, calibration efforts can be reduced significantly by optimising only a subset of all set of parameters using a limited number of properties.

For parameterisation of a force field there is a vast amount of data to be collected from the existing experimental results or the quantum mechanics calculations. These data include the structure properties, the geometry of the optimised molecular structure, conformation energies, and thermodynamic properties. However, often, there are missing data or sometimes difficult to obtain for some molecular systems. Therefore, the quantum mechanics calculations are often used to provide these data for optimisation of the force field parameters. Ab-initio approaches are used to reproduce experimental results, in particular, for small molecular systems. The validation of the force fields calibrated in this way is often a challenge when the comparison with experimental data is performed. In general, this comparison is in the following functional form:

$$U(\mathbf{R}) = \sum_t \sum_i U_t(\Psi_t(\mathbf{R}_i)) \tag{6.6}$$

where t runs over the all types of the interactions (such as those shown in Eq. 6.1), i runs over all particles representing the molecular structure, U_t is the model of the potential function, and Ψ_t is a scalar function of the Cartesian coordinates of each particle i, $\mathbf{R}_i$, which is a mathematical model of the particular type of interaction (described in Eq. 6.1). After the mathematical functional forms, U_t, are determined for each type of interactions, which are often chosen to be simple in their mathematical forms, the next stage includes the optimisation of the force field parameters.

There are two traditional optimisation approaches, such as the parameterisation by trial and error assessment and the least square fitting. A detailed discussion can be found in Refs. Leach (2001), Kamberaj (2020).

6.3.1 ForceBalance Approach

Molecular mechanics have many parameters; therefore, an accurate parametrisation is needed to obtain comparable results with experimental or quantum mechanics calculations data. For that a so-called ForceBalance initiative is created to allow scientific community to use and develop their molecular mechanics or classical force fields (Wang et al. 2013, 2014; Qiu et al. 2019). It is worthy mentioning that classical force fields do not involve solving the Schrödinger equation and the potential energy function is given in terms of elementary mathematical functions. A relatively few number of parameters may be obtained directly using experimental results; for instance, the covalent bonds can be determined fitting the experimental bond lengths and vibrational frequencies to a harmonic motion function. However, there are no experimental data used to calculate the non-bonded interaction parameters, such as Coulomb and van der Waals terms. Thus, the force field parametrisation is a difficult process.

Traditionally, the potential interaction terms for the parameterisation are classified into hard terms (stretching bond and angle bending) and soft terms (torsion angle, van der Waals and electrostatic). These terms are considered separately. In particular, bond stretching parameters for the same kind of atoms are almost identical among all the force fields; therefore, they are often omitted from the calibration, if there are already existing parameters. In contrast, the soft terms are parametrised together because they are coupled and hence they influence each other. Usually, the following steps may apply:

(i) A range of the parameters for the van der Waals interactions is determined.
(ii) The partial charges are assigned to atoms using electrostatic potential energy surface fitting procedure.
(iii) The torsion angle potential function is then determined as a function of the torsion angle using a fitting procedure.

The torsion angle terms are determined such that the heights of the barriers are reproduced together with relative energies of different conformations, iteratively until a desired accuracy achieves. For determination of the torsion potential energy barriers the quantum mechanics calculations are often employed. Often, the force field parameters are manually adjusted to reproduce the experimental results, such as protein nuclear magnetic resonance (NMR) structures.

The ForceBalance aims to generate force fields using a general systematic procedure with explicit input data and highly efficient optimisation methods. Their results are very accurate and with an improved reproducibility. Figure 6.8 presents the ForceBalance optimisation strategy (Wang et al. 2013, 2014; Qiu et al. 2019).

At a high level strategy, ForceBalance approach considers an empirical potential and uses a set of reference data for inputs. Then, it tunes the parameters such that computer simulations can reproduce the data at a desired level of accuracy. Usually, the reference data are the energy and forces from some high-level of accuracy QM calculations, molecular properties from experimental data (such as polarisation and multipole moments), or experimental bulk property data (such as density and dielectric constant).

The advantages of ForceBalance include the infrastructure that allows fitting any type of reference data using many existing empirical potentials and existing similar reference data and optimisation in an easily extensible framework.

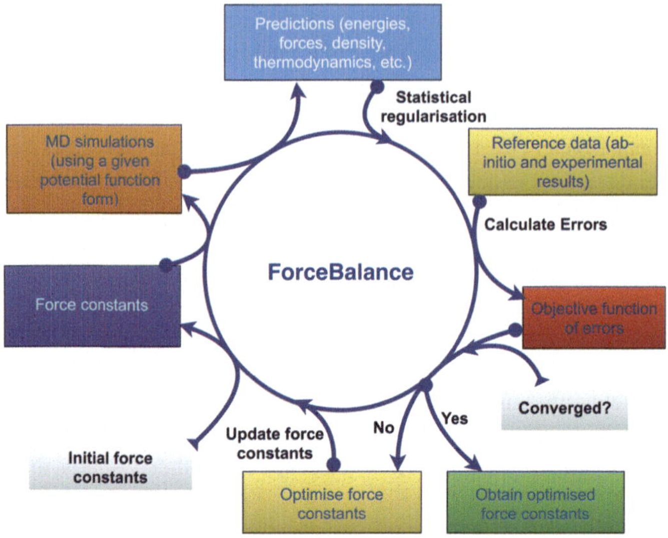

Fig. 6.8 ForceBalance approach strategy. Adapted from Refs. Wang et al. (2013), Wang et al. (2014), Qiu et al. (2019)

6.3.2 Machine Learning Approach

Recently, (Grimme 2014) automated parameterisation methods have shown a great interest, where the quantum mechanically produced input data are used in a machine learning represented as a black box producing in the output the classical force fields for specific types of the interaction terms. The input quantum mechanical data consist of optimised geometrical structures, atomic partial charges, and covalent bonds. More recently, (Metz et al. 2016) another automated method developed for calculation of the intermolecular potential energy surfaces using the symmetry-adapted perturbation theory calculation as reference data (Misquitta et al. 2005), which was applied to molecular systems up to 42 atoms. In Refs. Vleet et al. (2016), Vandenbrande et al. (2017) a new automated approach is introduced for calibration of the force field parameters using the properties of the atoms in a molecule as input parameters to characterise the chemical space of the molecules.

6.4 Force Field Problem

The force field issues relate to the calculations of the relatively small free energy differences driving various phenomena of interest in computer simulations of biological systems. Typically, they are of order of 10 kJ/mol (van Gunsteren et al. 2006), and thus the accuracy in calculations of non-bonded interactions, considered to drive these phenomena, has to be very high. However, the difficulty in achieving this accuracy becomes more evident for systems with a large number of atoms.

Furthermore, the calculations of the entropic contributions in the free energy analysis become a problem at the room temperature. This is because the free energy differences at constant T depend on the changes on the potential energy and entropy: $\Delta A = \Delta U - T\Delta S$.

Moreover, the development of the force fields for biological molecules with different chemical components is often difficult and it requires many computational resources because appropriate force field parameters may need to be derived. In some cases, the problem may be solvable, especially, when those parameters are transferable between the force fields; however, this is not always the case. Besides, the force field functional forms should be computationally simple and, at the same time, allow an efficient sampling of the configurational space.

Currently, open force field initiatives have been developed (Boothroyd et al. 2022), consisting of openly available sources of tools that can generate high-quality force field parameters for small molecules. The quality assessment of those force field parameters is made possible by using different condensed phase properties, curated by a huge amount of the accurate experimental data (Boothroyd et al. 2022).

6.5 Summary

We discussed the computational molecular modelling strategies and their problems. Besides, the CHARMM molecular mechanics force field of biomolecules is introduced and its parametrisation is discussed in details.

Furthermore, new initiatives in the force field development, such as ForceBalance, machine learning, and OpenFF approaches, were introduced. Moreover, the molecular mechanics force field problems were discussed.

References

M.P. Allen, D.J. Tildesley, *Computer Simulation of Liquids* (Oxford University Press, 1989)

N.L. Allinger, Conformational Analysis 130. MM2. A Hydrocarbon Force Field Utilizing V_1 and V_2 Torsional Terms. J. Am. Chem. Soc. **99**, 8127–8134 (1977)

N.L. Allinger, F. Li, L. Yan, J.C. Tai, Molecular mechanics. The MM3 force field for alkenes. J. Comput. Chem. **11**, 848–867 (1990a)

N.L. Allinger, F. Li, L. Yan, J.C. Tai, Molecular mechanics (MM3) calculations on conjugated hydrocarbons. J. Comput. Chem. **11**, 868–895 (1990)

N.L. Allinger, K. Chen, J.-H. Lii, An improved force field (MM4) for saturated hydrocarbons. J. Comput. Chem. **17**, 642–668 (1996)

N.L. Allinger, K. Chen, J.A. Katzenelenbogen, S.R. Wilson, G.M. Anstead, Hyperconjugative effects on carbon-carbon bond lengths in molecular mechanics (MM4). J. Comput. Chem. **17**, 747–755 (1996)

S. Boothroyd, L.P. Wang, D.L. Mobley, J.D. Chodera, M.R. Shirts, Open force field evaluator: an automated, efficient, scalable framework for the estimation of physical properties from molecular simulation. J. Chem. Theory Comput. **18**(6), 3566–3576 (2022)

W.D. Cornell, P. Cieplak, C.I. Bayly, I.R. Gould, K.M. Merz Jr., D.M. Ferguson, D.C. Spellmeyer, T. Fox, J.W. Caldwell, P.A. Kollman, A second generation force field for the simulation of proteins, nucleic acids and organic molecules. J. Am. Chem. Soc. **117**, 5179–5197 (1995)

N. Foloppe, A. MacKerell, All-atom empirical force field for nucleic acids: I. Parameter optimization based on small molecule and condensed phase macromolecular target data. J. Comput. Chem. **21**, 86–104 (2000)

S. Grimme, A general quantum mechanically derived force field (qmdff) for molecules and condensed phase simulations. J. Chem. Theory Comput. **10**(10), 4497–4514 (2014)

P.H. Hünenberger, W.F. van Gunsteren, *Computer Simulation of Biomolecular Systems Theoretical and Experimental Applications* (Kluwer, Dordrecht, The Netherlands, 1997)

W.L. Jorgensen, J. Tirado-Rives, The opls potential functions for proteins - energy minimizations for crystals of cyclic-peptides and crambin. J. Am. Chem. Soc. **110**, 1666–1671 (1988)

A.D. Jr MacKerell, D. Bashford, M. Bellot, R.L. Dunbrack Jr., J.D. Evanseck, M.J. Field, S. Fisher, J. Gao, H. Guo, S. Ha, D. Joseph-McCarthy, L. Kuchnir, K. Kuczera, F.T.K. Lau, C. Mattos, S. Michnick, T. Ngo, D.T. Nguyen, B. Prodhom, W.E. Reiher, B. Roux, M. Schlenkrich, J.C. Smith, R. Stote, J. Straub, M. Watanabe, J. Wiorkiewicz-Kuczera, D. Yin, M. Karplus. All-atom empirical potential for molecular modelling and dynamics studies of proteins. J. Phys. Chem. B **102**(18), 3586–3616 (1998)

H. Kamberaj, *Molecular Dynamics Simulations in Statistical Physics: Theory And Applications* Computational Series. (Springer Nature, Switzerland, 2020)

A.R. Leach, *Molecular Modelling, Principles And Applications*, 2nd edn. (Prentice Hall, Pearson Education Limited, Edingburgh Gate, 2001)

A. MacKerell, N. Banavali All-atom empirical force field for nucleic acids: II. Parameter optimization based on small molecule and condensed phase macromolecular target data. J. Comput. Chem. **21**, 105–120 (2000)

A. MacKerell, M. Feig, C. Brooks, Extending the treatment of backbone energetics in protein force fields: limitations of gas-phase quantum mechanics in reproducing protein conformational distributions in molecular dynamics simulations. J. Comput. Chem. **25**, 1400–1415 (2004)

M.P. Metz, K. Piszczatowski, K. Szalewicz, Automatic generation of intermolecular potential energy surfaces. J. Chem. Theory Comput. **12**(12), 5895–5919 (2016)

A.J. Misquitta, R. Podeszwa, B. Jeziorski, K. Szalewicz, Intermolecular potentials based on symmetry-adapted perturbation theory with dispersion energies from time-dependent density-functional calculations. J. Chem. Phys. **123**(21), 214103 (2005)

Y. Qiu, P.S. Nerenberg, T. Head-Gordon, L.P. Wang, Systematic optimization of water models using liquid/vapor surface tension data. J. Phys. Chem. B **123**, 7061–7073 (2019)

W. van Gunsteren, D. Bakowies, R. Baron, I. Chandrasekhar, M. Christen, X. Daura, P. Gee, D.P. Geerke, A. Glättli, P.H. Hünenberger, M.A. Kastenholz, C. Oostenbrink, M. Schenk, D. Trzesniak, N.F.A. can der Vegt, H.B. Yu, Biomolecular modeling: goals, problems, perspectives. Angew. Chem. Int. Ed. **45**(25), 4064–4092 (2006)

M.J. Van Vleet, A.J. Misquitta, A.J. Stone, J. Schmidt, Beyond Borm-Mayer: improved models for short-range repulsion in ab initio force fields. J. Chem. Theory Comput. **12**, 3851–3870 (2016)

S. Vandenbrande, M. Waroquier, V.V. Speybroeck, T. Verstraelen, The monomer electron density force field (MEDFF): a physically inspired model for noncovalent interactions. J. Chem. Theory Comput. **13**(1), 161–179 (2017)

L.P. Wang, J. Chen, T. Van Voorhis, Systematic parametrization of polarizable force fields from quantum chemistry data. J. Chem. Theory Comput. **9**, 452–460 (2013)

L.P. Wang, T.J. Martinez, V.S. Pande, Building force fields: an automatic, systematic, and reproducible approach. J. Phys. Chem. Lett. **5**, 1885–1891 (2014)

Chapter 7
Coarse-Grain Modelling Strategies

The chapter aims to introduce the computational coarse-grain modelling strategies.

In this chapter, the computational coarse-grain modelling strategies are discussed. Besides, different coarse-grained models (CGMs) will be described.

7.1 Coarse-Grained Models

The coarse-grained models are useful because allow to explore large (bio)molecular systems at longer time scales. However, CGMs have to have transferable potential, which should be suitable for different systems, and thus the main focus is on constructing CGM with a potential as transferable as the atomistic force fields (Emperador et al. 2008).

Often, the coarse-grained models are constructed from atomistic molecular dynamics trajectories for a given protein with a high level of precision (Izvekov and Voth 2005; Stepanova 2007; Kamberaj 2011), however, they will not be suitable for another system, and thus they are not transferable (Izvekov and Voth 2005; Voth 2008).

Coarse-grained force field models are efficient in treating the intrinsically disordered proteins, since there is no need to reproduce the native structure of system and they aggregate, and thus many proteins involve in the simulations. Many of current transferable CGM potentials work well as a protocol in simulations of intrinsically disordered proteins, although not for natively folded proteins (Emperador et al. 2008).

A common expectation is that the higher the resolution, the better reproduced the atomistic results, such as UNRES (Faraggi et al. 2018) or OPEP coarse-grained force fields (Sterpone et al. 2014; Kynast et al. 2016). However, the simulations become slower than with low resolution coarse-graining. Hydrogen bonding, which are

H. Kamberaj, *Computer Simulations in Molecular Biology*, Scientific Computation,
https://doi.org/10.1007/978-3-031-34839-6_7

essential in proteins, require high details of the biomolecular structures (such as dipole-dipole and other interactions) (Joseph et al. 2021); however, a high resolution means a large number of coarse-grained active sites, and thus the simulations become not so fast. On the other hand, the lipids can efficiently be coarse-grained (Marrink et al. 2004) without significant loss of the accuracy of atomistic resolution because they are mostly non-polar, and hence the hydrogen bonds do not play an important role.

Therefore, the main challenge in determining a coarse-grained model for a particular system is to determine the potential energy interactions between the sites of the CGM. In general, it should be expected that these interactions must capture the fluctuations effects of the atomistic details that are eliminated from the CGM. That is, the interactions in a CGM should describe the "correct physics" such that the model is highly efficient and at the same time provides detailed insights and quantitatively accurate results. For example, a CGM in implicit solvent of a folded protein should be such that the interactions describe correctly the hydrophobic effects provided by atomistic models of interactions with explicit solvent molecules.

There are different approaches that are being used to construct CGMs. Three important schemes in the field are the *top-down*, *bottom-up* and *knowledge-based* approaches to coarse-graining. The distinction between the first two approaches is becoming increasingly blurred as more models integrate aspects of both approaches. Schematically these three approaches are illustrated in Fig. 7.7.

A bottom-up CGM is a model of a system that is constructed on the basis of a more detailed model for the same system, which is a higher resolution representation of this system provided based on the fundamental principles for interpreting and predicting the experimental results of real materials. In particular, commonly, the classical atomistic models could be used to predict the experimental observables with quantitative accuracy and describe material properties and dynamics at any relevant length and time scale. This fundamental model may be considered itself empirical and may be represented from a more detailed model, for instance, from the quantum mechanics, which in practice may not be applied exactly due to its computational cost. Thus, atomistic models may be considered as highly optimised approximation of the fundamental model. Of course, if such high resolution model does not exist or it is not sufficiently accurate, then the bottom-up strategy may not be possible. Given that a high resolution atomistic model exist, then the statistical physics can be used as a rigorous framework for constructing lower resolution CGM that can reproduce all thermodynamic and structural properties of the atomistic model. The main problem in this framework is the many-body potential of mean force. In practice, the statistical physics framework can not be applied exactly, mainly, due to poor conformational sampling (Fig. 7.1).

On the other hand, top-down models do not need the information about more detailed systems, such as atomistic models, but they are related to the real system based on the observables that are directly measured by experiments in the length and time scale accessible by the CGM. Therefore, they are also called "fine-grained" (Uusitalo et al. 2015) models constructed based on the experimental measurements of physical macroscopic properties. In this sense, these macroscopic prop-

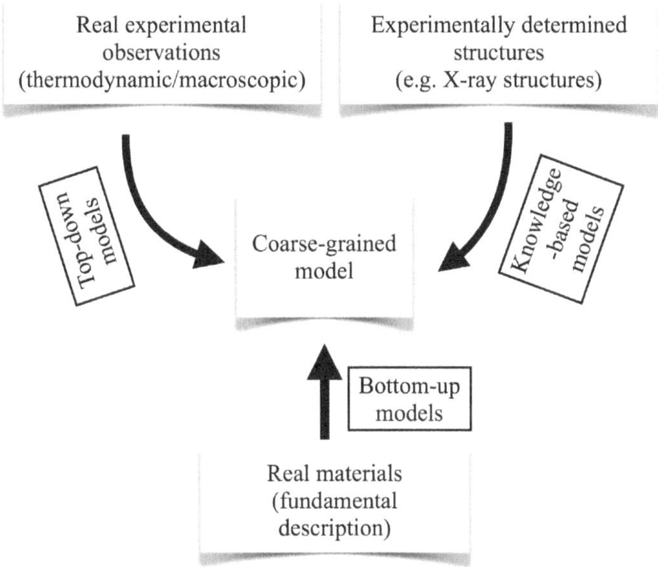

Fig. 7.1 A flowchart illustrating the distinctions between top-down, bottom-up, and knowledge-based strategies for a coarse-grained model. Adapted from Ref. Noid (2013)

erties may be considered even more coarse (Uusitalo et al. 2015). It can be distinguished two types of top-down models, the generic and chemically specific. The first ones are constructed for assessing the consequences of universal physical principles without specific consideration of any particular system. The last top-down models can be constructed using experimental observations for a particular system. Both models provide a useful framework for determining very important and emerging phenomena of the physical principles that were included in the model (Schmid 2009). However, these two cases of the top-down models only relate to the real system by addressing large-scale emerging phenomena, thus, it is not obvious that a top-down model can be related to any more detailed or more realistic model of the system (Noid 2013).

Another common classification of approaches for constructing CGMs of biomolecular systems includes physics-based and knowledge-based approaches. The physics-based approaches use the physical theories to model the interactions of coarse-grained system, for instance, via the bottom-up strategies by addressing more detailed observables or top-down strategies by addressing less detailed observables. On the other hand, knowledge-based models are constructed on the basis of information extracted from experimentally determined three-dimensional structures. These models typically employ empirical relations to determine the interactions in CGMs (or scoring functions for bioinformatics calculations), especially in the context of structure prediction and computational protein engineering.

7.2 The Many-Body Potential of Mean Force

Statistical physics provides a framework for determining the effective interactions that emerge as a coarse-grained manifestation of the interactions in a more detailed model (Zwanzig 1961). The many-body potential of mean force (PMF) is the main quantity to be determined in the approach for a CGM (Kirkwood 1935; Liwo et al. 1998; Lu and Voth 2011). In the bottom-up approach, which determines the interactions of the CGM sites on the basis of more detailed model, the many-body PMF, W, is typically determined by the underlying atomistic model and the coarse-grained mapping:

$$W(\mathbf{R}) = -k_B T \ln Z(\mathbf{R}) \tag{7.1}$$

with

$$Z(\mathbf{R}) = \int d\mathbf{r} \exp\left[-U(\mathbf{r})/k_B T\right] \delta\left(\mathbf{T}(\mathbf{r}) - \mathbf{R}\right) \tag{7.2}$$

where U is the potential of the atomistic model, which can be of any arbitrary complexity, $\mathbf{T}$ is the transformation matrix, and $\delta\left(\mathbf{T}(\mathbf{r}) - \mathbf{R}\right)$, is given as

$$\delta\left(\mathbf{T}(\mathbf{r}) - \mathbf{R}\right) = \prod_{I=1}^{N} \delta\left(\mathbf{T}_I(\mathbf{r}) - \mathbf{R}_I\right) \tag{7.3}$$

The factor W weighs every CGM configuration, $\mathbf{R}$, according to the total Boltzmann weight for the atomistic configurations, $\mathbf{r}$, that map to $\mathbf{R}$:

$$P_R(\mathbf{R}) \propto e^{-W(\mathbf{R})/k_B T} = \int d\mathbf{r} e^{-U(\mathbf{r})/k_B T} \delta\left(\mathbf{T}(\mathbf{r}) - \mathbf{R}\right) \tag{7.4}$$

That is, for a given temperature T and volume V, knowing $W(\mathbf{R})$ up to a constant value, it can be used to sample a canonical distribution of coarse-grained configurations for a given T and V, and this is equivalent to the one implied by the atomistic model and the coarse-grained mapping.

Note that the many-body PMF is not a conventional potential energy function, but it is a configurational-dependent free energy function that reflects both energetic and entropic contributions. The coarse-grained forces determined from the gradients of the PMF equal conditioned averages of atomistic forces, hence W is a potential that generates forces (Noid et al. 2008). Therefore, the PMF quantifies the reversible work associated with changing the coarse-grained configuration of the atomistic model for a molecular system (Chan and Dill 1991; Liwo et al. 2001). $W(\mathbf{R})$ is a function of the thermodynamic step and it depends on the molecular composition of the system, the volume V, and temperature T through the Boltzmann weight at a given $\mathbf{r}$. Thus, by knowing W one can in principle reconstruct both structural and thermodynamic properties of the atomistic model. On the other hand, U is assumed to be state-point independent, thus, the representation of thermodynamic quantities is fundamentally different in the CGM (Louis 2002; Stillinger et al. 2002).

Note that the calculation of $W(\mathbf{R})$ is quite impossible for most of the molecular systems, since the integral in Eq. (7.2) requires a free energy calculation as a function of the Cartesian coordinates for all the CGM sites (Chipot and Pohorille 2007; Christ et al. 2010).

7.3 Coarse-Grained Models of Proteins

Early works on reconstructing CGMs for proteins include the studies of Warshel (Levitt and Warshel 1975) and Levitt (Levitt 1976). In these CGMs, each amino acid is represented by one or a few sites having properties that are specified by the amino acid type (Tozzini 2005; Izvekov and Voth 2005; Tozzini et al. 2006). One of the sites is associated with C-α of the amino acid, which allows for the reconstruction of the protein backbone (Rotkiewicz and Skolnick 2008). The network-based and Gō-type models are often used to represent each amino acid with a single site by distinguishing each site on the basis of its position in the three-dimensional folded X-ray structure.

One of the most used databases for parametrisation of many CGMs (Levitt 1976; Tanaka and Scheraga 1976) is the protein databank (or PDB) (Berman et al. 2003). The most known CGMs using the PDB are the network models, native-centric models (also known as Gō, native-based, or structure-based models), and knowledge-based models. All these models are implicit solvent CGMs, and are treated as native structure based approaches, which provide various approximations to the many-body PMF. It is expected that these models will continue becoming increasingly important in the future as the quantity and quality of high resolution structures increases rapidly.

Native contacts is one of the characteristic features particularly used to construct native structure-based models. Given a folded structure (such as from PDB), the CGM can be reconstructed, represented by the coordinates $\mathbf{R}_0$, then a pair of amino acid residues (or a pair of coarse-grained sites) form a native contact if the distance between the residues is less than some predefined distance. Then, the interactions are classified as native and non-native, independent on the chemical properties of the residues. These models are found to be very useful in establishing experimentally determined three-dimensional structure, since the pair potentials governing the native interactions are minimised all for the folded structure. Therefore, these models are the most popular for studying the folding, fluctuations, and interactions of proteins, protein complexes and nucleic acids with well known equilibrium structures (Tozzini 2005).

Note that, in contrast, the knowledge-based approaches employ information obtained from the experimentally determined structures for multiple distinct proteins to reconstruct potentials that are intended to model proteins.

7.3.1 The Gō Model

The Gō model potential is given mathematically by Taketomi et al. (1975):

$$
\begin{aligned}
U = & \sum_{i=1}^{N-1} \sum_{j=i+1}^{N} k_b (r_{ij} - r_{ij}^0)^2 \\
& + \sum_{i=1}^{N-2} \sum_{j=i+1}^{N-1} \sum_{k=j+1}^{N} k_\theta (\theta_{ijk} - \theta_{ijk}^0)^2 \\
& + \sum_{n} \sum_{i=1}^{N-3} \sum_{j=i+1}^{N-2} \sum_{k=j+1}^{N-1} \sum_{l=k+1}^{N} k_\phi \left[1 + \cos\left(n\phi_{ijkl} - \phi_{ijkl}^0\right)\right] \\
& + \sum_{|i-j|>3}^{N_{\text{native}}} \varepsilon \left[5\left(\frac{r_{ij}^0}{r_{ij}}\right)^{12} - 6\left(\frac{r_{ij}^0}{r_{ij}}\right)^{10} \right] \\
& + \sum_{|i-j|>3}^{N_{\text{nonnative}}} \varepsilon \left(\frac{\sigma_0}{r_{ij}}\right)^{12} \\
& + \sum_{i<j}^{N} \frac{q_i q_j}{r_{ij}}
\end{aligned} \tag{7.5}
$$

where r_{ij}^0, θ_{ijk}^0 and ϕ_{ijkl}^0 are, respectively, the equilibrium distance, angle, and dihedral from the native structure between the sites i and j. The native contacts are defined from the crystal structure using a cutoff distance of 6 Å between the C_α atoms. k_b, k_θ and k_ϕ are the force field constants to be determined. The Gō-type model explicitly bias the coarse-grained potential to stabilise the native structure, thus, these models have proven to be useful for characterising protein fluctuations and folding (Jr. and III 2009).

7.3.2 Network Models

Network models are characterised by the simplest model of many-body interactions (Tirion 1996; Bahar et al. 1997; Haliloglu et al. 1997; Hinsen 1998; Atilgan et al. 2001) by approximating the potential energy function around the folded native structure, $\mathbf{R}_0$, as a quadratic function of either the distances between sites (elastic network model— ENM (Tirion 1996; Atilgan et al. 2001)), or the Cartesian displacements of sites (Gaussian network model—GNM (Bahar et al. 1997; Haliloglu et al. 1997)). The coefficients of the expansion are in general determined by free-energy calculations, which corresponds to the curvature of the PMF around $\mathbf{R}_0$. In contrast, in the network models, these coefficients are determined on the basis of the native contact map. For example, the ENM potential can be expressed as (Tirion 1996):

$$U_{ENM}(\mathbf{R}|\mathbf{R}_0) = \frac{1}{2}\sum_{i<j} K_{ij}\Delta_{ij}(\mathbf{R}_0) \mid \mathbf{R}_{ij} - \mathbf{R}_{ij;0} \mid^2 \tag{7.6}$$

where R_{ij} and $R_{ij;0}$ are the distances between coarse-grained sites i and j in the configuration $\mathbf{R}$ and in folded configuration, $\mathbf{R}_0$, respectively. In Eq. (7.6), K_{ij} is a spring constant and $\Delta_{ij}(\mathbf{R}_0)$ is the matrix of native contacts given by

$$\Delta_{ij}(\mathbf{R}_0) = \begin{cases} 1 & R_{ij;0} < R_c \\ 0 & R_{ij;0} \geq R_c \end{cases}$$

where R_c is a specified cutoff.

The GNM potential (Flory et al. 1976; Bahar et al. 1997) can be written as:

$$U_{GNM}(\mathbf{R}|\mathbf{R}_0) = \frac{1}{2}\sum_{i<j} K_{ij}\Delta_{ij}(\mathbf{R}_0) \mid \mathbf{R}_{ij} - \mathbf{R}_{ij;0} \mid^2 \tag{7.7}$$

where R_{ij} and $R_{ij;0}$ are the vector displacements from site i to site j in configurations $\mathbf{R}$ and $\mathbf{R}_0$, respectively. In both models, the folded structure corresponds to the minimum potential energy function.

Other CGMs include sites for both the peptide backbone and side-chain (Levitt and Warshel 1975; Rey and Skolnick 1991; Messer et al. 2010). For instance, the well-known UNRES model describes the amide peptide group with a point site and the side-chain with an ellipsoidal site (Liwo et al. 1997, 2001, 2011). Models with more detailed side-chains have proven useful for modelling the interactions of proteins with other proteins or with nucleic acids (Zacharias 2003; Basdevant et al. 2007). Furthermore, because models with three or more backbone sites per residue provide a better description of secondary structure and hydrogen bonding, they have considerable success in simulating peptide aggregation and in folding small proteins without relying upon explicit bias towards the native state (Takada et al. 1999; Irbäck et al. 2000; Guo et al. 2002; Nguyen and Hall 2004; van Giessen and Straub 2005; Ding et al. 2005; Chen et al. 2006; Mu and Gao 2007; Bereau and Deserno 2009; Cheon et al. 2010; Chebaro et al. 2012). On the other end of the spectrum, protein models with even lower resolution have provided insight into protein flexibility, functional fluctuations, biomolecular assemblies, cytosolic crowding, and protein-protein interactions (Zhang and Muthukumar 2009; Pfaendtner et al. 2009; Saunders and Voth 2012; Sinitskiy et al. 2012).

7.3.3 Martini Coarse-Grained Model of Proteins

The Martini CGM is one of the most used models in the molecular biological simulations due to its simple building block principle. In particular, Martini CGM uses a four-to-one bead mapping transformation; that is, on average every four heavy atoms (including the associated hydrogen atoms) are replaced by one coarse-grained bead. Besides, Martini CGM of parametrisation uses a top-down strategy where the thermodynamic data are used as the primary target (Monticelli et al. 2008; de Jong et al. 2013).

Figure 7.2 shows an overview of parameters and thermodynamic properties of amino acid side chain analogues, as reported in (de Jong et al. 2013).

Table 7.1 presents the overview of parameters and thermodynamic properties of amino acid side chain analogues and the experimental data, reported in de Jong et al. (2013).

The non-bonded interactions between non-charged beads in Martini coarse-grained force field are modelled as Lennard-Jones potentials and the interactions between the charged beads are modelled by Coulomb potentials. The Lennard-Jones interaction strength discriminates different levels of polarity of the CGM beads by introducing four classes of the CG types of beads, C, N, P, and Q, respectively, for non-polar, intermediately polar, polar, and charged chemical groups. Besides, within each class, sub-levels are defined to differentiate between groups with distinct degree of polarity or hydrogen donor and acceptor properties. In particular, almost all the beads have the same size, called regular beads (R-beads); however, small size beads have also been introduced to characterise the ring structures (S-beads). The tiny beads (the so-called T-beads) are modelled to reproduce correct stacking of the hydrogen bonds between nucleotides, discussed in the following. For a statistical mechanics interpretation of the origin of Martini force field, one can read the following review (Jarin et al. 2021).

It is shown that the size-dependent Lennard-Jones parameters in the Martini model can artificially increase the barrier in the free energy profiles of dimerisation

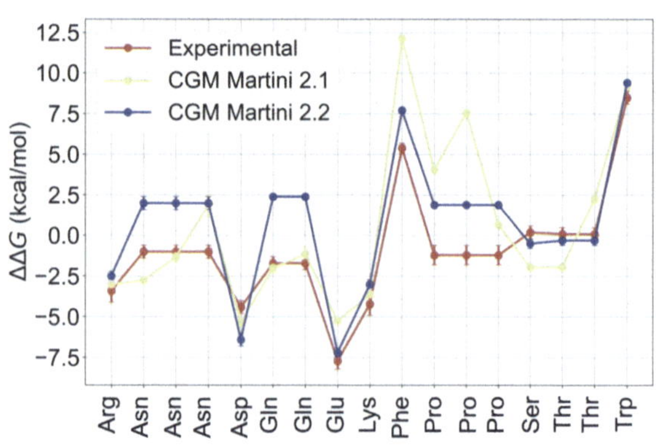

Fig. 7.2 Parameters and thermodynamic properties of amino acid side chain analogues (de Jong et al. 2013)

Table 7.1 Parameters and thermodynamic properties of amino acid side chain analogues (de Jong et al. 2013)

A.A	Type of charge	Exp.	CGM Martini 2.1	CGM Martini 2.2
Arg	N0-Qd; N0-Qd	-3.4 ± 0.7	-3.0 ± 0.1	-2.5 ± 0.3
Asn	P5; Nda (0.46)	-1.0 ± 0.4	-2.7±0.1	2.0±0.4
Asn	Nda (0.51); Nda (0.46)	-1.0 ± 0.4	1.9 ± 0.7	2.0 ± 0.4
Asn	N0 (0.54); Nda (0.46)	-1.0 ± 0.4	-1.3 ± 0.3	2.0 ± 0.4
Asp	Qa; Qa	-4.4 ± 0.4	-5.3 ± 0.1	-6.4 ± 0.4
Gln	P4; Nda (0.42)	-1.7 ± 0.4	-2.0 ± 0.1	2.4 ± 0.1
Gln	N0 (0.51); Nda (0.42)	-1.7 ± 0.4	-1.1 ± 0.5	2.4 ± 0.1
Glu	Qa; Qa	-7.7 ± 0.5	-5.2 ± 0.1	-7.2 ± 0.3
Lys	C3-Qd; C3-Qd	-4.2 ± 0.7	-3.6 ± 0.1	-3.0 ± 0.3
Phe	sc4-sc4-sc4; sc5-sc5-sc5	5.4 ± 0.3	12.2 ± 0.1	7.7 ± 0.1
Pro	Na-AC2; P4-C3	-1.2 ± 0.6	7.6 ± 0.1	1.9 ± 0.1
Pro	P4-AC2; P4-C3	-1.2 ± 0.6	4.1 ± 0.1	1.9 ± 0.1
Pro	P4-C5; P4-C3	-1.2 ± 0.6	0.7 ± 0.1	1.9 ± 0.1
Ser	P1; N0 (0.36)	0.2 ± 0.4	-1.9 ± 0.1	-0.5 ± 0.3
Thr	P1; N0 (0.36)	0.1 ± 0.4	-1.9 ± 0.1	-0.3 ± 0.3
Thr	Nda; N0 (0.36)	0.1 ± 0.4	2.3 ± 0.3	-0.3 ± 0.3
Trp	sc4-sp1-sc4-sc4; sc4-snd-sc5-sc5	8.5±0.4	9.2±0.1	9.4±0.1

(Alessandri et al. 2019). Besides, some molecules interact too strongly, for example, the proteins and carbohydrates in solution and membrane embedded proteins. Moreover, the sampling of the chemical space selectivity is observed for Martini 2. Therefore, a newer version, Martini 3 (Souza et al. 2021), has enabled more accurate simulations of molecular systems using the CGM by rebalancing the non-bonded interactions.

7.4 Coarse-Grained Models of Nucleic Acids

7.4.1 Worm-like Chain Model

The worm-like chain (WLC) model has been introduced initially by Fixman and Kovac (1973; 1982; 1994). The energy E of a stretched DNA molecule as a line integral of two terms: (1) $e_b(s)$ corresponding to the resistance of the chain to bending;

(2) $e_f(s)$ corresponding to the stretching energy due to the force F to the end of the molecular chain. The first term is proportional to the inverse square of the radius of the curvature, R:

$$e_b(s) = \frac{A}{2}\frac{1}{R^2} \tag{7.8}$$

where $1/R$ can be written as

$$\frac{1}{R} = \mid \frac{d\mathbf{t}(s)}{ds} \mid \tag{7.9}$$

where $\mathbf{t}(s)$ is the unit tangent vector and s is the arc length.

The second term is given as

$$e_f(s) = -F\cos\theta(s) \tag{7.10}$$

where $\theta(s)$ is the angle between the unit vector $\mathbf{t}(s)$ and the z-axis. Then, the expression for the total energy is

$$E_{WLC} = \int_0^L ds\left[\frac{A}{2} \mid \frac{d\mathbf{t}(s)}{ds} \mid^2 - F\cos\theta(s)\right] \tag{7.11}$$

where L is the contour length of the molecule.

The experimental data are analysed in terms of the persistence length L_p, which is related to A as:

$$A = k_B T L_p \tag{7.12}$$

The partition function, $Z(L, F, \mathbf{t}_0, \mathbf{t}_1)$ is given via the path integral

$$Z(L, F, \mathbf{t}_0, \mathbf{t}_1) = \int \mathcal{D}(\mathbf{t}) \exp\left(-E_{WLC}/k_B T\right) \tag{7.13}$$

Here, $\mathcal{D}(\mathbf{t})$ is the integration measure in functional space of the paths drawn on the unit sphere starting at the point $\mathbf{t}(s = 0) = \mathbf{t}_0$ and ending at $\mathbf{t}(s = L_0) = \mathbf{t}_1$. Following the discussion in Ref. Bouchiat et al. (1999), extending the integration along s in Eq. (7.13) in the imaginary axis, one obtains the action integral relative to a fictitious particle of unit charge moving upon a sphere under the influence of an electric field F with the time variable $t \propto \mathrm{Im}(s)$. This transforms the integral to the Feynman transition probability amplitude $\langle \mathbf{t}_1 t_f | t = 0 \mathbf{t}_0 \rangle$ for the fictitious particle, initially localised at the coordinate $\mathbf{t}_0$, and at the point of coordinate $\mathbf{t}_1$ at the final time t_f. The amplitude is calculated by the matrix element:

$$\langle \mathbf{t}_1 t_f | t = 0 \mathbf{t}_0 \rangle = \langle \mathbf{t}_1 | \exp\left(-\frac{i t_f}{\hbar}\hat{H}\right) | \mathbf{t}_0 \rangle \tag{7.14}$$

where $\hat{H}$ is the Hamiltonian operator, which in the case of $\hat{H}_{WLC}$ can be written as

$$\hat{H}_{WLC} = -\frac{1}{2\sin\theta}\frac{\partial}{\partial\theta}\sin\theta\frac{\partial}{\partial\theta} - \alpha\cos\theta \tag{7.15}$$

where

$$\alpha = \frac{L_p F}{k_B T} \tag{7.16}$$

is a dimensionless quantity. Then, the partition function can be transformed to Bouchiat et al. (1999):

$$Z(L_0, F, \mathbf{t}_0, \mathbf{t}_1) = \langle \mathbf{t}_1 | \exp\left(-\frac{L_0}{L_p}\hat{H}_{WLC}\right) | \mathbf{t}_0 \rangle \tag{7.17}$$

Denoting with ϵ_n the eigenvalues and $\Psi_n(\theta)$ the eigen-wave functions of the WLC Hamiltonian, it can be written that

$$\hat{H}_{WLC}\Psi_n(\theta) = \epsilon_n \Psi_n(\theta) \tag{7.18}$$

Then, the partition function can be expressed by a series of eigenstates, where each term is of order: $\exp(-L_0\epsilon_n/L_p)$. Usually, $L_0 \gg L_p$, thus only the term related to the ground state contributes, and hence the logarithm of the partition function is given by

$$\ln\left(Z(L_0, F, \mathbf{t}_0, \mathbf{t}_1)\right) = -\frac{L_0}{L_p}\epsilon_0(\alpha) + O(1) \tag{7.19}$$

with $\epsilon_0(\alpha)$ being the ground state energy of $\hat{H}_{WLC}$ and $O(1)$ includes all the terms of order unity associated with the pre-factor multiplying the ground state wave-function $\Psi_0(\theta)$.

The relative extension of the chain is:

$$\begin{aligned} \frac{\langle z \rangle}{L_0} = \frac{k_B T}{L_0}\frac{\partial \ln Z}{\partial F} &= \frac{k_B T}{L_0}\frac{\partial \ln Z}{\partial \alpha}\frac{\partial \alpha}{\partial F} \\ &= -\frac{\partial \epsilon_0(\alpha)}{\partial \alpha} \\ &= \langle \Psi_0 | \cos\theta | \Psi_0 \rangle \end{aligned} \tag{7.20}$$

Next, the following eigenvalue problem has to be solved

$$\hat{H}_{WLC}\Psi_0(\theta) = \epsilon_0 \Psi_0(\theta) \tag{7.21}$$

as suggested in Ref. Bouchiat et al. (1999) by solving the following differential equation

$$\frac{1}{2\sin\theta}\frac{\partial}{\partial\theta}\left(\sin\theta\frac{\partial\Psi_0(\theta)}{\partial\theta}\right)+(\alpha\cos\theta+\epsilon_0)\,\Psi_0(\theta)=0 \tag{7.22}$$

which can further be simplified by denoting $t=\cos\theta$, as:

$$(1-t^2)\frac{d^2\Psi_0(t)}{dt^2}-2t\frac{d\Psi_0(t)}{dt}+2\,(\alpha t+\epsilon_0)\,\Psi_0(t)=0 \tag{7.23}$$

The solution can be looked at the values satisfying the regularity conditions at $\theta=0$ and $\theta=\pi$ (or $t=\pm1$), in order for $\hat{H}_{WLC}$ to be a self-adjoint operator. It has been shown (Bouchiat et al. 1999) that this condition is satisfied if and only if the reduced energy ϵ has discrete values. The value of the ground state energy as given by the lowest order perturbation theory is:

$$\epsilon_0=-\frac{\alpha^2}{3} \tag{7.24}$$

Let $\Psi_a(\theta)$ and $\Psi_b(\theta)$ be the two regular solutions, respectively, for $\theta=0$ and $\theta=\pi$, constructed using a series expansion. Then, the energy eigenvalue is obtained by requiring that for some θ_0 satisfying the condition $0<\theta_0<\pi$, the two functions are equal, which also ensures the regularity of the eigenfunction for the all range of θ. Thus, requiring that the logarithmic derivatives of $\Psi_a(\theta)$ and $\Psi_b(\theta)$ are equal at $\theta=\theta_0$ (Bouchiat et al. 1999), can be written that

$$\frac{1}{\Psi_a}\frac{\partial\Psi_a}{\partial\theta}\,|_{\theta=\theta_0}-\frac{1}{\Psi_b}\frac{\partial\Psi_b}{\partial\theta}\,|_{\theta=\theta_0}=0 \tag{7.25}$$

Figure 7.4 shows the solution of the second order differential Eq. 7.23 for $\alpha=0.1$. Using the solution $\Psi_0(t)$ (or equivalently $\Psi_0(\theta)$ since $t=\cos\theta$) one can use the Eq. 7.20 to determine the relative length $\langle z\rangle/L_0$.

In Ref. (Bustamante et al. 1994), the following interpolated formula can be used as approximation to the WLC model:

$$\alpha=\frac{1}{4\left(1-\frac{\langle z\rangle}{L_0}\right)^2}-\frac{1}{4}+\frac{\langle z\rangle}{L_0} \tag{7.26}$$

This equation can be solved numerically, for example using the Steepest Descents method by minimising the value of the expression

$$V\left(\frac{\langle z\rangle}{L_0}\right)=\left(\alpha-\frac{1}{4\left(1-\frac{\langle z\rangle}{L_0}\right)^2}-\frac{1}{4}+\frac{\langle z\rangle}{L_0}\right)^2 \tag{7.27}$$

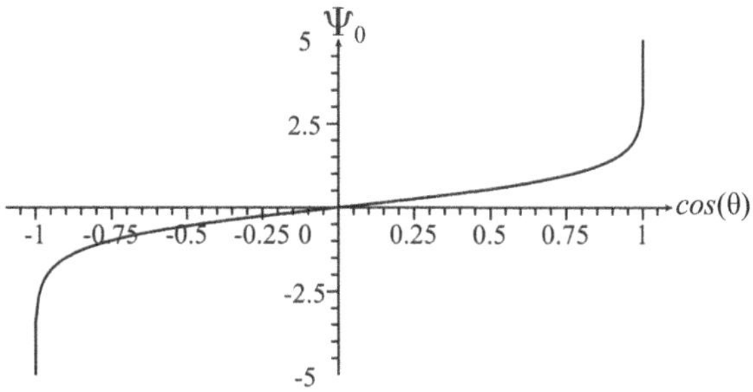

Fig. 7.3 The values of $\alpha = L_p F/k_B T$ versus the relative persistence length for $\epsilon_0 = -\alpha^2/3$ calculated using the exact analytical expression and interpolated approximate formula Eq. 7.26

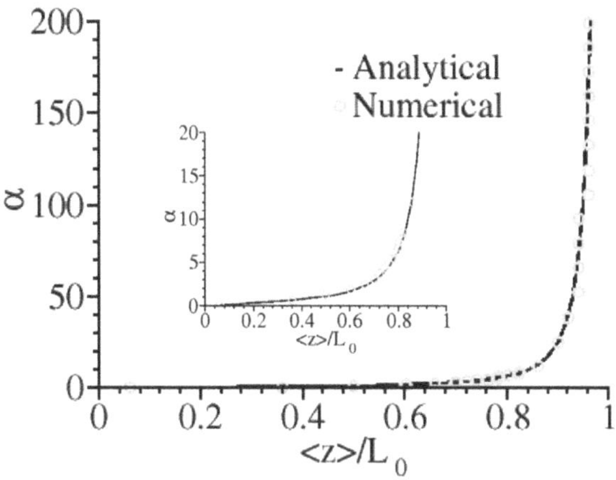

Fig. 7.4 The eigenwave-function $\Psi_0(\theta)$ of the ground state energy versus the $\cos\theta$ for $\alpha = 0.1$ and $\epsilon_0 = -\alpha^2/3$

In Fig. 7.4, it is shown a plot of the values of $\alpha = L_p F/k_B T$ versus the relative persistence length for $\epsilon_0 = -\alpha^2/3$ calculated using both the analytical expression and numerically. It can be seen that the expression Eq. 7.26 has nice property of reducing the exact solution for $\langle z \rangle / L_0$ close to zero and one. For values in between, the expression Eq. 7.26 produces well the general trend of the exact solution.

In Ref. Bouchiat et al. (1999) has been suggested another approximate formula, which is expressed as a seventh-order polynomial with all correction terms included:

$$\alpha = \frac{1}{4\left(1 - \frac{\langle z \rangle}{L_0}\right)^2} - \frac{1}{4} + \sum_{i=1}^{7} a_i \left(\frac{\langle z \rangle}{L_0}\right)^i \tag{7.28}$$

where $a_1 = 1$, $a_2 = -0.5164228$, $a_3 = -2.737418$, $a_4 = 16.07497$, $a_5 = -38.87607$, $a_6 = 39.49944$, and $a_7 = -14.17718$. In Fig. 7.5, it is presented a plot of both the exact analytical values of α and interpolated approximated values from formula 7.28 versus relative persistence length $\langle z \rangle / L_0$. It can be seen that above approximate formula reproduces very well all the properties as the analytical function (Fig. 7.5).

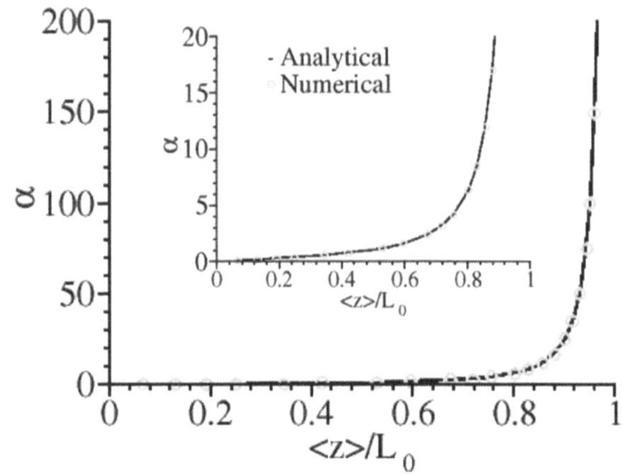

Fig. 7.5 The values of $\alpha = L_p F / k_B T$ versus the relative persistence length for $\epsilon_0 = -\alpha^2/3$ calculated using the exact analytical expression and interpolated approximate formula 7.28

It has been argued (Smith et al. 1996; Wang et al. 1997) that the B-DNA three-dimensional structure exhibit a linear stretching behaviour for high values of the external force F, which is due to modification of the chemical bonds, slightly, by strong stretching of high forces. This also increases the length of the module, and it is in particular specific for values of the external forces in the range $5 - 50$ pN. Above 60 pN significant structure changes occur leading to a so-called *over-stretching transition*. In Refs. Odijk (1995), Wang et al. (1997) it has been proposed that for high external forces but below the over-stretching transition an approximate formula is

$$\alpha = \frac{1}{4\,(1-\ell)^2} - \frac{1}{4} + \sum_{i=1}^{7} a_i \ell^i \tag{7.29}$$

where $\ell = \langle z \rangle / L_0 - F/K_0$, with F being the external force and K_0 is ranging between 500 pN to 1500 pN.

7.4.2 *Elasticity Theory Framework*

For studying the DNA double helix and their interactions with proteins, single-molecule stretching technique is an important tool. Many experimental works have followed from the DNA-stretching studies (Smith et al. 1992). Theoretical models used to describe these experimental results (Bustamante et al. 1994; Vologodskii 1994; Odijk 1995; Marko and Siggia 1995) have as their basis the model of semi-flexible polymer. According to this model, the DNA fluctuations are described in terms of only the persistence length, which for DNA double helix structure has a value of 50 nm, from both the base-pair scale (0.34 nm) and the total molecular length scale (>1000 nm nm for >3 kb molecules; 1 nm $= 10^{-9}$ m). Homogeneous bending with equal flexibility in all directions, is also assumed as the first approximation. Although, current computational models include local twist using the body-centred coordinate frames along the helical axis using Euler transformations.

Elastic deformation energy can be written as a sum of two terms: bending and twisting, with force constants A and C, respectively, which are determined from the experimental measurements of DNA bending and twisting, as describe in Ref. Hagerman (1988). The bending term of the potential energy is proportional to the square of the curvature κ, and the twisting term is proportional to the twist deformation:

$$E = E_B + E_T = \frac{A}{2} \oint_{\mathcal{L}} \kappa^2(s) ds + \frac{C}{2} \oint_{\mathcal{L}} (\omega - \omega_0)^2 ds \tag{7.30}$$

In Eq. (7.30), s is the arc-length, and the integrals are computed over the entire closed DNA curve of length $\mathcal{L}$. ω_0 denotes the intrinsic twist rate of the DNA, for example, its value is $2\pi/10.5$ radians between two successive base-pair steps.

The elastic bending constant A of DNA can be linearly related to the bending persistence length of DNA as:

$$A = k_B T p_b \tag{7.31}$$

where p_b is the persistence length of the DNA. Moreover, A can also be related to the root-mean-square bending angle $\langle \theta_b^2 \rangle^{1/2}$ of a semi-flexible chain with a preferred axis of bending, perpendicular to the helix axis, for which the angular fluctuations are independent from each other (Landau and Lifshitz 1980):

$$A = \frac{2 k_B T h}{\langle \theta_b^2 \rangle} \tag{7.32}$$

where h is the base-pair separation distance in DNA, which is about 3.4 Å (Schellman 1974; Hagerman 1988).

The torsional rigidity constant can also be related to the persistence length of twisting p_{tw} and the root-mean-square twist angle Ω as (Bustamante et al. 1994):

$$C = \frac{p_{tw} k_B T}{2} \tag{7.33}$$

and

$$C = \frac{k_B T h}{\langle \Omega^2 \rangle} \tag{7.34}$$

Denoting the bending to torsional rigidity constant ratio

$$r = \frac{A}{C} \tag{7.35}$$

the following relationship can be found:

$$\langle \Omega^2 \rangle^{1/2} = \langle \theta^2 \rangle^{1/2} \sqrt{\frac{r}{2}}. \tag{7.36}$$

The common values of $A = 2.0 \times 10^{-19}$ erg cm and $C = 3.0 \times 10^{-19}$ erg cm for DNA will produce the persistence lengths for bending and twisting of $p_b = 500$ Å (Hagerman 1988), $p_{tw} = 750$ Å, and $r = 2/3$. These values correspond to root-mean-square bend and twist-angle values of $\langle \theta^2 \rangle^{1/2} = 6.7°$ and $\langle \Omega^2 \rangle^{1/2} = 3.9°$ at room temperature, in the context of isotropic bending. The bending constant A can also be related to measured local angular variations in solved DNA structures. The torsional rigidity constant has recently been estimated from pulling experiments on single molecules of DNA to be 40 % higher than generally accepted values (Bryant et al. 2003).

Recently, CGMs have also been introduced showing greater details of the DNA structure (Drukker et al. 2001; Sales-Pardo et al. 2005; Starr and Sciortino 2006; Trovato and Tozzini 2008; Savelyev and Papoian 2009, 2010; Sayar et al. 2010), which are used to characterise DNA melting (Savelyev and Papoian 2010), renaturation (Sambriski et al. 2009) or nano-tweezers (Ouldridge et al. 2010).

7.4.3 *Three-Bead DNA Model*

The three-bead model of DNA is developed by Knotts et. al. (Knotts et al. 2007), in which the beads that are used to describe sugar and phosphate are placed at the centres of mass of these groups. The adenine (A) and guanine (G) base beads are placed in the position of their N1 atoms and the thymine (T) and cytosine (C) beads in the position of their N3 atoms, as shown in Fig. 7.6. The DNA backbone are represented with two beads placed at the P and C3′ atom positions, which are necessary for reproducing the deformations of the grooves for characterising the DNA-protein interactions and for backward projection from the CGM to fully atomistic representation of the DNA.

The internal potential energy interaction is written as (Knotts et al. 2007):

$$U_{intra} = U_{bond} + U_{angle} + U_{dihedral} + U_{stack} \tag{7.37}$$

where U_{bond} is the pseudo-bond energy term, U_{angle} is the pseudo-angle energy term, and $U_{dihedral}$ is the pseudo-torsional energy term implemented as cosine potential. These three terms are standard quadratic terms. U_{stack} energy term characterises the stacking interactions, and it is modelled only to base beads belonging to one strand with native structure positioned within 9 Å cutoff distance. It counts for both $i : i+1$ and $i : i+2$ interactions.

The non-bonded interaction $U_{non-bonded}$ are modelled as Lennard-Jones potential using the 12–10 powers:

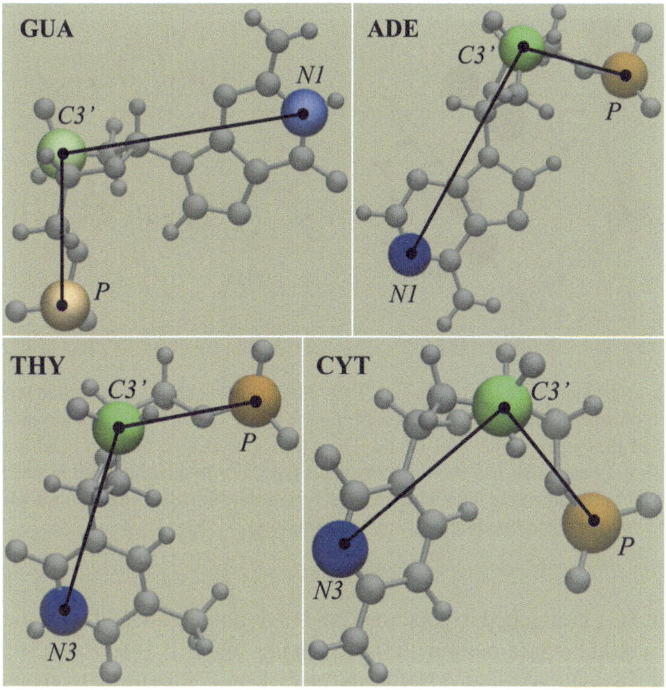

Fig. 7.6 The three-bead models of adenine and guanine centred at N1, C3′ and P atoms, and thymine and cytosine centred at N3, C3′ and P atoms

$$U_{non-bonded}(r_{ij}) = 4\varepsilon_{ij}\left[\left(\frac{\sigma_{ij}}{r_{ij}}\right)^{12} - \left(\frac{\sigma_{ij}}{r_{ij}}\right)^{10}\right] \tag{7.38}$$

where the summations is over all G-C and A-T base pairs that are not already considered in U_{stack}.

7.4.4 Martini Coarse-Grained Model of DNA

In Martini DNA model (Uusitalo et al. 2015), each nucleotide is mapped to six or seven coarse-grained beads. Three beads are used to model the backbone by mapping the phosphate to one bead and the sugar to two beads. The cytosine and thymine are modelled as three-bead rings and adenine and guanine as four bead rings.

Figure 7.7 shows a mapping of each base to a CGM characterised by backbone beads (BB1, BB2, and BB3) and side chain beads (SC1, SC2, SC3 and SC4) for adenine and guanine. The first backbone bead (BB1) is the phosphate and the last one (BB3) at 3′ end of the sugar. For the side chains, the beads are defined in cyclical

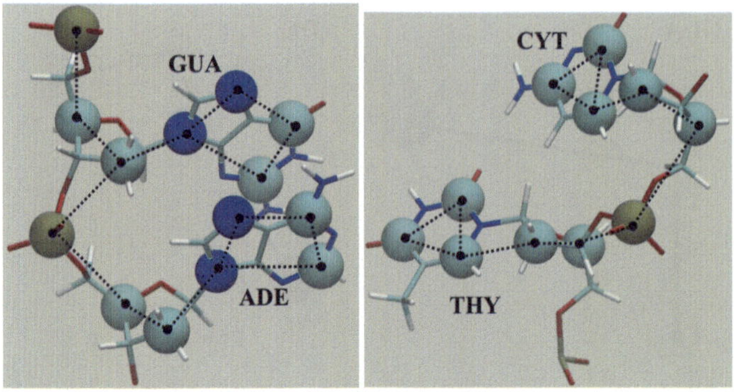

Fig. 7.7 Coarse-grained mapping of Martini DNA: The DNA backbone is modelled with one bead describing the phosphate and two beads describing the sugar. The pyrimidines are modelled with three beads and the purines with four beads. The Martini bead type of each bead is shown; the T-prefix marks the beads that use the new tiny bead type. For hydrogen bonding beads, the new special bead types are shown together with the bead type describing their interactions with all beads except the special hydrogen bonding beads (in parentheses)

order so that SC1 is attached to backbone, and in double-stranded DNA, the SC2 and SC3 beads would be base-pairs with the opposing strand.

The bases in double-stranded DNA are stacked very close to each other, typically, with a distance between bases of 0.34 nm, which lead to problems in a CGM that describes the underlaying atomistic structure with large isotropic beads. In Martini model, four atoms are characterised approximately with a single bead, where spherical model used is part of the simplicity of the model and to describe uniformity of the distribution of atom inside a sphere. However, this model fails for thin and planar molecules such as nucleobases, because beads are too large in the dimension perpendicular to the plane of the base.

The standard Martini CGM uses a Lennard-Jones parameter $\sigma = 0.47$ nm, and smaller ring type beads (S type) use $\sigma = 0.43$ nm, which is too large to model base stacking (Uusitalo et al. 2015). Therefore, smaller bead types have been modelled in order to fit the bases at the correct distance in the coarse-grained DNA structure with $\sigma = 0.32$ nm (Uusitalo et al. 2015), which are denoted with T (for tiny). These particles interact with each other with reduced σ, but they behave as S type particles toward standard and S type particles ($\sigma = 0.43$ nm and epsilon reduced to 75% with S particles and $\sigma = 0.47$ nm with regular particles).

Compare to the standard Martini, other additional are made to describe the hydrogen bonding between the bases, which are crucial for the formation of the double-stranded DNA and specificity of base pairing. A specific bead type has been introduced to describe the hydrogen bonding interaction by adding eight special beads that are meant to describe these interactions. These beads interact with all the other beads based on their underlying chemical group. For example, TA2 based on TN0 interacts like TN0 with all the other bead types except for the eight specific hydrogen bonding bead types. Energy levels are also modified to describe the interactions between these specific eight beads.

7.5 Summary

The computational coarse-grain modelling strategies were discussed for use in computer simulations of biological molecules. Besides, different coarse-grained models were described in details, in particular, for proteins and DNAs.

Furthermore, the Gō coarse-graining potential energy function for proteins was given. Besides, the network and Martini coarse-grained models of proteins were discussed.

Moreover, the coarse-grained models of nucleic acids were described in details, such as worm-like chain model, elasticity theory framework, three-bead DNA model, and Martini coarse-grained models.

Note that the Martini coarse-grained force fields have also been developed for small molecules (Alessandri et al. 2022) and membranes (Marrink et al. 2019). Besides, the Gō-Martini coarse-grained force fields are optimised for computer simulations of proteins and lipid membranes systems (Mahmood et al. 2021).

References

R. Alessandri, P.C.T. Souza, S. Thallmair, M.N. Melo, A.H. de Vries, S.J. Marrink, Pitfalls of the Martini model. J. Chem. Theory Comput. **15**(10), 5448–5460 (2019)

R. Alessandri, J. Barnoud, A.S. Gertsen, I. Patmanidis, A.H. de Vries, P.C.T. Souza, S.J. Marrink, Martini 3 coarse-grained force field: small molecules. Adv. Theory Simul. **5**, 2100391 (2022)

A. Atilgan, S. Durell, R. Jernigan, M. Demirel, O. Keskin, I. Bahar, Anisotropy of fluctuation dynamics of proteins with an elastic network model. Biophys. J . **80**(1), 505–515 (2001)

I. Bahar, A.R. Atilgan, B. Erman, Direct evaluation of thermal fluctuations in protein using a single parameter harmonic potential. Folding Des. **2**, 173–181 (1997)

N. Basdevant, D. Borgis, T. Ha-Duong, A coarse-grained protein-protein potential derived from an all-atom force field. J. Phys. Chem. B **111**, 9390 (2007)

T. Bereau, M. Deserno, Generic coarse-grained model for protein folding and aggregation. J. Chem. Phys. **130**, 235106 (2009)

H. Berman, K. Henrick, H. Nakamura, Announcing the worldwide Protein Data Bank. Nat. Struct. Mol. Biol. **10**(12), 980 (2003)

C. Bouchiat, M.D. Wang, J.F. Allemand, T. Strick, S.M. Block, V. Croquette, Estimating the persistence length of a worm-like chain molecule from force-extension measurements. Biophys. J. **76**, 409–413 (1999)

Z. Bryant, M.D. Stone, J. Gore, S.B. Smith, N.R. Cozzarelli, C. Bustamante, Structural transitions and elasticity from torque measurements on DNA. Nature **424**, 338–341 (2003)

C. Bustamante, J.F. Marko, E.D. Siggia, S. Smith, Entropic elasticity of λ-phage DNA. Science **265**, 1599–1600 (1994)

H.S. Chan, K.A. Dill, Polymer principles in protein structure and stability. Annu. Rev. Biophys. Biophys. Chem. **20**(1), 447–490 (1991)

Y. Chebaro, S. Pasquali, P. Derreumaux, The coarse-grained OPEP force field for non-amyloid and amyloid proteins. J. Phys. Chem. B **116**(30), 8741–8752 (2012)

N.Y. Chen, Z.Y. Su, C.Y. Mou, Effective potentials for folding proteins. Phys. Rev. Lett. **96**, 078103 (2006)

M. Cheon, I. Chang, C.K. Hall, Extending the PRIME model for protein aggregation to all twenty amino acids. Proteins **78**(14), 2950–2960 (2010)

C. Chipot, A. Pohorille (eds) *Free energy calculations: Theory and Applications in Chemistry and Biology* (Springer, 2007)

C.D. Christ, A.E. Mark, W.F. van Gunsteren, Basic ingredients of free energy calculations: a review. J. Comput. Chem. **31**(8), 1569–1582 (2010)

D.H. de Jong, G. Singh, W.F.D. Bennett, C. Arnarez, T.A. Wassenaar, L.V. Schäfer, X. Periole, D.P. Tieleman, S.J. Marrink, Improved parameters for the martini coarse-grained protein force field. J. Chem. Theory Comput. **9**, 687–697 (2013)

F. Ding, S.V. Buldyrev, N.V. Dokholyan, Folding Trp-cage to NMR resolution native structure using a coarse-grained protein model. Biophys. J . **88**(1), 147–155 (2005)

K. Drukker, G. Wu, G.C. Schatz, Model simulations of DNA denaturation dynamics. J. Chem. Phys. **114**, 579–590 (2001)

A. Emperador, O. Carrillo, M. Rueda, M. Orozco, Exploring the suitability of coarse-grained techniques for the representation of protein dynamics. Biophys. J . **95**, 2127–2138 (2008)

E. Faraggi, P. Krupa, M.A. Mozolewska, A. Liwo, A. Kloczkowski, Reoptimized UNRES potential for protein model quality assessment. Genes **9**, 601 (2018)

M. Fixman, J. Kovac, Modified gaussian model for rubber elasticity. 2 the wormlike chain. Macromolecules **15**, 537–541 (1982)

M. Fixman, J. Kovac, Polymer conformational statistics iii: Modified gaussian models of the stiff chains. J. Chem. Phys. **58**, 1564–1568 (1973)

P.J. Flory, M. Gordon, N.G. McCrum, Statistical thermodynamics of random networks. Proc. R. Soc. Lond. Ser. A **351**(1666), 351–380 (1976)

C. Guo, M.S. Cheung, H. Levine, D.A. Kessler, Mechanisms of co-operativity underlying sequence-independent β-sheet formation. J. Chem. Phys. **116**, 4353–4365 (2002)

P.J. Hagerman, Flexibility of DNA. Ann. Rev. Biophys. Biophys. Chem. **17**, 265–286 (1988)

T. Haliloglu, I. Bahar, B. Erman, Gaussian dynamics of folded proteins. Phys. Rev. Lett. **79**(16), 3090 (1997)

R.D. Hills Jr., C.L. Brooks III., Insights from coarse-grained Gō models for protein folding and dynamics. Int. J. Mol. Sci. **10**(3), 889–905 (2009)

K. Hinsen, Analysis of domain motions by approximate normal mode calculations. Proteins **33**, 417–429 (1998)

A. Irbäck, F. Sjunnesson, S. Wallin, Hydrogen bonds, hydrophobicity forces and the character of the collapse transition. Proc. Natl. Acad. Sci. U.S.A. **97**, 13614 (2000)

S. Izvekov, G.A. Voth, A multiscale coarse-graining method for biomolecular systems. J. Phys. Chem. B **109**(7), 2469–73 (2005)

Z. Jarin, J. Newhouse, G.A. Voth, Coarse-grained force fields from the perspective of statistical mechanics: better understanding of the origins of a MARTINI hangover. J. Chem. Theory Comput. **17**, 1170–1180 (2021)

J.A. Joseph, A. Reinhardt, A. Aguirre, P.Y. Chew, K.O. Russell, J.R. Espinosa, A. Garaizar, R.C. Guevara, Physics-driven coarse-grained model for biomolecular phase separation with near-quantitative accuracy. Nat. Comput. Sci. **1**(11), 732–743 (2021)

H. Kamberaj, A theoretical model for the collective motion of proteins by means of principal component analysis. Cent. Eur. J. Phys. **9**(1), 96–109 (2011)

J.G. Kirkwood, Statistical mechanics of fluid mixtures. J. Chem. Phys. **3**, 300 (1935)

T.A. Knotts, N. Rathore, D.C. Schwartz, J.J. de Pablo, A coarse grain model for DNA. J. Chem. Phys. **126**(27), 084901 (2007)

P. Kynast, P. Derreumaux, B. Strodel, Evaluation of the coarse-grained OPEN force field for protein-protein docking. BMC Biophys. **9**, 4 (2016)
L.D. Landau, E.M. Lifshitz (eds.), *Course of Theoretical Physics,* vol. 5, 3rd edn. (Pergamon Press, Oxford, 1980)
M. Levitt, A simplified representation of protein conformations for rapid simulation of protein folding. J. Mol. Biol. **104**(1), 59–107 (1976)
M. Levitt, A. Warshel, Computer simulation of protein folding. Nature **253**, 694–698 (1975)
A. Liwo, R. Kaźmierkiewicz, C. Czaplewski, M. Groth, S. Oldziej, R.J. Wawak, S. Rackovsky, M.R. Pincus, H.A. Scheraga, A united-residue force field for off-lattice protein-structure simulations. III. Origin of backbone hydrogen-bonding cooperativity in united-residue potentials. J. Comput. Chem. **19**, 259–276 (1998)
A. Liwo, S. Oldziej, M.R. Pincus, R.J. Wawak, S. Rackovsky, H.A. Scheraga, A united-residue force field for off-lattice protein-structure simulations. I. Functional forms and parameters of long-range side-chain interaction potentials from protein crystal data. J. Comput. Chem. **18**(7), 849–873 (1997)
A. Liwo, C. Czaplewski, J. Pillardy, H.A. Scheraga, Cumulant-based expressions for the multibody terms for the correlation between local and electrostatic interactions in the united-residue force field. J. Chem. Phys. **115**, 2323 (2001)
A. Liwo, Y. He, H.A. Scheraga, Coarse-grained force-field: general folding theory. Phys. Chem. Chem. Phys. **13**, 16890–16901 (2011)
A.A. Louis, Beware of density dependent pair potentials. J. Phys.: Condens. Matter **14**, 9187 (2002)
L. Lu, G.A. Voth, The multiscale coarse-graining method. VII. Free energy decomposition of coarse-grained effective potentials. J. Chem. Phys. **134**, 224107 (2011)
J.F. Marko, E.D. Siggia, Stretching DNA. Macromolecules **28**, 8759–8770 (1995)
S.J. Marrink, A.H. de Vries, A.E. Mark, Coarse grained model for semiquantitative lipid simulations. J. Phys. Chem. B **108**(2), 750–760 (2004)
S.J. Marrink, V. Corradi, P.C.T. Souza, H.I. Ingólfsson, D.P. Tieleman, M.S.P. Sansom, Computational modeling of realistic cell membranes. Chem. Rev. **119**, 6184–6226 (2019)
Md I. Mahmood, A.B. Poma, K. Okazaki, Optimizing Gō-MARTINI Coarse-Grained Model for F-BAR Protein on Lipid Membrane. Front. Mol. Biosci. **8**, 619381 (2021)
B.M. Messer, M. Roca, Z.T. Chu, S. Vicatos, A.V. Kilshtain, A. Warshel, Multiscale simulations of protein landscapes: using coarse-grained models as reference potentials to full explicit models. Proteins **78**(5), 1212–1227 (2010)
L. Monticelli, S.K. Kandasamy, X. Periole, R.G. Larson, D.P. Tieleman, S.J. Marrink, The MARTINI coarse-grained force field: extension to proteins. J. Chem. Theory Comput. **4**, 819–834 (2008)
Y. Mu, Y.Q. Gao, Effects of hydrophobic and dipole-dipole interactions on the conformational transitions of a model polypeptide. J. Chem. Phys. **127**(10), 105102 (2007)
H.D. Nguyen, C.K. Hall, Molecular dynamics simulations of spontaneous fabril formation by random-coil peptides. Proc. Natl. Acad. Sci. U.S.A. **101**(46), 16180–16185 (2004)
W.G. Noid, Perspective: coarse-grained models for biomolecular systems. J. Chem. Phys. **139**, 090901 (2013)
W.G. Noid, J.W. Chu, G.S. Ayton, V. Krishna, S. Izvekov, G.A. Voth, A. Das, H.C. Andersen, A rigorous bridge between atomistic and coarse-grained models. J. Chem. Phys. **128**, 244114 (2008)
T. Odijk, Stiff chains and filaments under tension. Macromolecules **28**(20), 7016–7018 (1995)
T.E. Ouldridge, A.A. Louis, J.P.K. Doye, DNA Nanotweezers studied with a coarse-grained model of DNA. Phys. Rev. Lett. **104**, 178101 (2010)
J. Pfaendtner, D. Brabduardi, M. Parrinello, T.D. Pollard, G.A. Voth, Nucleotide-dependent conformational states of actin. Proc. Natl. Acad. Sci. U.S.A. **106**, 12723–12728 (2009)
A. Rey, J. Skolnick, Comparison of lattice Monte Carlo dynamics and Brownian dynamics folding pathways of α-helical harpins. Chem. Phys. **158**(2–3), 199–219 (1991)
P. Rotkiewicz, J. Skolnick, Fast procedure for reconstruction of full-atom protein models from reduced representations. J. Comput. Chem. **29**(9), 1460–1465 (2008)

M. Sales-Pardo, R. Guimerá, A.A. Moreira, J. Widow, L.A.N. Amaral, Mesoscopic modelling for nucleic acid chain dynamics. Phys. Rev. E **71**, 051902 (2005)

E. Sambriski, D. Schwartz, J. de Pablo, A mesoscale model of DNA and its renaturation. Biophys. J. **96**, 1675–1690 (2009)

M.G. Saunders, G.A. Voth, Coarse-graining of multiprotein ensemblies. Curr. Opin. Struct. Biol. **22**(2), 144–150 (2012)

A. Savelyev, G.A. Papoian, Molecular renormalisation group coarse-graining of polymer chains: application to double-stranded DNA. Biophys. J. **96**(10), 4044–4052 (2009)

A. Savelyev, G.A. Papoian, Chemically accurate coarse graining of double-stranded DNA. Proc. Natl. Acad. Sci. U.S.A. **107**(47), 20340–5 (2010)

M. Sayar, B.C. Avşaroğlu, A. Kabakçioğlu, Twist-writhe partitioning in a coarse-grained DNA mini circle model. Phys. Rev. E **81**, 041916 (2010)

J.A. Schellman, Flexibility of DNA. Biopolymers **13**, 217–226 (1974)

F. Schmid, Toy amphiphiles on the computer: what can we learn from generic models? Macromol. Rapid Commun. **30**, 741 (2009)

A.V. Sinitskiy, M.G. Saunders, G.A. Voth, Optimal number of coarse-grained sites in different components of large biomolecular complexes. J. Phys. Chem. B **116**, 8363–8374 (2012)

S. Smith, L. Finzi, C. Bustamante, DNA molecules by using magnetic beads. Science **258**, 1122–1126 (1992)

S. Smith, Y. Cui, C. Bustamante, Over-stretching B-DNA: the elastic response of individual double-stranded and single-stranded DNA molecules. Science **271**, 795–799 (1996)

P.C.T. Souza, R. Alessandri, J. Barnoud, S. Thallmair, I. Faustino, F. Grünewald, I. Patmandis, H. Abdizadeh, B.M.H. Bruininks, T.A. Wassenaar, P.C. Kroon, J. Melcr, V. Nieto, V. Corradi, H.M. Khan, J. Domanski, M. Javanainen, H. Martinez-Seara, N. Reuter, R.R. Best, I. Vattulainen, L. Monticelli, X. Periole, D.P. Tieleman, A.H. de Vries, S.J. Marrink, Martini 3: a general purpose force field for coarse-grained molecular dynamics. Nat. Methods **18**, 382–388 (2021)

F.W. Starr, F. Sciortino, Model of assembly and gelation of four-armed DNA dendrimers. J. Phys.: Condens. Matter **18**, L347 (2006)

M. Stepanova, Dynamics of essential collective motions in proteins: Theory. Phys. Rev. E **76**(5), 051918 (2007)

F. Sterpone, S. Melchionna, P. Tuffery, S. Pasquali, N. Mousseau, T. Cragnolini, Y. Chebaro, J.F. Saint-Pierre, M. Kalimeri, A. Barducci, Y. Laurin, A. Tex, M. Baaden, P.H. Nguyen, P. Derreumaux, The OPEN coarse-grained protein model: from single molecules, amyloid formation, role of macromolecular crowding and hydrodynamics to RNA/DNA complexes. Chem. Soc. Rev. **43**(13), 4871–4893 (2014)

F.H. Stillinger, H. Sakai, S. Torquato, Statistical mechanical models with effective potentials: definitions, applications, and thermodynamic consequences. J. Chem. Phys. **117**, 288 (2002)

S. Takada, Z. Luthey-Schulten, P.G. Wolynes, Folding dynamics with nonadditive forces: a simulation study of a designed helical protein and a random heteropolymer. J. Chem. Phys. **110**, 11616 (1999)

H. Taketomi, Y. Ueda, N. Go, Studies on protein folding, unfolding and fluctuations by computer simulations. I. The effect of specific amino acid sequence represented by specific inter-unit interactions. Int. J. Pept. Protein Res. **7**, 445–459 (1975)

S. Tanaka, H.A. Scheraga, Medium- and long-range interaction parameters between amino acids for predicting three-dimensional structures of proteins. Macromolecules **9**(6), 945–950 (1976)

M.M. Tirion, Large amplitude elastic motions in proteins from a single-parameter. Atomic Anal. Phys. Rev. Lett. **77**, 1905–1908 (1996)

V. Tozzini, Coarse grained models for proteins. Curr. Opin. Struc. Biol. **15**, 144–150 (2005)

V. Tozzini, W. Rocchia, J.A. McCammon, Mapping all-atom models onto one-bead coarse grained models: general properties and applications to a minimal polypeptide model. J. Chem. Theory Comput. **2**(3), 667–673 (2006)

F. Trovato, V. Tozzini, Supercooling and local denaturation of plasmids with a minimalist DNA model. J. Phys. Chem. B **112**(42), 13197–200 (2008)

J.J. Uusitalo, H.I. Ingólesson, P. Akhshi, D.P. Tielman, S. Marrink, Martini coarse-grained force field: extension to DNA. J. Chem. Theory Comput. **11**, 393–3945 (2015)

A.E. van Giessen, J.E. Straub, Monte Carlo simulations of polyalanine using a reduced model and statistics-based interaction potentials. J. Chem. Phys. **122**, 024904 (2005)

A.V. Vologodskii, Exploring polymer dynamics with single DNA molecules. Macromolecules **27**, 5623–5625 (1994)

G.A. Voth (ed.), *Coarse-Graining of Condensed Phase and Biomolecular Systems* (CRC Press, Boca Raton, FL, 2008)

M.D. Wang, H. Yin, R. Landick, J. Gelles, S.M. Block, Stretching DNA with optical tweezers. Biophys. J . **72**, 1335–1346 (1997)

M. Zacharias, Protein-protein docking with a reduced protein model accounting for side-chain flexibility. Protein Sci. **12**, 1271 (2003)

J. Zhang, M. Muthukumar, Simulations of nucleation and elongation of amyloid fibrils. J. Chem. Phys. **130**, 035102 (2009)

R. Zwanzig, Memory effects in irreversible thermodynamics. Phys. Rev. **124**, 983 (1961)

Chapter 8
Implementation of Molecular Dynamics in Simulations

The chapter aims to describe some practical aspects to implementation of molecular dynamics simulations.

In this chapter, we will briefly describe the practical aspects of molecular dynamics simulations, such as designing the constraints, periodic boundary conditions, spherical cutoffs, treatment of the long-range interactions (in particular, electrostatic interactions), and identifying the equilibrium states of the simulations. For further details, one can refer to the following literatures (Allen and Tildesley 1989; Leach 2001; Kamberaj 2020).

8.1 Designing Constraints for Molecular Dynamics Simulations

Running a molecular dynamics simulation has to account for both, the physical nature of the system under study and the available computational resources. System size, time step, and total time duration must be selected so that the simulations finish within a reasonable period. Besides, the simulation time has to be long enough to study properties of interest at the timescales of natural processes. Moreover, statistically, the simulations' timescales should match the kinetics of the natural process. Most of scientific studies about the dynamics of proteins, DNA/RNA, lipids, and their complexes use simulations ranging from nanoseconds to microseconds. In general, several days to years of central processing units (CPUs), depending on the system size, is necessary to obtain the simulations at these timescales. Multiple CPU simulations in parallel using specific algorithms allow sharing the computation workload among CPUs, and special algorithms for running in the computer's graphical processing unit (GPU) (Stone et al. 2011; Phillips et al. 2014).

H. Kamberaj, *Computer Simulations in Molecular Biology*, Scientific Computation,
https://doi.org/10.1007/978-3-031-34839-6_8

In a classical MD simulation, the most expensive computational task is evaluating the potential energy and forces on particles as a function of their internal coordinates, dominated by the evaluation of the non-bonded or non-covalent terms (such as the electrostatic and van der Waals interactions). Typically, the molecular dynamics simulations scale as $\mathcal{O}(N^2)$, with N being the number of particles, if all pair-wise electrostatic and van der Waals interactions are included. However, that computation cost reduces by employing electrostatics methods, such as Particle Mesh Ewald ($\mathcal{O}(N \log(N))$) or good spherical cutoff techniques ($\mathcal{O}(N)$).

The integration time step Δt equals the time between evaluations of the potential and forces of a system. Since the total CPU time of a simulation is proportional to the number of these evaluations, then it depends on Δt. The time step Δt is of order 2 femtoseconds, if covalent bond constraints are applied, to avoid numerical errors. In general, Δt is smaller than the fastest vibrational degree of freedom in the system. Classical MD simulation uses a time step about 1 femtosecond, which increases to 2 femtoseconds by using SHAKE algorithm (Ryckaert et al. 1977) fixing the vibrations of the fastest atoms (such as the covalent bonds involving hydrogen atom). Multiple time scale methods allow using larger time steps between updates of slower long-range forces (Martyna et al. 1992).

For performing the simulations of the macromolecules in a solvent environment, often a choice should be made between explicit solvent and implicit solvent. Usually, the explicit solvent models of the force fields, such as the TIP3P and SPC/E waters, are computationally expensive, while implicit solvents use a mean-field approach, and hence they are usually less costly. Typically, explicit solvent models require the inclusion of about ten times more particles in the simulation. Besides, the explicit models are such that they are allowed to reproduce specific properties of the solute macromolecules their kinetics.

In the molecular dynamics simulations of the macromolecules, the simulation box size must be large enough to avoid boundary condition artefacts. Usually, the boundary conditions are treated by either having fixed values at the edges or by applying periodic boundary conditions in which one side of the simulation loops back to the opposite side. In practice, this mimics a bulk phase.

8.2 Initial Configuration

The initial configuration of a system is chosen before the MD simulation starts. The stability of the numerical integration depends on the initial arrangement of the particles, and hence it is very important step. Usually, the initial configuration for a system at equilibrium is close to the desired state. Also, it is recommended that the energy of the initial configuration does not contain terms with high-energy interactions. Often, to remove such terms, energy minimisation can be performed before the start of MD simulations to minimise the potential energy of system to the closest local minimum.

If the system's experimental structure exists, this is used. However, if it is not available, then other biomolecular structures with similar amino acids sequence can be chosen as a starting configuration using structure alignment techniques. For that, the homology modelling approach is often used to construct three-dimensional atomic level model of a protein based on its amino acid sequence and an experimental three-dimensional structure of a known related homologous template protein (Bordoli et al. 2009).

The solvent molecules' coordinates are added at a given density in normal conditions. For some solvent molecules the coordinates of the X-Ray structures can also be used. In practice, when the solute immerses in a solvent, which was previously equilibrated, we remove all the solvent molecules in close contact with the solute. Then, some minimisation steps may be required priory of MD simulations to equilibrate the system to the nearest local minimum of the potential energy function.

8.3 Periodic Boundary Conditions

In computer simulations, the maximum size of the system is limited by the available storage on the computer or by the speed of execution of the program. As an example, if we consider that one mole of a liquid will contain approximately 10^{23} particles we are far from being able to study this system. Mainly if we are interested in studying the properties of the bulk fluid, a small number of particles we can simulate, if not all of them, would be within the influence of the walls of the boundary. *Periodic boundary conditions* enable a simulation to be performed using a relatively small number of particles, in such a way that every particle experiences a force as if it was in bulk fluid phase. In the two-dimensional each box is surrounded by eight neighbours as is shown in Fig. 8.1. Note that in three dimensions, each box has 26 nearest neighbours.

The coordinates of the particles in the image boxes can be computed merely by adding or subtracting integral multiples of the box sides. If a particle leaves the box during the simulation, then it is replaced by an imaged particle that enters from the opposite side, as illustrated in Fig. 8.2.

The number of particles within the central box thus remains constant after applying the periodic boundary conditions. The cubical cell and its close relation, the parallelepiped, are the simplest periodic systems to use in the computer programs and visualise them. However, a cell of a different shape might be more appropriate for a given simulation. It is often sensible to choose a periodic cell that reflects the specified property of the system. For example, allowing the cell to change the shape during a molecular dynamics or Monte Carlo simulation in the smectic phase might affect the packing. In simulations of biological molecules in explicit solvent, often the octahedral simulation box is chosen to reduce the number of water molecules, and hence reducing the computation costs.

Periodic boundaries are widely used in computer simulations, but they do have some drawbacks (Allen and Tildesley 1989). A limitation of the periodic cell is that it is not possible to achieve fluctuations that have a wavelength larger than the length

Fig. 8.1 Illustration of the periodic boundary conditions in two dimensions

of the cell. The range of the interactions present in the system is also important; if the scale over which the interactions act is larger than the cell size then there should be no problems.

8.4 Potential Cutoffs and the Minimum Image Convention

The most time-consuming part of the computer simulations is the calculation of the energies and forces. For example, for a pairwise model of the potential, such as the Lennard-Jones, the time for calculation of the forces is of order N^2 for an N particles system. In principle, the interactions are calculated for each pair of particles in the system. In practice, for many of the interaction models, this is not necessary since the potentials fall off very rapidly with distance, as it is shown in Fig. 8.3.

In such cases, a cut-off applies to the interaction potential with the minimum image convention (Allen and Tildesley 1989). In the minimum image convention, each particle interacts at the most with just one image of every other particle in the

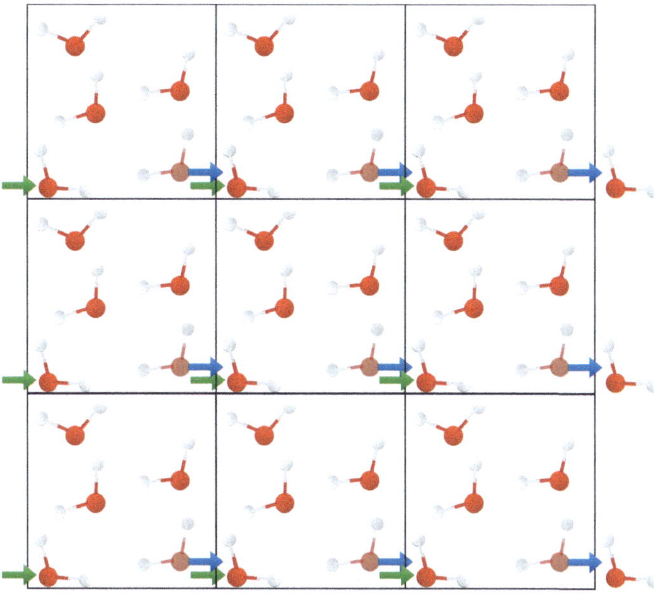

Fig. 8.2 Demonstration of changes occurring when a particle leaves the simulation box from the right and entering to the left

Fig. 8.3 Lennard-Jones potential energy (in kcal/mol) as a function of atoms' distance r_{ij} (in Å) for the pairs CA−HA and CA−CB, using CHARMM36 force field

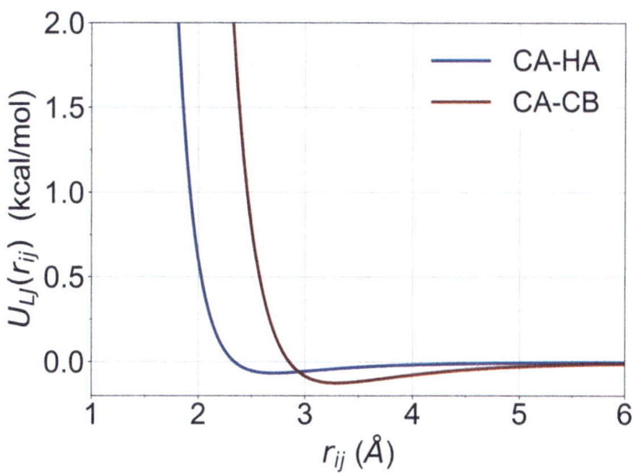

system, which is repeated infinitely via periodic boundary conditions. When the cut-off is employed the interactions between all pairs of particles that are further apart than the cut-off value is modified using some function $f(r)$, taking into account the closest image.

Lennard-Jones and electrostatic interactions between every pair of atoms i and j take the forms:

$$U_{\mathrm{LJ}}(r_{ij}) = 4\varepsilon_{ij}\gamma_{ij}^{\mathrm{LJ}} f(r_{ij}) \left[\left(\frac{\sigma_{ij}}{r_{ij}}\right)^{12} - \left(\frac{\sigma_{ij}}{r_{ij}}\right)^{6} \right] \tag{8.1}$$
$$U_{\mathrm{Elec}}(r_{ij}) = k_e \gamma_{ij}^{\mathrm{Elec}} f(r_{ij}) \frac{q_i q_j}{\varepsilon r_{ij}}$$

In Eq. (8.1), $f(r_{ij})$ is either the switch or shift function, $\gamma_{ij}^{\mathrm{LJ}}$ and $\gamma_{ij}^{\mathrm{Elec}}$ are scaling factors, which are constants defined *a priory* of simulations. They are often chosen such that $\gamma_{ij}^{\mathrm{LJ}} = \gamma_{ij}^{\mathrm{Elec}} = 0$ for atoms separated by less than three bonds, and they are either one or less than one, depending on the force field, for atoms separated by three bonds, the so-called 1-4 interactions, and moreover, they are both one for atoms separated by more than three bonds, the so-called non-bonded atoms.

If a simple truncation of the potential energy function is used, then the switch function is given as:

$$f(r_{ij}) = \begin{cases} 1, & r_{ij} < R_c \\ 0, & r_{ij} \geq R_c \end{cases} \tag{8.2}$$

where R_c is the cutoff distance. It can be seen that a function given by Eq. (8.2) introduces a discontinuity in the force, and hence it does not guarantee the conservation of the energy. Therefore, in practice, different molecular dynamics simulation engines have introduced the so-called switch and/or shift function in either potential energy or force calculations. In this approach, the distance-dependent function $f(r_{ij})$ is determined in terms of the cutoff distance R_c and the so-called switch distance R_s, such as

$$f(r_{ij}) = \begin{cases} 1, & r_{ij} < R_c \\ S(r_{ij}), & R_c \leq r_{ij} \leq R_s \\ 0, & r_{ij} > R_s \end{cases} \tag{8.3}$$

where, now, $S(r_{ij})$ is a smooth function decaying to zero between R_c and R_s. In general, $S(r_{ij})$ can be designed different for the Lennard-Jones and the electrostatic interactions, therefore, Eq. (8.1) can also be written as:

$$\begin{aligned} V_{\mathrm{LJ}}(r_{ij}) + V_{\mathrm{Elec}}(r_{ij}) = \Bigg[& 4\varepsilon_{ij} f^{\mathrm{LJ}}(r_{ij}) \gamma_{ij}^{\mathrm{LJ}} \left(\frac{\sigma_{ij}^{12}}{r_{ij}^{12}} - \frac{\sigma_{ij}^{6}}{r_{ij}^{6}} \right) \\ & + f^{\mathrm{Elec}}(r_{ij}) \gamma_{ij}^{\mathrm{Elec}} k_e \frac{q_i q_j}{\varepsilon r_{ij}} \Bigg] \end{aligned} \tag{8.4}$$

We can design the so-called *potential switch* functions, $S(r_{ij})$, to be a polynomial function of the distance such that potential energy of non-bonded interactions slowly decays to zero in the interval $[R_c,\ R_s]$, where $V(r)$, for $r \leq R_c$, is unchanged and $V(R_s) = 0$. In this case, the polynomial function is chosen with such degree that energy function and its gradient are continuous functions. Such a function is given in the following as designed in CHARMM program (Brooks et al. 2009) (based on the CHARMM's manual):

$$f(r_{ij}) = \begin{cases} 1, & r_{ij} < R_c \\ \dfrac{(R_s^2 - r_{ij}^2)^2(R_s^2 + 2r_{ij}^2 - 3R_c^2)}{(R_s^2 - R_c^2)^3}, & R_c \leq r_{ij} \leq R_s \\ 0, & r_{ij} > R_s \end{cases} \tag{8.5}$$

From Eq. (8.5), it can easily be seen that $S(r_{ij})$ decays smoothly to zero as r_{ij} increases from R_c to R_s. In addition, the derivative of $f(r_{ij})$ with respect to the separation r_{ij} is given as:

$$\frac{\partial f(r_{ij})}{\partial r_{ij}} = \begin{cases} 0, & r_{ij} < R_c \\ \dfrac{-12r_{ij}(R_s^2 - r_{ij}^2)(r_{ij}^2 - R_c^2)}{(R_s^2 - R_c^2)^3}, & R_c \leq r_{ij} \leq R_s \\ 0, & r_{ij} > R_s \end{cases} \tag{8.6}$$

which indicates that at R_c and R_s the derivative is equal to zero. Hence, the potential function and its derivative are continuous functions.

Other function modifications that can be used are the so-called *potential shift* function, which are used to avoid the sudden changes on the potential energy function or its gradient. For example, typical potential shift function has the form:

$$f(r_{ij}) = \begin{cases} \left(1 - \left(\dfrac{r_{ij}}{R_s}\right)^2\right)^2, & r_{ij} \leq R_s \\ 0, & r_{ij} > R_s \end{cases} \tag{8.7}$$

and its gradient with respect to r_{ij} is given as the following:

$$\frac{\partial f(r_{ij})}{\partial r_{ij}} = \begin{cases} -4\dfrac{r_{ij}}{R_s^2}\left(1 - \left(\dfrac{r_{ij}}{R_s}\right)^2\right), & r_{ij} \leq R_s \\ 0, & r_{ij} > R_s \end{cases} \tag{8.8}$$

It can be seen that potential function and its gradient decay smoothly to zero at $r_{ij} = R_s$, and hence they are continuous functions of the distance.

Another form of the potential shift function is given as the following:

$$f(r_{ij}) = \begin{cases} \left(1 - \dfrac{r_{ij}}{R_s}\right)^2, & r_{ij} \leq R_s \\ 0, & r_{ij} > R_s \end{cases} \tag{8.9}$$

The gradient with respect to r_{ij} is given as the following:

$$\frac{\partial f(r_{ij})}{\partial r_{ij}} = \begin{cases} -\dfrac{2}{R_s}\left(1 - \dfrac{r_{ij}}{R_s}\right), & r_{ij} \leq R_s \\ 0, & r_{ij} > R_s \end{cases} \tag{8.10}$$

These two potential shift functions are better suited to electrostatic potential interactions when used without the Ewald summation to correct for long-range interactions. For Lennard-Jones potential interactions, it is recommended to use the potential switch functions of the form given by Eq. (8.5). Note that the potential shift function of the form as in Eq. (8.9) used for the electrostatic interactions as follows:

$$V'_{\text{Elec}}(r_{ij}) = k_e \frac{q_i q_j}{\varepsilon r_{ij}} f(r_{ij})$$

yield the following gradient of the potential

$$\begin{aligned} \frac{\partial V'_{\text{Elec}}(r_{ij})}{\partial r_{ij}} &= -k_e \frac{q_i q_j}{\varepsilon r_{ij}^2} \left(1 - \frac{r_{ij}}{R_s}\right)^2 - k_e \frac{q_i q_j}{\varepsilon r_{ij}} \frac{2}{R_s} \left(1 - \frac{r_{ij}}{R_s}\right) \\ &= -k_e \frac{q_i q_j}{\varepsilon r_{ij}^2} + k_e \frac{q_i q_j}{\varepsilon R_s^2} \\ &= \frac{\partial V_{\text{Elec}}(r_{ij})}{\partial r_{ij}} + k_e \frac{q_i q_j}{\varepsilon R_s^2} \end{aligned} \tag{8.11}$$

where $\frac{\partial V_{\text{Elec}}(r_{ij})}{\partial r_{ij}}$ represents the gradient of unmodified potential energy function, and the second term is a constant term, which represents the force shift. That is why sometimes this method is also called *force shift*.

To provide a smooth separation between the fast and slow forces, in particular in multiple-time step integration schemes, the so-called *force switching* functions are also used. The force switching function is applied as the following (Schlick 2010:

$$\begin{aligned} F_{\text{fast}}^{\text{LJ, Elec}}(r_{ij}) &= f(r_{ij}) F^{\text{LJ, Elec}}(r_{ij}) \\ F_{\text{slow}}^{\text{LJ, Elec}}(r_{ij}) &= \left(1 - f(r_{ij})\right) F^{\text{LJ, Elec}}(r_{ij}) \end{aligned} \tag{8.12}$$

where the switching function is defined as follows:

$$f(r_{ij}) = \begin{cases} 1, & r_{ij} < R_c \\ 1 + \left(S(r_{ij})\right)^2 \left(2S(r_{ij}) - 3\right), & R_c \le r_{ij} \le R_s \\ 0, & r_{ij} > R_s \end{cases} \tag{8.13}$$

where

$$S(r) = \frac{r^2 - R_c^2}{R_s^2 - R_c^2}$$

When periodic boundary conditions are used, the cut-off must not be so large that a particle sees its image or the same particle twice: the cut-off is limited to no more than half of the length of the cell in the case of the cubical cell; for rectangular cells, it must be smaller than half of the length of the shortest side.

8.5 Neighbour Lists

The use of the cut-off may not significantly reduce the time required to compute the number of interactions. That is because to decide if every pair of particles is close enough to calculate their interaction energy, we need to figure out the distance between the particles in each pair in the system. Verlet (Allen and Tildesley 1989) proposed a method that would significantly decrease the time of calculation of the energy and forces, by keeping a stored list of the nearest neighbours of a selected molecule, which are all particles within the cut-off distance, R_C, together with all particles that are slightly further away than the cut-off distance (see Fig. 8.4), $R_L > R_C$. For the simulations of the multi-sites molecules, the upper limit on the cut-off may also be affected by the size of the molecules. That is the most efficiently done using a large pointer array P of neighbour lists as described in Allen and Tildesley (1989), Leach (2001) and illustrated in Fig. 8.4.

After, the neighbour lists create, the program does not loop through all atoms but only those in the list for atom i and the computation time is of order N (number of atoms in the system). The nearest neighbouring lists to atoms are updated once a specified criterion satisfies, such as one of the atoms displaces more than $R_\mathrm{L} - R_\mathrm{C}$. The computation time required to update it is of order N^2. Usually, the difference $R_\mathrm{L} - R_\mathrm{C}$ is in the range $1 - 2$ Å. It is worth noting that large values of $R_\mathrm{L} - R_\mathrm{C}$ yield a less often update of the neighbour lists; however, it increases the length of the neighbour lists for the atoms. On the other hand, small values of $R_\mathrm{L} - R_\mathrm{C}$ produces short neighbour lists for the atoms, however, more often requires to update the neighbour lists. Therefore, practically, it is a trade off the choice of R_L and R_C.

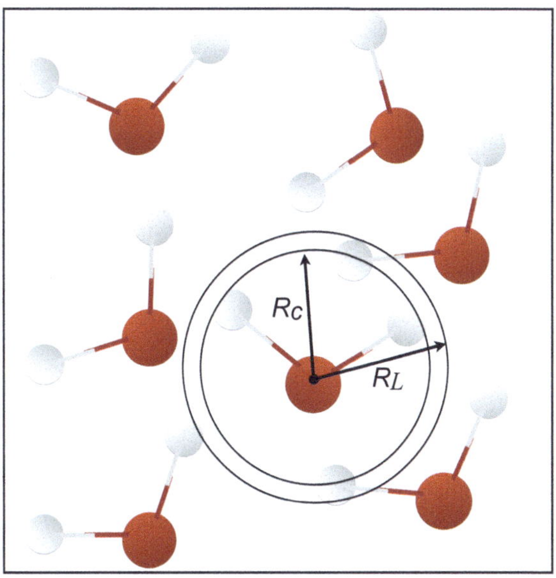

Fig. 8.4 Illustration of the cutoff radius R_C and neighbour list radius R_l used to implement the neighbour list

8.6 Long-range Forces

Electrostatic interactions are critical, especially, in biomolecular systems. Compared to van der Waals and covalent interactions, their range is relatively long, because of electrostatic interactions between molecules or parts of molecules at a distance r from each other decrease only slowly with increasing value of r. In particular, interaction energy, U, between two charged molecules is proportional to r^{-1}, while the corresponding force, $\mathbf{F} = -\nabla U$, is proportional to r^{-2}. The interactions between a neutral molecule with a dipole moment and a charged molecule is proportional to r^{-2}, and the corresponding force is proportional to r^{-3}. Furthermore, the interaction energy between two neutral molecules with dipole moments is proportional to r^{-3}, while the corresponding force is proportional to r^{-4}.

By considering in this multipole expansion, the quadrupole moments or higher moments, it can be shown that even the interaction between two neutral molecules without dipole moments, but with quadrupole moments is proportional to r^{-5}, so it has a longer range than the van der Waals interactions, which are proportional to r^{-6}. If we consider the electrostatic interaction energy of either a single charge, dipole, or quadrupole with all the charges, dipoles, and quadrupoles surrounding it, we have to integrate the electrostatic interaction $U_{\text{elec}}(r)4\pi r^2$ from r to infinity, where $4\pi r^2 dr$ is the volume of the spherical shell between r and $r + dr$ surrounding the central charge, dipole, or quadrupole:

$$\int_0^{\infty} U_{\text{elec}}(r)4\pi r^2 dr \tag{8.14}$$

which converges if the following condition is satisfied:

$$U_{\text{elec}}(r) \propto r^{-n}, \quad n > 3$$

Note that that the total electrostatic energy of ionic systems depends on the spatial boundary conditions that restrict the range of the integral Eq. (8.14) in practical calculations. That indicates that decrease of the pair interaction energies and forces with interatomic separation r will affect how the long-range interactions are treated in the force and energy calculations during simulations.

8.6.1 Ewald Summation Method

The Ewald summation method was introduced by Ewald (1921). Based on this method, the sum of the long-range interactions between particles and all their infinite periodic images are calculated efficiently. The total electrostatic interaction energy of a system of N particles in a cubic box of size L and infinite replicas in periodic boundary condition is given by Toukmaji and Board (1996)

$$U = \frac{1}{2} \sum_{n=0}^{\infty}{}' \sum_{i=1}^{N} \sum_{j=1}^{N} \frac{q_i q_j}{4\pi\epsilon_0 \mid \mathbf{r}_i - \mathbf{r}_j + \mathbf{n}L \mid} \tag{8.15}$$

where q_i is the charge of the i-th particle and $\mathbf{n}$ is the cell-coordinate vector

$$\mathbf{n} = n_1 L\mathbf{i} + n_2 L\mathbf{j} + n_3 L\mathbf{k}$$

where $\mathbf{i}, \mathbf{j}, \mathbf{k}$ are cartesian coordinate unit vectors. The origin cell is located at $\mathbf{n} = (0, 0, 0)$ with image cells located at $L\mathbf{n}$ intervals in all three dimensions as $\mathbf{n}$ goes to infinity.

The sign (′) in the first sum indicates that the terms with $i = j$ are not included for $\mathbf{n} = 0$. According to the Ewald method, the potential energy of Eq. (8.15) can be expresses as a sum of three terms

$$U_{\text{Ewald}} = U_r + U_m + U_0 \tag{8.16}$$

where U_r is the real space sum, U_m is the reciprocal sum, and U_0 is the self-term a constant term. These three terms are given, respectively, as:

$$U_r = \frac{1}{2} \sum_{i,j}^{N'} \sum_{n=0}^{\infty} \frac{q_i q_j}{4\pi\epsilon_0} \frac{\text{erfc}(\alpha r_{ij,n})}{r_{ij,n}} \tag{8.17}$$

$$U_m = \frac{1}{2\pi V} \sum_{i,j}^{N} \frac{q_i q_j}{4\pi\epsilon_0} \sum_{m \neq 0} \frac{\exp\left(-(\pi m/\alpha)^2\right) \cos\left(2\pi \mathbf{m} \cdot (\mathbf{r}_i - \mathbf{r}_j)\right)}{m^2} \tag{8.18}$$

$$U_0 = \frac{-\alpha}{\sqrt{\pi}} \sum_{i=1}^{N} \frac{q_i^2}{4\pi\epsilon_0} \tag{8.19}$$

where

$$r_{ij,n} = \mid \mathbf{r}_i - \mathbf{r}_j + \mathbf{n}L \mid$$

V denotes the volume of the simulation box, $\mathbf{m} = (l, j, k)$ is the reciprocal-space vector, and erfc is the complementary error function, given as

$$\text{erfc}(x) = \frac{2}{\sqrt{\pi}} \int_x^{\infty} \exp(-t^2) dt$$

It should be noted that this summation involving the error function converges very fast and for distances larger than a cutoff value its value can be omitted. Specifically, α is chosen such that the only terms in the series of U_r are those for which $\mid \mathbf{n} \mid = 0$.

The reciprocal sum U_m also converges much faster than the first term U_r. But the number of terms increases with the width α. Hence, there is a need for a balance between the real-space and reciprocal-space sums. Here, the first term U_r converges more rapidly for large α, whereas the second U_m converges more rapidly for small α. Following the discussion in Toukmaji and Board (1996), α is large for small systems such that the real-space sum extends up to the nearest neighbours of the original cell. In Toukmaji and Board (1996) (and the reference therein), a value of $\alpha = \sqrt{\pi}/L$ is suggested, which provides an equal rate of convergence for both direct and reciprocal-space terms. However, for large systems (typically, $N > 10^4$ Toukmaji and Board 1996), using minimum-image convention with a cutoff radius smaller than $L/2$, larger α are used. Based on our experience, for simulations of macromolecular systems in explicit solvent using CHARMM program (Brooks et al. 2009) with a cutoff radius in the range $10 - 12$ Å, typical value of α is $\alpha = 5/L$.

The self-term U_0 is a correction term that cancels out the interaction of each of the introduced artificial counter-charges with itself.

Another term is also added depending on the medium that surrounds the sphere of simulation box. If the surrounding medium is a vacuum, i.e. with a relative permittivity $\epsilon = 1$, then the following term is added

$$U_{corr} = \frac{2\pi}{3V} \mid \sum_{i=1}^{N} \frac{q_i}{4\pi\epsilon_0} \mathbf{r}_i \mid^2 \tag{8.20}$$

This term is zero, if the surrounding medium is a conductor, i.e. with an infinite relative permittivity.

The Ewald sum is an accurate way for how to treat the long-range forces in an MD simulation. It has been used in simulations of highly charged systems and to other systems as well, such as protein, DNA and lipid bilayers where the electrostatic effects are significant. The Ewald summation is computationally very expensive to implement. If α is allowed to vary, the algorithm can be made to scale as $N^{3/2}$. However, the problem can be that for these α, the range of electrostatic interactions are incompatible with the scope of the van der Waals interactions. There have been other methods introduced to speed up computational demands for calculation of the reciprocal space terms. For example, using the polynomial approximations, which, however, do not solve unfavourable N^2 scaling. Another proposed method is to use Fast Fourier Transform (FFT) for calculation of the reciprocal space summation, which scales as $N \ln N$. This method combined with using a large value of α, such that interatomic interactions are negligible for r_{ij} greater than some cutoff distance (typically, 9 Å), can reduce the real-space summation to order of N and the order of the entire algorithm is $N \ln N$.

A known approach for implementation of FFT method is particle-mesh method (Hockney and Eastwood 1988; Darden et al. 1993) or alternative variants, such as particle-particle-particle-mesh method (Luty et al. 1994, 1995). Unified version of these algorithms has also been discussed as an alternative for increasing the accuracy (Deserno and Holm 1998a, b).

The Ewald method has been widely used to simulate highly polar or charged systems, and it is often used to study different types of solid-state materials. Of particular interest, are the applications of the method for analysing large molecular systems, such as proteins, DNA, and their complexes (York et al. 1994; Darden et al. 1999). The nature of DNA systems makes the use of the particle-mesh Ewald method particularly crucial for dealing with long-range electrostatic interactions during the simulations, producing more stable trajectories (Cheatham et al. 1995).

8.7 Equilibration

A trajectory during MD simulations is a sequence of configurations for a given system saved at regular time steps, let say every N_{steps}. For example, for a system of N particles, every configuration of system is a vector of $3N$ dimensions

$$\mathbf{r}(t) \equiv (\mathbf{r}_1(t), \mathbf{r}_2(t), \cdots, \mathbf{r}_N(t))^T$$

where $\mathbf{r}_i(t) = (x_i(t), y_i(t), z_i(t))$ are the coordinates of the i-th atom. If we consider a path through the configurational phase space, then $\mathbf{r}(t)$ is the vector of coordinates of all atoms at the time step t in this path. The path $\mathbf{r}(t)$ is called a *trajectory*.

In general, the values of any physical property, such as temperature, pressure, energy, and so on, fluctuate as we move along the trajectory. The average value of a physical property, $\mathcal{P}(\mathbf{r})$, overall the configurations visited by the system during a trajectory with T steps is

$$\langle \mathcal{P} \rangle_T = \sum_{t=1}^{T} \mathcal{P}(\mathbf{r}(t))$$

Assuming that the trajectory is ergodic, then the time averages are equal to ensemble averages, that is:

$$\langle \mathcal{P} \rangle = \lim_{T \to \infty} \langle \mathcal{P} \rangle_T$$

In practice, MD simulations run for a finite time, typically, of order hundred of nanoseconds, hence the average over configurations provides only an estimate of $\langle \mathcal{P} \rangle$.

As an example, Fig. 8.5 shows the running averages versus T along a trajectory. The magnitude of the fluctuations along the average $\langle \mathcal{P} \rangle$ equals the statistical uncertainty in the average value. For this, we can divide the trajectory into blocks of the length L. Assuming that L is long enough, then the averages on each block can be considered statistically independent. Let us denote with $\langle \mathcal{P} \rangle_m$ the average of $\mathcal{P}(\mathbf{r}(t))$ over the configurations of the block m, then

$$\langle \mathcal{P} \rangle_T = \frac{1}{M} \sum_{m=1}^{M} \langle \mathcal{P} \rangle_m$$

where $M = T/L$ is the total number of blocks.

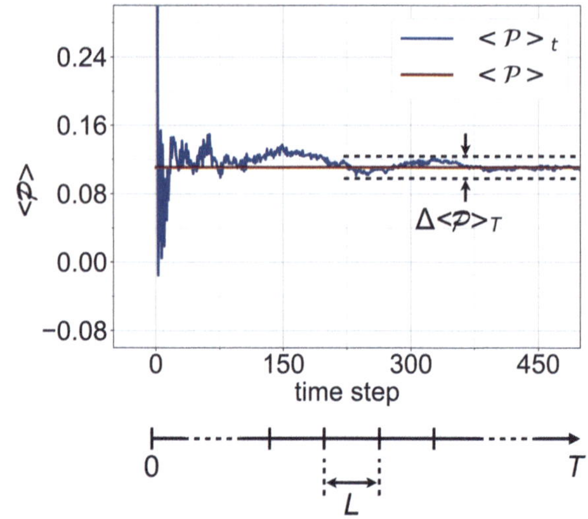

Fig. 8.5 Running averages along a MD trajectory. Adapted from Kamberaj (2020)

The standard deviation gives the statistical error of the average value as

$$\Delta \langle \mathcal{P} \rangle_T = \frac{1}{M} \sqrt{\sum_{m=1}^{M} \left(\langle \mathcal{P} \rangle_m - \langle \mathcal{P} \rangle_T \right)^2}$$

which for $T \to \infty$ tends to zero.

The prove of the *ergodic hypothesis* is difficult for any system. However, there can be a simple evaluation of the ergodic hypothesis using MD simulation results from different independent trajectories. In that case, averages of a macroscopic property, for example, total energy or pressure of an ergodic system, over two different independent trajectories should be equal. For that criteria to be satisfied, averages over a trajectory should be identical to the statistical path over conformation space.

It can be argued that correlations in low-frequency atomic displacements of order one nanosecond are not sampled efficiently due in part to the fact that trajectory remains constrained only in some region of the phase space, especially for complex molecular systems. Practically, the following quantity is used as an estimation of the simulation length needed to guarantee that the ergodic hypothesis is satisfied:

$$E(t) = \frac{1}{N_{\text{bins}}} \sum_{i=1}^{N_{\text{bins}}} \left(\mathcal{P}_i^{(1)}(t) - \mathcal{P}_i^{(2)}(t) \right)^2 \tag{8.21}$$

where $\mathcal{P}_i^{(j)}$ is the value of the quantity measured from the j-th trajectory at the bin i and N_{bins} is the total number of histogram bins. In general, the quantity $\mathcal{P}$ can be any macroscopic property of the system, for instance, root mean square deviation, energy, and so on, and E is presented as a function of the simulation time t. If the system is ergodic, then E should decay to zero as $1/D\tau$, where D is the generalised diffusion constant and τ is a timescale for self-averaging in the simulation. The decay of the ergodic measure, $E(t)$, to zero in the limit of the long times is considered a necessary condition for the system's average properties to correspond to equilibrium thermodynamic averages.

8.8 Summary

Here, we briefly described some practical aspects of molecular dynamics simulations of the biological systems, such as designing the constraints, periodic boundary conditions, spherical cutoffs, treatment of the long-range interactions, choosing the initial structure, and identifying the equilibrium states of the simulations.

Furthermore, the standard statistical analysis methodologies were introduced, such as block analysis approach and convergence of the averages.

References

M.P. Allen, D.J. Tildesley, *Computer Simulation of Liquids*. Oxford University Press (1989)

L. Bordoli, F. Kiefer, K. Arnold, P. Benkert, J. Battey, T. Schwede, Protein structure homology modeling using SWISS-MODEL workspace. Nat. Protocols **4**, 1–13 (2009)

B.R. Brooks, C.L. Brooks, A.D. MacKerell, L. Nilsson, R.J. Petrella, B. Roux, Y. Won, G. Archontis, C. Bartels, S. Boresch, A. Caflisch, L. Caves, Q. Cui, A.R. Dinner, M. Feig, S. Fischer, J. Gao, M. Hodoscek, W. Im, K. Kuczera, T. Lazaridis, J. Ma, V. Ovchinnikov, E. Paci, R.W. Pastor, C.B. Post, J.Z. Pu, M. Schaefer, B. Tidor, R.M. Venable, H.L. Woodcock, X. Wu, W. Yang, D.M. York, M. Karplus, Charmm: the bimolecular simulation program. J. Comput. Chem. **30**(10), 1545–1614 (2009)

T.E. Cheatham III, J.L. Miller, T. Fox, T.A. Darden, P.A. Kollman, Molecular dynamics simulations on solvated biomolecular systems: The particle mesh Ewald method leads to stable trajectories of DNA, RNA and proteins. J. Am. Chem. Soc. **117**, 4193–4194 (1995)

T. Darden, D. York, L. Perdersen, Particle-mesh Ewald: an $n \log n$ method for Ewald sums in large systems. J. Chem. Phys. **98**, 10089–10092 (1993)

T.A. Darden, L. Perera, L. Li, L. Pedersen, New tricks for modellers from the crystallography toolkit: the Particle mesh Ewald algorithm and its use in nucleic acid simulations. Struct Fold. Des. **7**, R55–R60 (1999)

M. Deserno, C. Holm, How to mesh up Ewald sums. I A theoretical and numerical comparison of various particle mesh routines. J. Chem. Phys. **109**, 7678–7693 (1998a)

M. Deserno, C. Holm, How to mesh up Ewald sums. II An accurate error estimate for the particle-particle-particle-mesh algorithm. J. Chem. Phys. **109**, 7694–7701 (1998b)

P. Ewald, Die Berechnung optischer und elektrostatischer Gitter potenciale. Annalen der Physik **64**, 253–287 (1921)

R.W. Hockney, J.W. Eastwood, *Computer Simulation using Particles* (Adam Hilger, Bristol, 1988)

H. Kamberaj, *Molecular Dynamics Simulations in Statistical Physics: Theory and Applications* Computational Series. (Springer Nature, Switzerland, 2020)

A.R. Leach, *Molecular Modelling. Principles and Applications*, 2nd edn. (Prentice Hall, Pearson Education Limited, Edingburgh Gate, 2001)

B.A. Luty, M.E. David, I.G. Tironi, W.F. van Gunsteren, A comparison of particle-particle, particle-mesh and Ewald methods for calculating electrostatic interactions in periodic molecular systems. Mol. Simul. **14**, 11–20 (1994)

B.A. Luty, I.G. Tironi, W.F. van Gunsteren, Lattice-sum methods for calculating electrostatic interactions in molecular simulations. J. Chem. Phys. **103**, 3014–3021 (1995)

G.J. Martyna, M.E. Tuckerman, M.L. Klein, Nosé-hoover chains. The canonical ensemble via continuous dynamics. J. Chem. Phys. **97**, 2635–2643 (1992)

J.C. Phillips, J.E. Stone, K.L. Vandivort, T.G. Armstrong, J.M. Wozniak, M. Wilde, K. Schulten, Petascale Tcl with NAMD, VMD, and Swift/T, in *SC'14 workshop on High Performance Technical Computing in Dynamic Languages* (IEEE Press, 2014)

J.P. Ryckaert, G. Ciccotti, H.J.C. Berendsen, Numerical integration of the Cartesian equations of motion of a system with constraints: molecular dynamics of n-alkanes. J. Comput. Phys. **23**, 327–341 (1977)

T. Schlick, *Molecular Modeling and Simulation*, An Interdisciplinary Guide, 2nd edn. (Springer, New York, NY, 2010)

J.E. Stone, D.J. Hardy, B. Isralewitz, K. Schulten, Gpu algorithms for molecular modelling. In: D.A. Bader, J. Kurzak (eds.), *Scientific Computing with Multicore and Accelerators* (Chapman & Hall, 2011), pp. 351–371

A.Y. Toukmaji, J.A. Board Jr., Ewald summation techniques in perspective: a survey. Comput. Phys. Commun. **95**, 73–92 (1996)

D.M. York, A. Wlodawer, L. Pedersen, T.A. Darden, Atomic-level accuracy in simulations of large protein crystals. Proc. Natl. Acad. Sci. USA **91**, 8715–8718 (1994)

Chapter 9
Python Interactive GUI for CHARMM Software Package

The chapter aims to describe a Python Interactive Graphical User Interface for CHARMM software package.

In this chapter, we will briefly describe a Python interactive graphical user interface (GUI) for CHARMM package for performing MD simulations of a biological molecular system and standard analysis of *in silico* data produced from the simulations.

9.1 Introduction

The biological molecular systems are characterised by complex dynamics involving different time- and size-scales. Only the most experienced biologist, chemist or physicist with a lot of imagination and thinking may arrive at qualitative microscopic understanding of phase transitions, thermal equilibrium, irreversible behaviour and other specific phenomena.

Further analysis are necessary to describe quantitatively the phenomena occurring at the microscopic scale in terms of thermodynamics, statistics, kinetic theory, and others. A non-expert in the field may not be able to automatically correctly imagine and think about these phenomena. For instance, the students will find it difficult to imagine the motions involved in a system of atoms moving under the laws of classical mechanics, as discussed in Chaps. 5 and 6. (Note that the laws of dynamics for a classical system of N particles are well described in any classical mechanics textbook.)

Physics's students, but also the students of other fields, gradually learn quantitative approaches to statistical physics, thermodynamic, classical mechanics, and electrodynamics; however, this quantitative understanding of any of related phenomena does not go much behind some equations and problem solving techniques.

H. Kamberaj, *Computer Simulations in Molecular Biology*, Scientific Computation,
https://doi.org/10.1007/978-3-031-34839-6_9

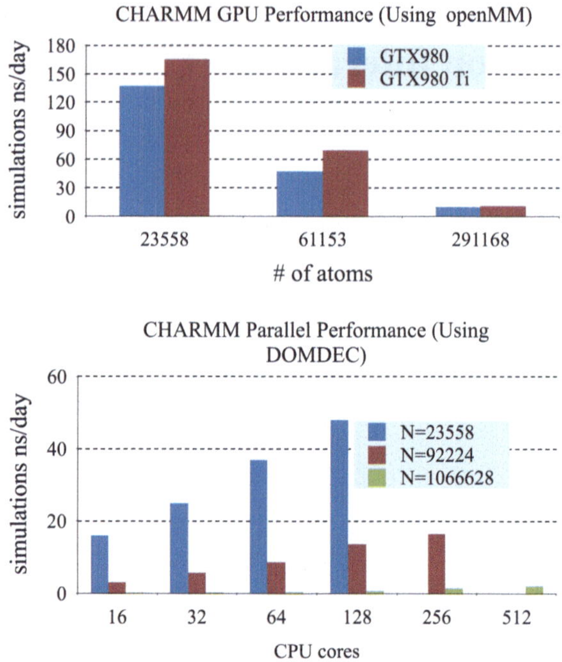

Fig. 9.1 Performance of the CHARMM program Brooks et al. (2009) at different computer architectures and for various system sizes

The interactive computer simulations software packages are made available to the students and scientists to model, visualise, and animate the motions of particles Schroeder (2015) and biological systems Humphrey et al. (1996). The primary aim of the interactive computer simulations is to encourage more widespread use of molecular dynamics simulations in biophysics, biology, chemistry, engineering at all levels, enabling the students and scientists a better understanding of the atomic view of matter and learn to apply it, both qualitatively and quantitatively, to a wider variety of phenomena in science and technology.

In the following, we will describe an interactive GUI for the CHARMM software package. Figure 9.1 shows the performance of the CHARMM program Brooks et al. (2009) at different computer architectures and various system sizes[1]

9.2 Designing GUI

Figure 9.2 shows the front-view of the GUI program.

The interactive interface consists of different modules, namely Energy Minimisation (EM), System Setup (SS), System Equilibration (SE), Classical Molecular

[1] Source of data: www.charmm.org.

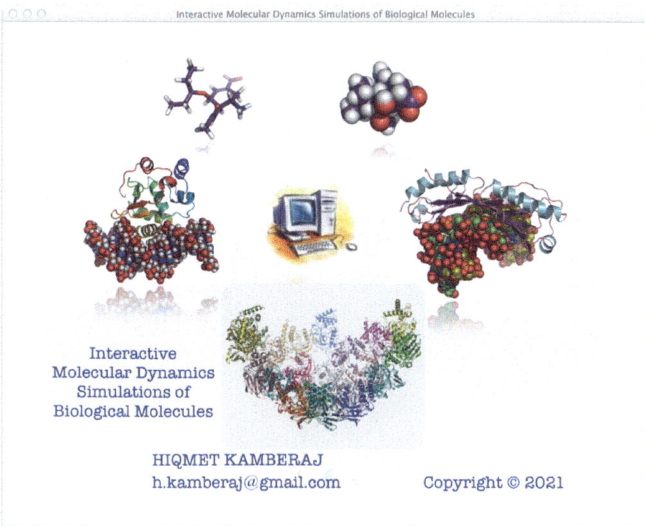

Fig. 9.2 Front-view of the GUI program

Fig. 9.3 Interactive modules of the GUI program

Dynamics (CMD), Quantum Mechanics (QM), Free Energy (FE), and *in silico* data analysis (or Trajectory Analysis) (TA), as presented in Fig. 9.3.

The modules will be introduced in the following.

9.2.1 Energy Minimisation Module

The energy minimisation module contains different energy minimisation methods (see also Fig. 9.4), such as Steepest Descent (SD), Conjugate Gradient (CONJ), Adopted Basis Newton-Raphson (ABNR), Newton-Raphson (NRAP), Powell (POWE), and Truncated Newton Method (TNPACK).

Besides, a Help menu about all the minimisation functions is created in the form of a documentation, as shown in Fig. 9.5. Here, the users can find a detailed information about the use of each energy minimisation method and the parameters' setup.

Clicking the menu of each method takes the user to the interactive design dialog, which consists of a set of parameters to be initialised for an energy minimisation run. Figure 9.6 presents a dialog form for performing Steepest Descent energy minimisation of a system. The values shown in the dialog form text boxes are the default ones. Checking the checkboxes indicates that the current variable is in use and the value entered in the text box will be read. The button *Run* should be clicked to call

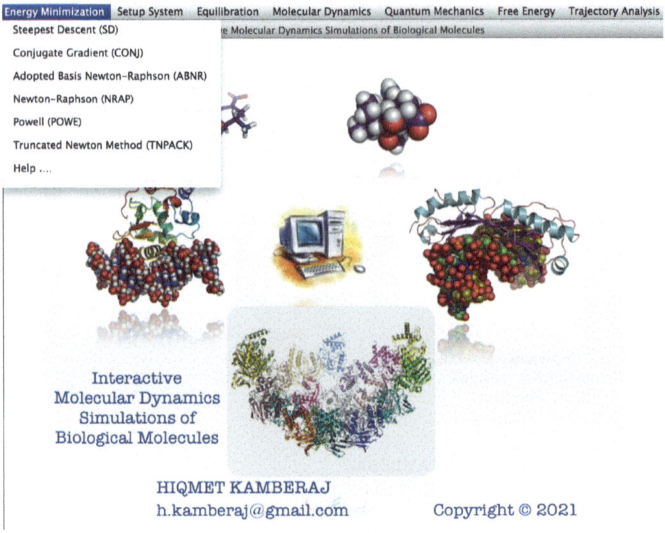

Fig. 9.4 Energy minimisation functions used by EM module

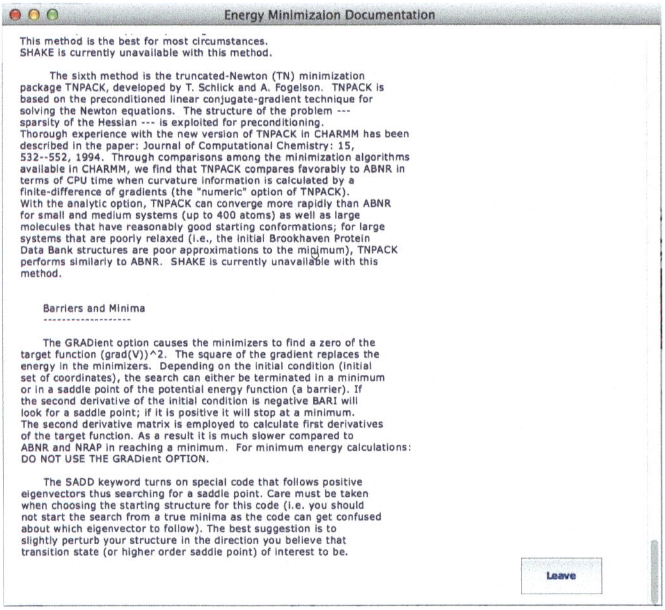

Fig. 9.5 Energy minimisation documentation

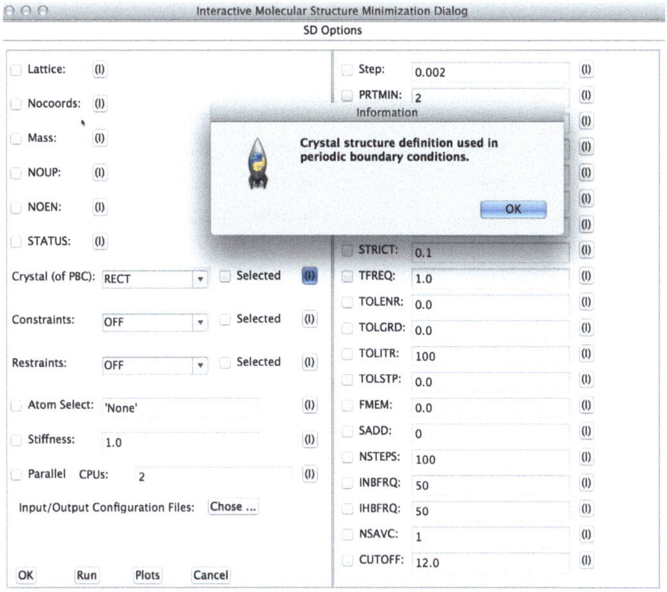

Fig. 9.6 Interactive dialog for performing energy minimisation using SD method

energy minimisation external program, which runs in the background independently on the state of the current dialog form after it is called. Next to each variable, the user can see an information button, namely the button *(I)*, (see also Fig. 9.6), which, if clicked, presents the additional information about that variable.

It is interesting to mention about the functionality of common button *chose*, which open the input-output dialog form used to select the input and output files options, as shown in Fig. 9.7.

Another interesting functionality includes the analysis of the results of a simulation run using the button *Plots*, which accesses an output logo file (which will be discussed in details in the Case Studies chapter). The energy terms printed out from energy minimisation run will be summarised in a table, as shown in Fig. 9.8.

The button marked using an oval shape can be used to obtain a plot view of the data, as presented in Fig. 9.9.

In the plotting view, the user has many selection options for creating nice plots for publications, such as styles, fonts, and many others, as illustrated in Fig. 9.10. Besides, the plots can be saved in various high compressed data formats, such as JPEG, PNG, and TIFF.

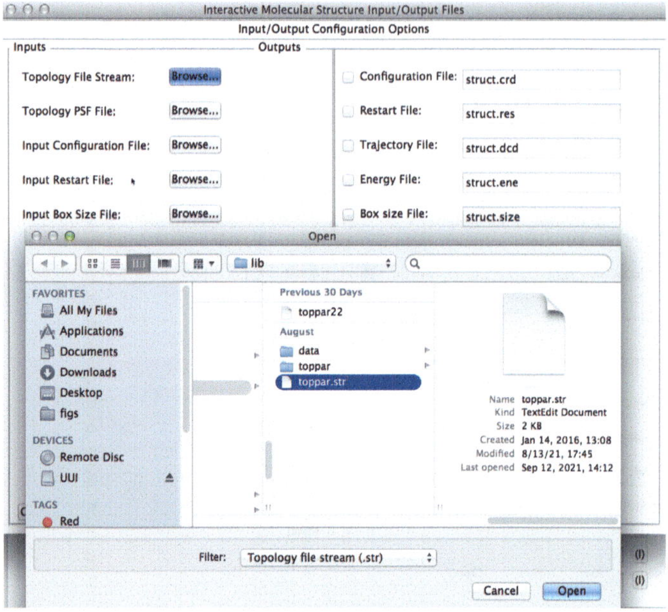

Fig. 9.7 Interactive dialog for choosing the input-output information

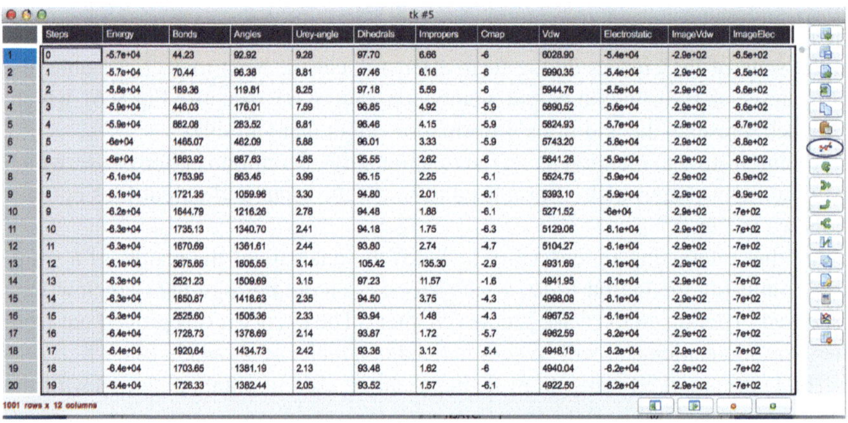

	Steps	Energy	Bonds	Angles	Urey-angle	Dihedrals	Impropers	Cmap	Vdw	Electrostatic	ImageVdw	ImageElec
1	0	-5.7e+04	44.23	92.92	9.28	97.70	6.66	-6	6028.90	-5.4e+04	-2.9e+02	-6.5e+02
2	1	-5.7e+04	70.44	96.38	8.81	97.46	6.16	-6	5990.35	-5.4e+04	-2.9e+02	-6.5e+02
3	2	-5.8e+04	189.36	119.81	8.25	97.18	5.59	-6	5944.76	-5.5e+04	-2.9e+02	-6.6e+02
4	3	-5.9e+04	446.03	176.01	7.59	96.85	4.92	-5.9	5890.52	-5.6e+04	-2.9e+02	-6.6e+02
5	4	-5.9e+04	882.08	283.52	6.81	96.46	4.15	-5.9	5824.93	-5.7e+04	-2.9e+02	-6.7e+02
6	5	-6e+04	1465.07	462.09	5.88	96.01	3.33	-5.9	5743.20	-5.8e+04	-2.9e+02	-6.8e+02
7	6	-6e+04	1863.92	687.63	4.85	95.55	2.62	-6	5641.26	-5.9e+04	-2.9e+02	-6.9e+02
8	7	-6.1e+04	1753.95	863.45	3.99	95.15	2.25	-6.1	5524.75	-5.9e+04	-2.9e+02	-6.9e+02
9	8	-6.1e+04	1721.35	1059.96	3.30	94.80	2.01	-6.1	5393.10	-5.9e+04	-2.9e+02	-6.9e+02
10	9	-6.2e+04	1644.79	1216.26	2.78	94.48	1.88	-6.1	5271.52	-6e+04	-2.9e+02	-7e+02
11	10	-6.3e+04	1735.13	1340.70	2.41	94.18	1.75	-6.3	5129.06	-6.1e+04	-2.9e+02	-7e+02
12	11	-6.3e+04	1670.69	1361.61	2.44	93.80	2.74	-4.7	5104.27	-6.1e+04	-2.9e+02	-7e+02
13	12	-6.1e+04	3675.66	1805.55	3.14	105.42	135.30	-2.9	4931.69	-6.1e+04	-2.9e+02	-7e+02
14	13	-6.3e+04	2521.23	1509.69	3.15	97.23	11.57	-1.6	4941.95	-6.1e+04	-2.9e+02	-7e+02
15	14	-6.3e+04	1850.87	1418.63	2.35	94.50	3.75	-4.3	4998.08	-6.1e+04	-2.9e+02	-7e+02
16	15	-6.3e+04	2525.60	1505.36	2.33	93.94	1.48	-4.3	4967.52	-6.1e+04	-2.9e+02	-7e+02
17	16	-6.4e+04	1728.73	1378.69	2.14	93.87	1.72	-5.7	4962.59	-6.2e+04	-2.9e+02	-7e+02
18	17	-6.4e+04	1920.64	1434.73	2.42	93.36	3.12	-5.4	4948.18	-6.2e+04	-2.9e+02	-7e+02
19	18	-6.4e+04	1703.65	1381.19	2.13	93.48	1.62	-6	4940.04	-6.2e+04	-2.9e+02	-7e+02
20	19	-6.4e+04	1726.33	1382.44	2.05	93.52	1.57	-6.1	4922.50	-6.2e+04	-2.9e+02	-7e+02

1001 rows x 12 columns

Fig. 9.8 Energy minimisation decomposition summarised in tabular form

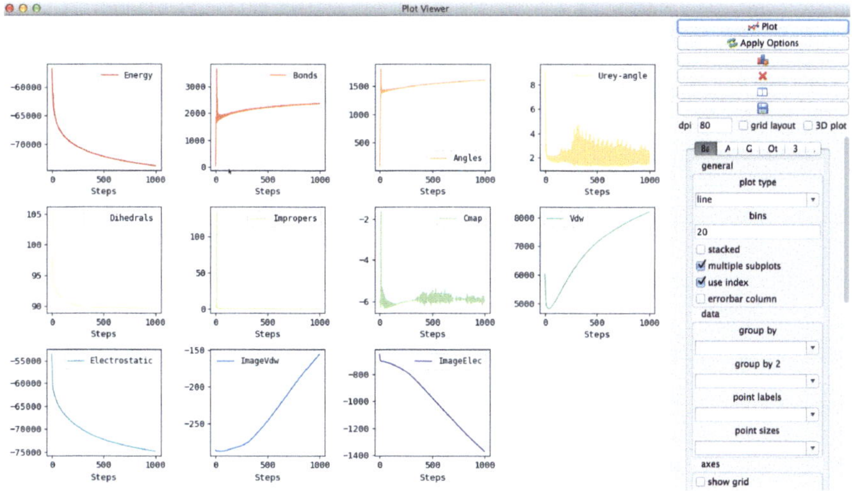

Fig. 9.9 Plot view of energy minimisation decomposition terms

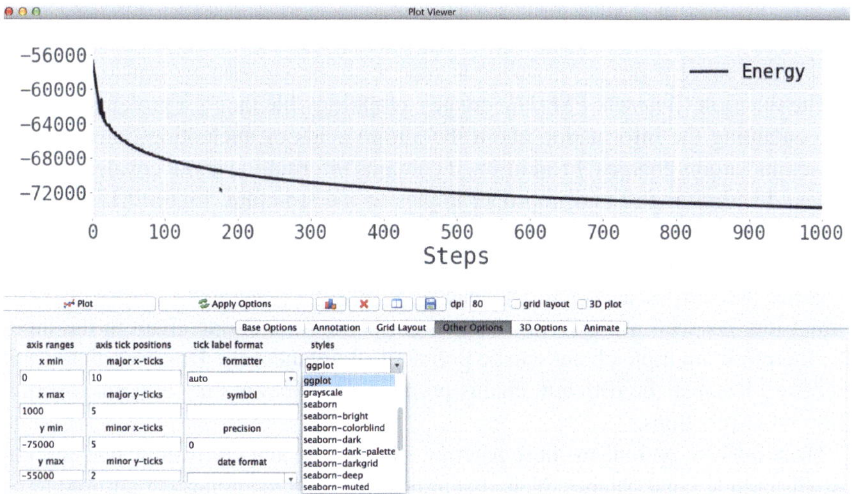

Fig. 9.10 Formatting plots using different graph styles

9.2.2 System Setup Module

System Setup module contains five different functions, such as the function for generating the topology PSF file, joining two PSF files, performing amino acid mutation in the structure, adding a water box around the solute, and adding counter ions and salt in the solution. Figure 9.11 presents SS module.

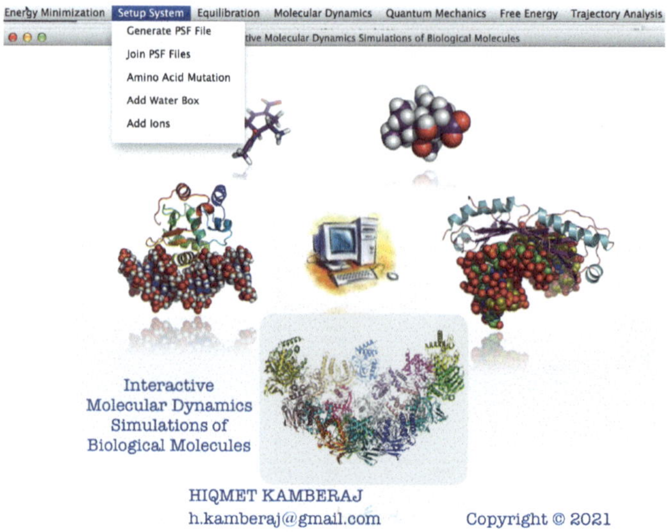

Fig. 9.11 System setup module

The function *Generate PSF File* creates a topology file that is a protein structure file containing the information about the amino acids or nucleotides, atom types, masses and atoms charges. In addition, it contains information about covalent bonds, angles, dihedral angles and improper angles in the structure. Figure 9.12 presents an interactive dialog for generating a molecular structure PSF file. The inputs are the topology and parameters force field stream file (e.g., CHARMM force field) and PDB file that can be a Protein Databank file or a pre-generated CHARMM PDB format file. As seen in Fig. 9.12, there might be more than one chain in the input file; therefore, multiple chains can be prepared simultaneously. However, that is not necessary because the different chains prepared in that way may be joint together using other functions.

There are also options to chose patches, such as first and last residue patches, but also disulphide bond patches. Note that multiple disulphide bonds are allowed to be chosen simultaneously.

After clicking on button *Generate PSF*, three output files will be generated, namely the PSF file, CHARMM configuration file, and simulation cell dimensions file. All three files, including the output logo file will be stored in the default folder: *scratch/setup/psf/*. Note that this folder is automatically created even if it does not exist.

The functions of the other two buttons, namely *OK* and *Cancel*, are the usual ones; in particular, clicking the button *OK* with destroy the dialog and keep the input parameters for the next operation, and, in contrast, by clicking on the button *Cancel*, the dialog window form is destroyed and the input data will be lost.

The next function within SS module is the *Join PSF Files* function, used to join two different input molecular structures, such as two segments. For that, see also

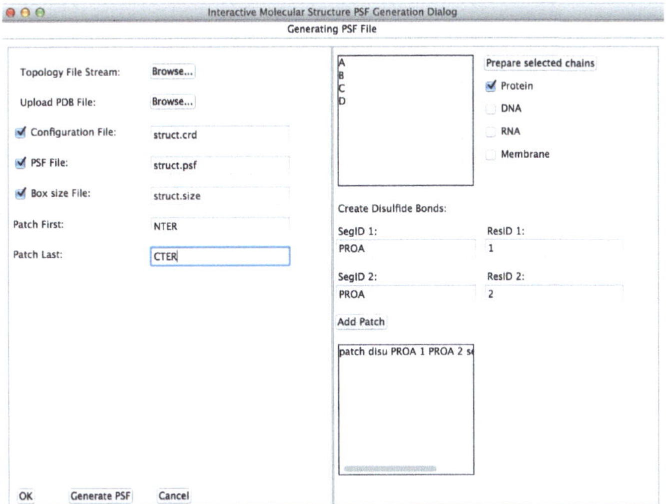

Fig. 9.12 Generate PSF file function

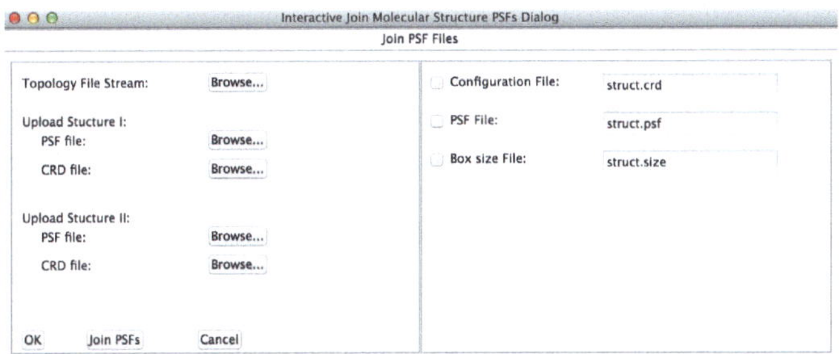

Fig. 9.13 Joining two PSF files

Fig. 9.13, the topology file stream is needed and the two molecular structure files (PSF and configuration files). After clicking on the button *Join PSFs*, three files (such as *Configuration File*, *PSF File*, and *Box size File* are stored under the folder *scratch/setup/join/*, which is created automatically. The role of the buttons *OK* and *Cancel* is as described above. Note that if the *Configuration File*, *PSF File*, and *Box size File* already exist under the same name in that folder, then the files will be replaced; therefore, backup or renaming the files may be needed.

Another available function is the *Amino Acids Mutation*. The interactive dialog form is shown in Fig. 9.14. As it can be seen, for that two specifications are needed, namely the wild-type amino acids information (such as segment ID as in PSF file, residue name, and residue ID) and the mutation residue information (such as the new name). Other input information includes the topology structure PSF file and

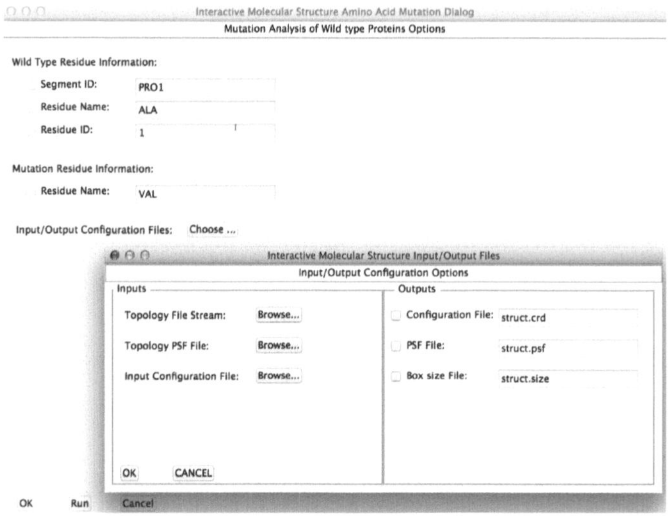

Fig. 9.14 Amino acids mutation

configuration coordinates file. The output information (as shown in interactive dialog form in Fig. 9.14) is automatically stored in the folder *scratch/aamutation/*.

For computer simulations of biological molecules an important element is the environment. Often, water is chosen as the main ingredient of that. Figure 9.15 shows an interactive dialog form for solvating a biological molecule.

The first input information that is needed includes the macromolecule topology file and configuration coordinates. Besides, the water box configuration coordinates are needed as CHARMM coordinates format. Other additional information that is needed includes the specification of the number of water molecules for output, distance between the edge of protein size and water box, which must not be smaller than the non-bonded long-range interactions cutoff, and the shape of the simulation box for determining the periodic boundary conditions. Three type of periodic boundary conditions are currently supported, namely *rectangular* (RECT), *cubical* (CUBIC), and *octahedral* (OCTA), as shown Fig. 9.15. The output information, such as PSF file, configuration file, and box size file, are automatically stored under the folder *scratch/setup/solvate/*.

The last function of the SS module includes *Add Ions*, which adds counter ions (if the biological molecule is charged) and salt at the specified concentration. The interactive dialog form is depicted in Fig. 9.16.

The input information includes the topology stream files, PSF file, and configuration coordinates file of the biological molecule in water. In addition, the user has to specify the salt concentration (in Mol/L), the minimum distance between a pair of ions and between ions and biological molecule. Typical values of that minimum distance are $5 - 5.5$ Å. The ions are placed at the water positions (by replacing them) and then their positions are optimised by performing Monte Carlo moves trying to

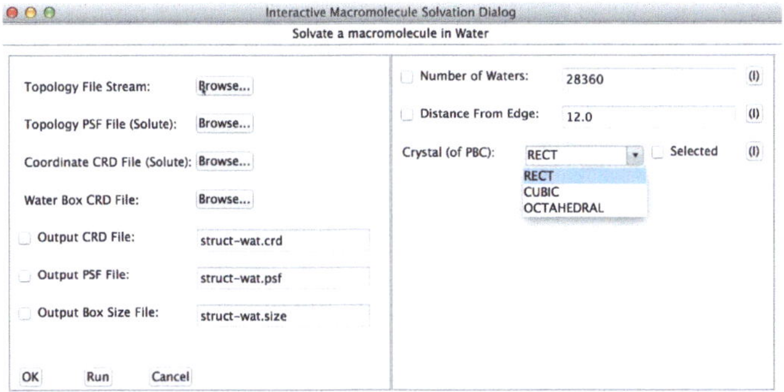

Fig. 9.15 Water box

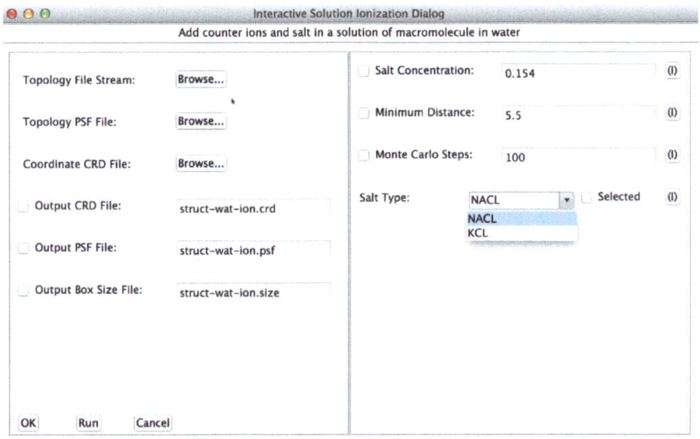

Fig. 9.16 Adding ions in solvent

minimise the energy function. Typically, 100 steps might be enough, depending on the system size and the electrostatic properties; however, this is not a crucial decision because the system will be further equilibrated.

Furthermore, the user needs to specify the type of salt. Two type of salts are supported, namely sodium chloride (NaCl) and potassium chloride (KCl). After clicking on the *Run* button, the output information, such as PSF file, configuration file, and box size file, are automatically stored under the folder *scratch/setup/ionize/*.

9.2.3 System Equilibration Module

The *System Equilibration* module consists of three functions, namely *Annealing Simulations*, *Constraining Simulations*, and *Restraining Simulations*, as presented in Fig. 9.17.

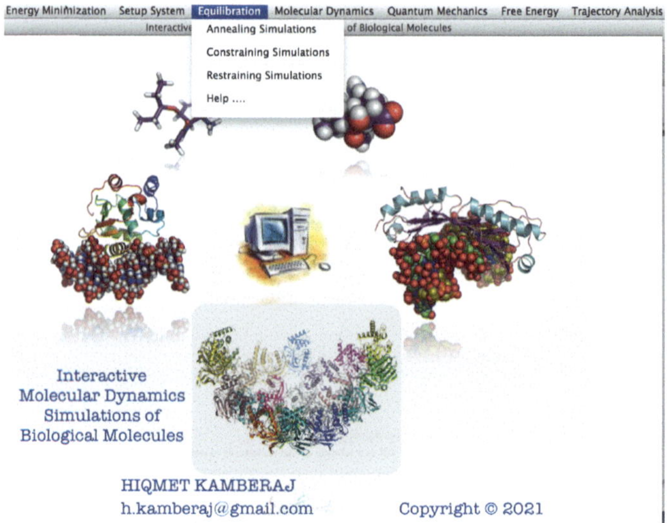

Fig. 9.17 System equilibration module

The annealing simulation function performs heating and cooling molecular dynamics simulations using constant NVT ensemble by increasing the temperature at regular intervals with a constant amount. An interactive dialog form is presented in Fig. 9.18. On the lefthand side, the parameters that control the constraints can be identified, which select the atoms to be fixed at their initial positions. Usually, the heavy atoms of biological molecule are held fixed at this stage and solvent atoms and hydrogen atoms of solute are allowed to move freely.

Next, the integrator to control the dynamics can be chosen (such as VVER standing for velocity-Verlet algorithm, LANGEVIN standing for Langevin dynamics, and LEAP standing for Leap-Frog algorithm). The parameter *Status* indicates if the simulation starts from previous state (RESTART option) or from a new state (START option). The parameter *EWALD* determines if the Ewald method is used to handle long-range electrostatic interactions. The important parameters that control the temperature of the system are the initial temperature, temperature increment, and the final temperature.

From the parameters on the righthand side, the most important parameters include the IHTFRQ that is the step frequency for heating (or cooling) the molecule in increments of *Increment Temperature* degrees in the heating (or cooling) portion of a dynamics run; zero means do no heating (or cooling). Besides, the IEQFRQ gives the step frequency for assigning or scaling velocities to *Final Temperature* during the equilibration stage of the dynamics run.

Clicking the button *Run*, the MD annealing simulations will run and the output are stored in the folder *scratch/equilibration/anneal/*. Besides, the button *Plots* provides the table view and plot view facilities of the simulation run.

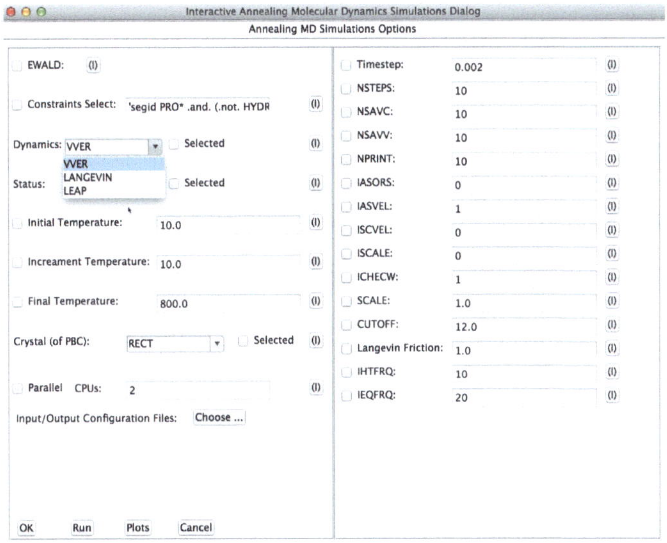

Fig. 9.18 Interactive annealing simulation dialog form

Next function of the SE module includes the *Constraining MD simulations*, as shown in Fig. 9.19. Also here, the parameter that control the constraints is specified by selecting the atoms to be fixed at their initial positions. Usually, the heavy atoms of biological molecule are held fixed at this stage and solvent atoms and hydrogen atoms of solute are allowed to move freely. The constrained MD simulations can be performed at constant NVE, NVT, and NPT ensembles.

Furthermore, several algorithms can be selected to control dynamics, such as Berendsen, Langevin, Nosé, Nosé-Hoover, and Glein-Martyna-Tuckerman (or the so-called VV2) algorithms, as depicted in Fig. 9.19.

Moreover, Langevin friction coefficient is given in units of 1/ps, which is also in VV2 algorithm the friction constant (in 1/ps) to damp the motion of the centre of mass of the system. Any nonzero value will create a separate thermostat, which will actually be a heat sink instead of a heat bath coupled to the three degrees of freedom associated with the motion of the centre of mass of the whole system. The thermostat τ is the characteristic response time for each thermostat (in ps). It is also used as input for the temperature coupling constant in BERENDSEN dynamics (taking an input value of 5.0 ps or more). The mass of the pressure piston (in amu) is used to control the pressure in Nosé-Hoover dynamics. Note that its default value is infinity. The mass of the thermal piston (in $\text{kcal} \cdot \text{mol}^{-1} \cdot \text{ps}^2$) has its default value of infinity. Note that it is also used to define *QREF* in Nosé dynamics, which should be between 10 and 1000.

The button *Run* will allow constrained MD simulations run, and produce a logo of that output and other output files. The outputs are stored in the folder

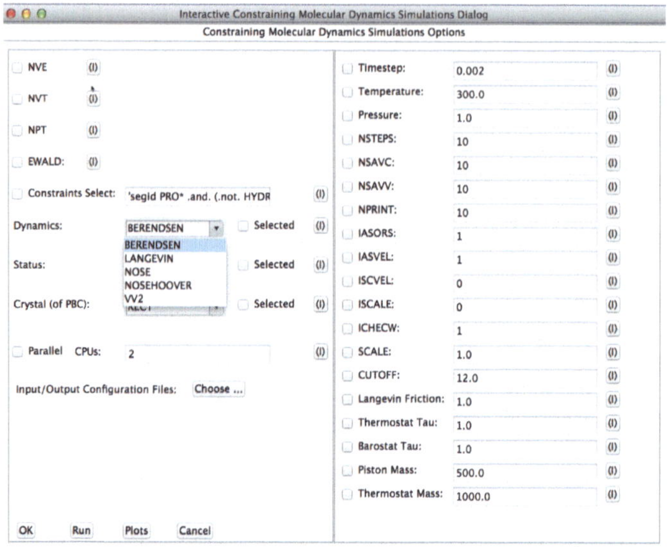

Fig. 9.19 Interactive constraining simulation dialog form

scratch/equilibration/constraints/. Besides, there is an option to present the results as a table and plot views by pressing the button *Plots*.

Figure 9.20 presents an interactive dialog form for running restrained molecular dynamics simulations in different statistical ensembles (such as NVE, NVT, and NPT). Compared to constrained MD simulations there are two changes made; the first is on the definition of the restraints atoms, and the second on determining the harmonic force constant of restrained dynamics. The restraint force constant (such as 1 unit = 2 · M · kcal/mol/Å^2, where M is mass of atom) can be set to any positive value (specified by the *FORCE* keyword followed by the desired value). The force constants may also be obtained from the weight array, in which case the *FORCE* keyword is not read. When using this option, a negative values may be used for some atoms, however, the total weight must be positive.

Clicking on the button *Run* will allow restrained MD simulations run and produce a logo of that output on the folder *scratch/equilibration/restraints/*. Besides, there is an option to present the results as the table and plot views by pressing the button *Plots*.

Furthermore, the user can find a detailed documentation on how to use constraints or restraints in a typical MD simulation. That can be obtained in a viewing form by clicking on *Help* menu shown in Fig. 9.17.

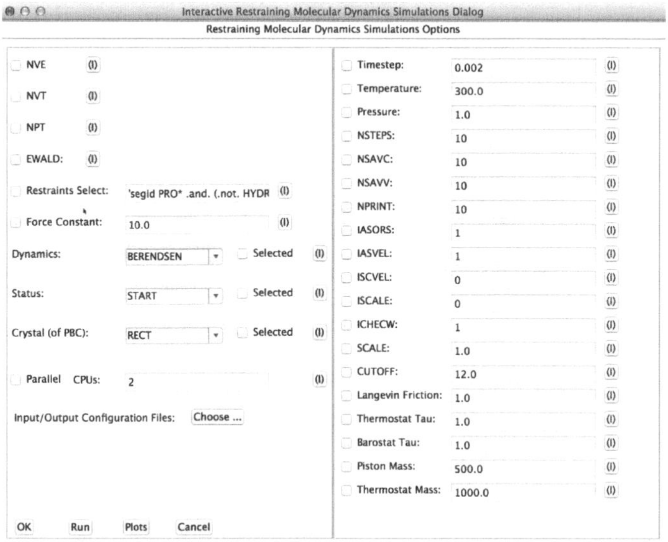

Fig. 9.20 Interactive restraining simulation dialog form

9.2.4 Molecular Dynamics Simulations Module

Figure 9.21 presents the *Molecular Dynamics* menu. It contains three functions: *Standard MD* used to perform standard classical MD simulations in one of the statistical ensembles (such as constant NVE, NVT, NPT, or pH), *Augmented MD* enabling running augmented MD simulations (such as Tsallis statistics and Swarm Particle Intelligence) at different statistical ensembles (such as constant NVT and NPT), and *Replica Exchange MD* that performs temperature replica exchange MD simulations at constant NVT and NPT ensembles.

Standard MD function performs classical MD simulations with an interactive dialog form given in Fig. 9.22. The button *(I)* next to parameter gives the information about that parameter and its possible value. The output logo file along with all output files will be stored in the folder *scratch/md/standard/*. The button *Run* can be pressed to perform a run. The button *Plots* can be clicked to view the output logo file in a tabular form or plot view.

Figure 9.23 shows the interactive dialog form of the *Augmented MD* function. That function enables augmented dynamics using Tsallis statistics or Swarm Particle Intelligence approach (see also Fig. 9.23). Three important parameters can be distinguished. *Tsallis q* factor such that for $q = 1$ Boltzmann distribution is sampled, and for $q > 1$ Tsallis statistics is sampled with longer tails. *Gamma 1* gives the strength of the bias toward the best local positions, and *Gamma 2* gives the strength of the bias toward the best global (target) positions. In the *Atom selection*, the user chooses the subset of atoms of the system to be biased. In particular, the

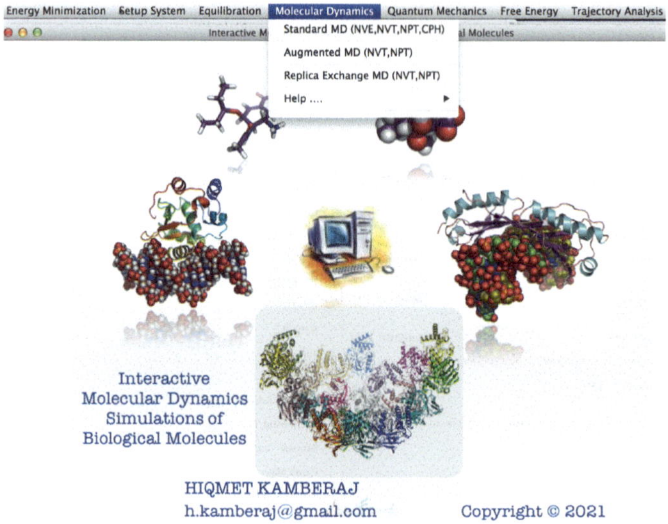

Fig. 9.21 Molecular dynamics module

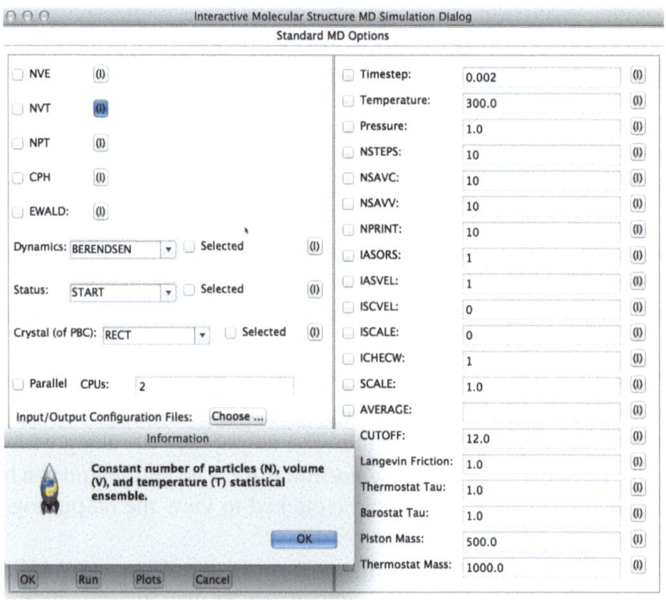

Fig. 9.22 Standard molecular dynamics function

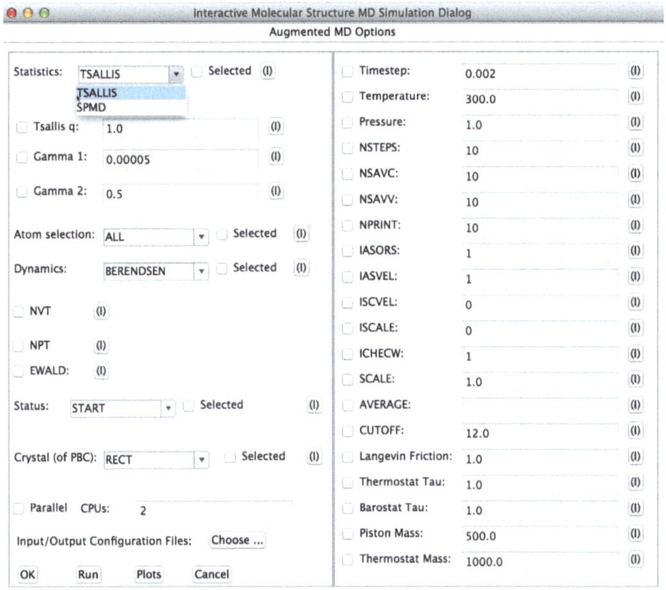

Fig. 9.23 Augmented molecular dynamics function

user may choose between *ALL* (all atoms), *CA* (C_α atoms), *BACKBONE* (backbone atom), and *TORSIONS* (all the dihedral and backbone angles) that is chosen only in the case of Tsallis dynamics. After making a selection of the input parameters and pressing the button *Run*, the output files and logo file will be saved in the folder *scratch/md/augmented/*. The output logo file properties, such as the energy components, temperature, pressure, and volume can be viewed as a table format or as plot view by clicking on the button *Plots*.

Figure 9.24 presents an interactive dialog form of the *Replica Exchange MD* function. That function sets up temperature replica exchange MD simulations using either standard dynamics or augmented dynamics using Tsallis statistics or Swarm Particle Intelligence approaches. The same parameters should be set up for augmented dynamics, namely *Tsallis q* factor, *Gamma 1*, or *Gamma 2*. If the Boltzmann statistics is chosen, then their values are ignored. Another important parameter is the *Exchange Frequency*, which gives the number of MD steps for swapping replicas. After making a selection of the input parameters and pressing the button *Run*, the output files and logo file will be saved in the folder *scratch/md/remd/*. The output logo file properties, such as the energy components, temperature, pressure, and volume can be viewed as a table format or as plot view by clicking on the button *Plots*.

In the MD module, there is a detailed documentation about the dynamics and replica exchange dynamics, as shown in Fig. 9.25. Besides, the user can find a detailed documentation about the control of the temperature and pressure, or both simultaneously. Furthermore, the user can find documentation about the non-bonded interactions options and Ewald summation (see also Fig. 9.25).

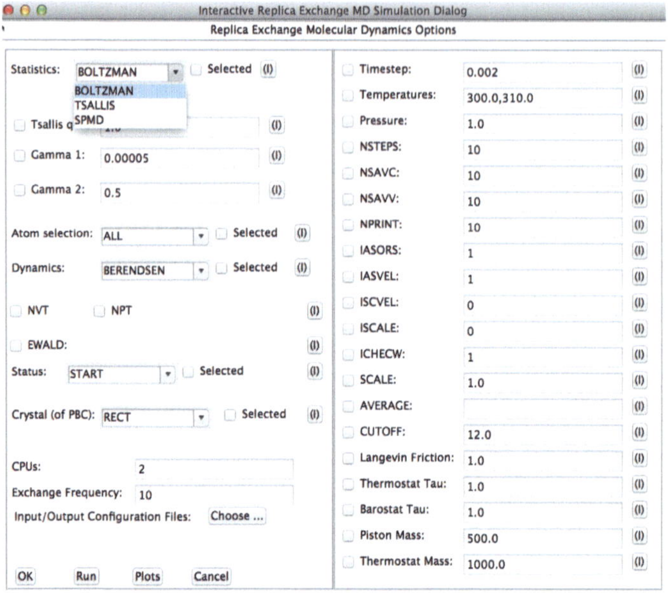

Fig. 9.24 Replica exchange molecular dynamics function

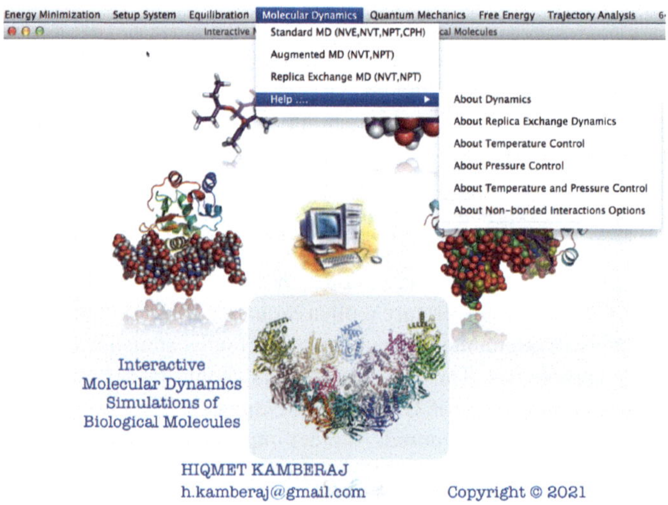

Fig. 9.25 Molecular dynamics functions helping documentation

9.2.5 *Quantum Mechanics Simulations Module*

Figure 9.26 shows the *Quantum Mechanics* module menu. It contains the *QM/MM Hamiltonian MD simulations* function and *Help* menu.

The *QM/MM Hamiltonian MD simulations* function is an interactive dialog form for performing QM/MM Hamiltonian MD simulations, as shown in Fig. 9.27. From the interactive dialog form, the user can determine the QM region and the rest of the system is identified as MM region. Currently, the QM region is modelled using the AM1, PM3, and MNDO semi-empirical quantum models (as depicted from Fig. 9.27). Different statistical ensembles can be sampled, such as NVE, NVT, and NPT, along with EWALD option for treating the long-range electrostatic interactions. Besides, the dynamics can be performed using the Langevin Dynamics, BERENDSEN (which provides NVT and NPT dynamics for controlling temperature and pressure), Nosé dynamics are used only in NVT ensemble, Nosé-Hoover dynamics sample NVT and NPT ensembles. Note that Leap-Frog integrator provides an increased in accuracy, and it allows Langevin dynamics and the so-called Constant Temperature and Pressure (CPT) (based on Berendsen's method). Moreover, velocity-Verlet integrator provides an increase in accuracy, and it allows Constant Temperature (Nosé) (Nosé-Hoover method) Multiple Time Step (MTS). Glen-Martyna-Tuckerman (or the so-called VV2) is a velocity-Verlet integrator based on the operator-splitting technique; It works in conjunction with the TPCONTROL command, allowing temperature and pressure control, and provides an efficient integration algorithm for polarisable force fields based on Drude oscillators.

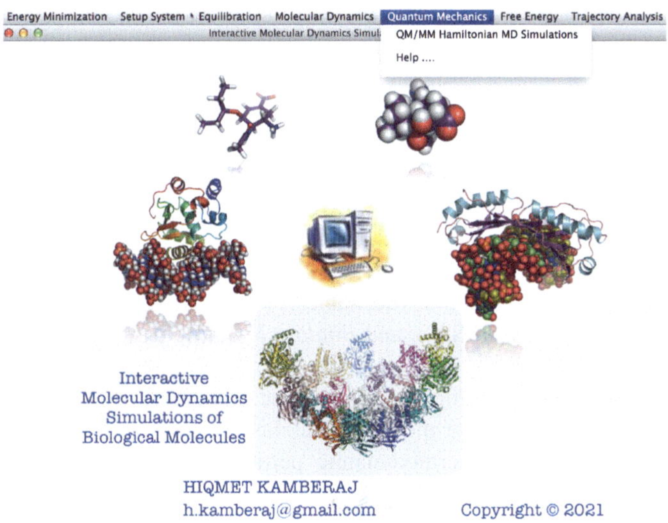

Fig. 9.26 Quantum mechanics module

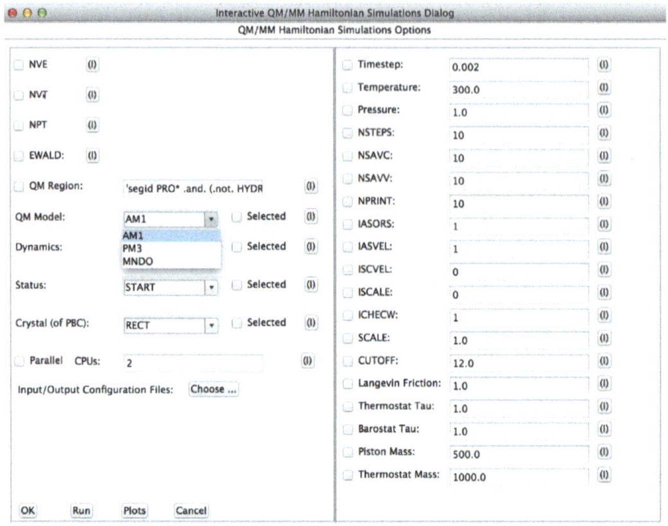

Fig. 9.27 QM/MM hamiltonian MD simulations function

After the parameter selections, the button *Run* can be pressed to perform QM/MM Hamiltonian MD simulations. The output logo file and all other configuration files are stored in the folder *scratch/qmmm/*. The results of the output logo file can be presented in a table view or plot view by clicking on the button *Plots*.

Besides, there is a detailed documentation of performing QM/MM simulations, which can be obtained by clicking on the menu *Help*, as shown in Fig. 9.28.

9.2.6 Free Energy Calculations Module

Free Energy module is shown in Fig. 9.29. The module contains several menus, such as *Thermodynamics Perturbation* functions (using MM and QM/MM Hamiltonians), *pKa Calculations* function (using Poisson-Boltzmann approach), *Absolute Free Energy* function (using Molecular Mechanics Poisson-Boltzmann (MMPB) Solute Accessible Surface Area (SASA)), and *Binding Free Energy* (using MMPB-SASA approach).

The first two models perform perturbation free energy calculations using explicit solvent models. Each of these modules contains different functions, as depicted in Fig. 9.29. In particular, thermodynamic perturbation using MM Hamiltonian approach contains the following interactive dialog forms:

1. Chemical Perturbation: This function performs chemical perturbation free energy calculations using forward-backward perturbation approach and thermodynamic integration method.

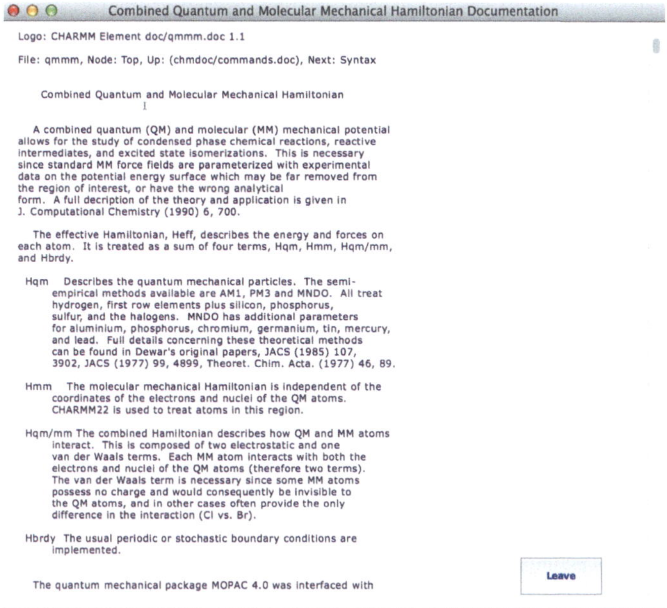

Combined Quantum and Molecular Mechanical Hamiltonian Documentation

Logo: CHARMM Element doc/qmmm.doc 1.1

File: qmmm, Node: Top, Up: (chmdoc/commands.doc), Next: Syntax

Combined Quantum and Molecular Mechanical Hamiltonian

A combined quantum (QM) and molecular (MM) mechanical potential allows for the study of condensed phase chemical reactions, reactive intermediates, and excited state isomerizations. This is necessary since standard MM force fields are parameterized with experimental data on the potential energy surface which may be far removed from the region of interest, or have the wrong analytical form. A full decription of the theory and application is given in J. Computational Chemistry (1990) 6, 700.

The effective Hamiltonian, Heff, describes the energy and forces on each atom. It is treated as a sum of four terms, Hqm, Hmm, Hqm/mm, and Hbrdy.

Hqm Describes the quantum mechanical particles. The semi-empirical methods available are AM1, PM3 and MNDO. All treat hydrogen, first row elements plus silicon, phosphorus, sulfur, and the halogens. MNDO has additional parameters for aluminium, phosphorus, chromium, germanium, tin, mercury, and lead. Full details concerning these theoretical methods can be found in Dewar's original papers, JACS (1985) 107, 3902, JACS (1977) 99, 4899, Theoret. Chim. Acta. (1977) 46, 89.

Hmm The molecular mechanical Hamiltonian is independent of the coordinates of the electrons and nuclei of the QM atoms. CHARMM22 is used to treat atoms in this region.

Hqm/mm The combined Hamiltonian describes how QM and MM atoms interact. This is composed of two electrostatic and one van der Waals terms. Each MM atom interacts with both the electrons and nuclei of the QM atoms (therefore two terms). The van der Waals term is necessary since some MM atoms possess no charge and would consequently be invisible to the QM atoms, and in other cases often provide the only difference in the interaction (Cl vs. Br).

Hbrdy The usual periodic or stochastic boundary conditions are implemented.

The quantum mechanical package MOPAC 4.0 was interfaced with

Leave

Fig. 9.28 QM/MM hamiltonian MD simulations helping documentation

Fig. 9.29 Free energy simulations module

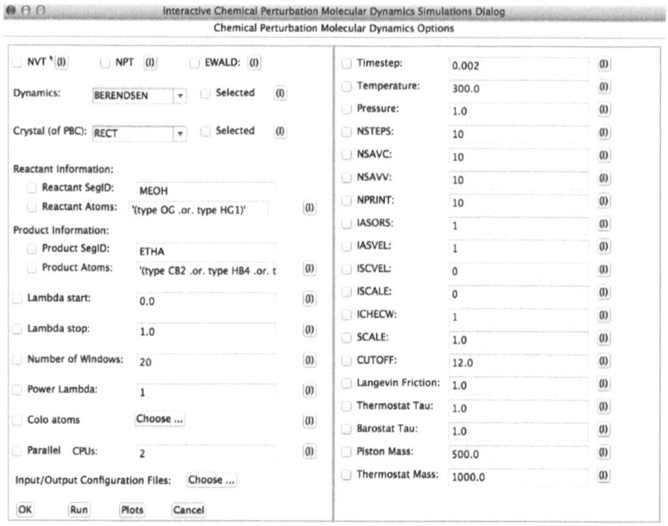

Fig. 9.30 Chemical perturbation free energy simulations function

2. Internal Coordinate Perturbation: This function performs free energy calculations along an internal coordinate (such as dihedral angle, bond, angle, distance, and backbone angle) using forward-backward perturbation approach and thermodynamic integration method.
3. Soft Core Potentials Free Energy Perturbation: This function performs chemical perturbation free energy calculations using soft core potentials.
4. pKa Free Energy Perturbation: This function performs pKa calculation using chemical perturbation free energy approach.
5. Mutation Free Energy Perturbation: Here, free energy changes due to a mutation in a wild-type biological molecule is calculated using chemical perturbation approach.

Figure 9.30 shows an interactive dialog form for chemical perturbation free energy simulations. For that the user should specify the reactant information and product information, as depicted in Fig. 9.30. Besides, range of the values for λ (such as initial value λ_0 and final value λ_f) and the number of windows for λ.

Furthermore, the power of λ scaling of the potential should be given.

It is important to note that sometimes the van der Waals parameters and the identity of an atom remains the same while the charge and hence the electrostatic interactions with this atom are perturbed from a reactant value to a product value. Therefore, the so-called *colo* atoms should be added as patches, as depicted in Fig. 9.31.

The button *Run* can be clicked to perform chemical perturbation free energy calculations. The output logo file and all other configuration files are stored in the folder *scratch/pert/chem/npt/* (for constant NPT dynamics) or *scratch/pert/chem/nvt/*

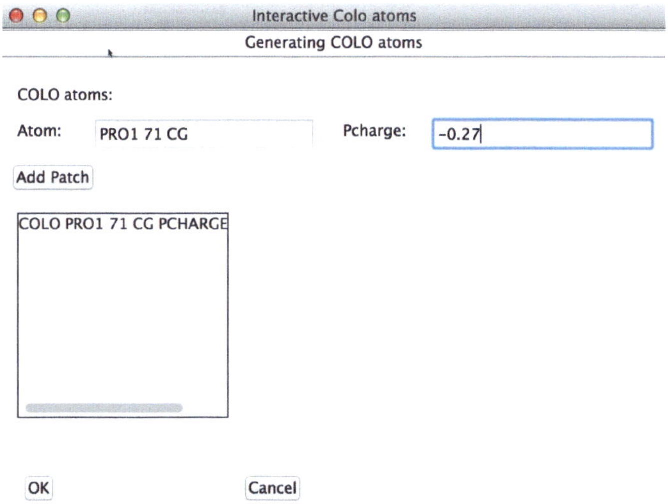

Fig. 9.31 *Colo* interactive dialog form

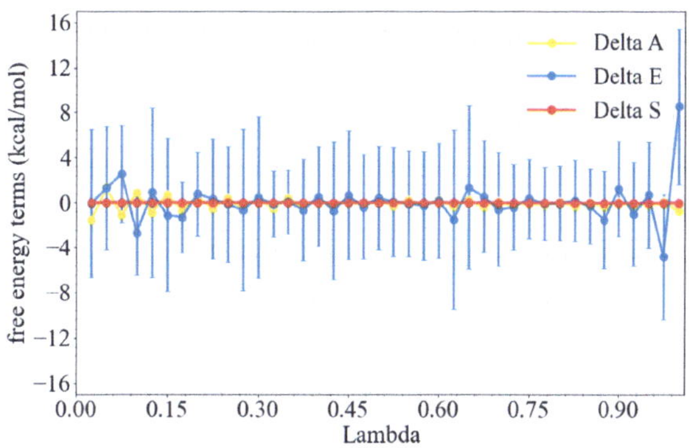

Fig. 9.32 Plot view of the chemical perturbation free energy run in short constant NPT MD simulations

(for constant NVT dynamics). The results of the output logo file can be presented in a table view or plot view by clicking on the button *Plots*, as shown in Fig. 9.32.

Furthermore, a detailed documentation of performing perturbation free energy calculations can be obtained by clicking on the menu *Help*, as shown in Fig. 9.29.

9.2.7 Trajectory Analysis Module

Fig. 9.33 shows the functions within the module *Trajectory Analysis* used to analyse MD simulations data. These functions can be classified, based on the analyses, into functions enabling manipulations of the trajectory inputs (such as orienting the solute

Fig. 9.33 Trajectory analysis module

and re-centring the solute in a solvent box), functions analysing the solvent, functions analysing solvent-solute correlations, and the functions analysing solute.

Furthermore, the user can have an interface to the PyMol and VMD software for visualisation of the MD simulations data (see also Fig. 9.33). Moreover, there are detailed documentations about the data analysis functions as depicted in Fig. 9.33.

9.3 Front-end Programming

The GUI dialog forms are designed using Python 3.8 programming language. The required packages for running the GUI include:

1. Tkinter (version 3.8), a wrapper depository of functions for Tcl/Tk. Tkinter contains classes that allow to display, position, and control the widgets. The Toplevel widgets are Tk and Toplevel. Other widgets include Frame, Label, Entry, Text, Canvas, Button, Radiobutton, Checkbutton, Scale, Listbox, Scrollbar, Menu, Spinbox, LabelFrame, and PanedWindow.
2. Pandas (version 1.3.0), which is a Python package for manipulation of data. It provided flexible, fast, and expressive data structures for working with relational tabular data. Pandas is a high-level object-oriented doing data analysis using Python. It also includes many open source tools for manipulation of data, such as DataFrame, data alignment, aggregation and transformation of the data, intelligent label-based slicing, fancy indexing, large data splitting, hierarchical labelling of axes, input/output tools for loading data (such as CSV format), and time-series functionality tools.

9.4 Back-end Programming

As the back-end programming language, CHARMM program Brooks et al. (2009) scripting language is used. For calculation of the free energy surfaces, weighted-histogram analysis method (the so-called WHAM) Kamberaj and van der Vaart (2007) is applied, which is written in Fortran 90. The computations of the information-theoretic measures, such as mutual information and transfer entropy, are done using the Fortran 90 code in Nebiu and Kamberaj (2020) and the method described in Kamberaj and van der Vaart (2009b). The augmented dynamics are according to method introduced in Kamberaj and van der Vaart (2007), Kamberaj (2015), Kamberaj (2018) implemented in CHARMM program Brooks et al. (2009). The network analyses are performed using the methods described in Ref. Kamberaj (2021), written in Python.

9.5 Summary

We described a Python interactive graphical user interface of CHARMM software package for performing MD simulations of a biological molecular system and standard analysis of trajectories produced from these simulations.

This interactive GUI software package will enable the students and scientists to model, visualise, animate, and analyse the motions of biological systems. The focus is on encouraging a widespread use of computer simulations in biophysics, biology, chemistry, engineering at all levels, enabling the students and scientists a better understanding of the atomistic simulations.

References

B.R. Brooks, C.L. Brooks, A.D. MacKerell, L. Nilsson, R.J. Petrella, B. Roux, Y. Won, G. Archontis, C. Bartels, S. Boresch, A. Caflisch, L. Caves, Q. Cui, A.R. Dinner, M. Feig, S. Fischer, J. Gao, M. Hodoscek, W. Im, K. Kuczera, T. Lazaridis, J. Ma, V. Ovchinnikov, E. Paci, R.W. Pastor, C.B. Post, J.Z. Pu, M. Schaefer, B. Tidor, R.M. Venable, H.L. Woodcock, X. Wu, W. Yang, D.M. York, M. Karplus, Charmm: the bimolecular simulation program. J. Comput. Chem. **30**(10), 1545–1614 (2009)

W. Humphrey, A. Dalke, K. Schulten, VMD-visual molecular dynamics. J. Mol. Graph. **14**, 33–38 (1996)

H. Kamberaj, Conformational sampling enhancement of replica exchange molecular dynamics simulations using swarm particle intelligence. J. Chem. Phys. **143**, 124105–8 (2015)

H. Kamberaj, Faster protein folding using enhanced conformational sampling of molecular dynamics simulation. J. Mol. Graph. Model. **81**, 32–49 (2018)

H. Kamberaj, Heat flow random walks in biomolecular systems using symbolic transfer entropy and graph theory. J. Mol. Graph. Model. **104**, 107838 (2021)

H. Kamberaj, A. van der Vaart, Multiple scaling replica exchange for the conformational sampling of biomolecules in explicit water. J. Chem. Phys. **127**, 234102–234109 (2007)

H. Kamberaj, A. van der Vaart, Extracting the causality of correlated motions from molecular dynamic simulations. Biophys. J. **97**, 1747–1755 (2009)

D. Nebiu, H. Kamberaj, Symbolic information flow measurement (SIFM): a software for measurement of information flow using symbolic analysis. SoftwareX **11**, 100470 (2020)

D.V. Schroeder, Interactive molecular dynamics. Am. J. Phys. **83**(3), 210–219 (2015)

Chapter 10
Equilibration of a Molecular Biological System Simulation

The chapter aims to introduce the setup of a molecular biological system simulation in explicit solvent and the protocol for equilibration of solvent. This aims to demonstrate how to equilibrate a complex system of a solute immersed in solvent.

This chapter introduces the setup protocol for preparing a molecular biological system simulation in explicit solvent. The energy minimisation methods are discussed in details. Furthermore, the complete protocol of the equilibration of solvent is shown for a protein system in explicit solvent.

10.1 Input Files

10.1.1 Residue Topology File (RTF)

This library file specifies how the atoms are connected to form amino acids and other small molecules. It also contains definitions of all atom types and partial charges associated with each atom.

In the following, the topology of the de-protonated aspartate (total charge $-1.00e$, where $e = 1.6 \times 10^{-19}$ C) is shown.

Aspartate—states: D, 1, 2

```
PRES ASPD -1.00 ! deprotonated ASP, dummy protons @OD1/OD2
GROUP
ATOM CB CT2A -0.28
ATOM HB1 HA2 0.09  !   HB1   OD1..HD1
```


H. Kamberaj, *Computer Simulations in Molecular Biology*, Scientific Computation,
https://doi.org/10.1007/978-3-031-34839-6_10

```
ATOM HB2 HA2 0.09 !   |      //
ATOM CG CC 0.62 !  --CB--CG  (-)
ATOM OD1 OC -0.76 !   |      \
ATOM HD1 HD 0.00 !   HB2   OD2..HD2
ATOM OD2 OC -0.76
ATOM HD2 HD 0.00
BOND OD1 HD1 OD2 HD2
IC CB CG OD1 HD1 1.5218 117.9900 180.0000 115.0000 0.9600
IC CB CG OD2 HD2 1.5218 117.7000 180.0000 115.0000 0.9600
```

10.1.2 Parameter Files

Parameter file contains the necessary parameters for calculating the energy of the system. This library contains force constants and equilibrium distances for bond stretching, angle bending, torsional angle, improper angle, CMAP, and van der Waals parameters (such as ε_i and $R_{min,i}/2 = \sigma_i 2^{-5/6}$ of the atom i) for a Lennard-Jones 6–12 potential.

An example of the parameter file is shown in the following:

```
ATOMS
MASS 131 HD 1.00800 ! polar H dummy
...
BONDS
! carboxylates
OC HD 545.000 0.9600 ! same as OH1-H
...
ANGLES
! carboxylates
HD OC CC 55.000 115.00 ! same as H-OH1-CD
...
DIHEDRALS
! carboxylates
CT2 CD OB HD 0.8800 2 180.00 ! from ACEH/TIP3P PMF of HO1/HO2
...
IMPROPERS
! carboxylates
OH1 X X CD 100.0000 0 0.0000 ! same as OB-X-X-CD
...
CMAP
! 2D grid correction data.
! alanine map
C NH1 CT1 C NH1 CT1 C NH1 24
! phi = -180.0
0.126790 0.768700 0.971260 1.250970 2.121010
2.720430 2.089440 1.789790 0.780870 -0.688474
1.001130 -2.200520 -4.827670 -4.821447 -4.913223
-3.591106 -2.766446 -2.784200 -2.454589 -2.346991
```

```
-2.335350 -1.522656 -0.951542 -0.036650
...
NONBONDED
HA 0.000000 -0.022000 1.320000 ! ALLOW PEP ALI POL SUL ARO PRO ALC
```

10.1.3 Protein Structure File

Specific molecules are described in the form of a Protein Structure File (PSF). This file is generated using the information contained in topology file (that is, RTF file). It contains the information about the composition and connectivity of the system of interest. PSF can include several molecular entities, defined as segments, which can range from a single macromolecular chain to a collection of water molecules or ions (such as counter ions and a salt concentration). PSF must be specified before any calculation can be performed on the molecular biological system.

Note that the PSF file does not contain any information related to the force fields parameters, such as the equilibrium bond lengths, angles, dihedrals, CMAP, and van der Waals, but only about the topology of the molecule; however, it does contain the partial charges of atoms and their masses. Therefore, it is necessary to add the parameters and coordinates information for any computation to take place.

The following paragraph presents a typical example of a PSF file of a protein system, including the header of the file and the author information.

Example of PSF File

```
PSF CMAP CHEQ

5 !NTITLE
* GENERATED PSF, CRD FOR SEGMENTS
* READ CRD, ADD PATCHES (IF NEEDED) AND GENERATE TOPOLOGY FILE
* AUTHOR: HIQMET KAMBERAJ (H.KAMBERAJ.AT.GMAIL.COM)
* COPYRIGHT (C) 2021
* DATE: 3/18/22 12:25:54 CREATED BY USER: hiqmetkamber

21300 !NATOM
1 MEM1 1 POPC N 153 -0.600000 14.0070 0 0.00000 -0.301140E-02
2 MEM1 1 POPC C12 147 -0.100000 12.0110 0 0.00000 -0.301140E-02
3 MEM1 1 POPC H12A 136 0.250000 1.00800 0 0.00000 -0.301140E-02
4 MEM1 1 POPC H12B 136 0.250000 1.00800 0 0.00000 -0.301140E-02
...
21300 MEM4 50 LPPC H12Z 141 0.900000E-01 1.00800 0 0.00000 -0.301140E-02

21100 !NBOND: bonds
1 5 1 9 1 13 5 6
5 7 5 8 9 10 9 11
...
```

10.1.4 Coordinates File

CHARMM uses Cartesian coordinates as its primary representation system, although the internal coordinates (bond lengths, bond angles, and dihedral angles) can also be used. Cartesian coordinates are often taken from crystal or NMR structures, deposited in the PDB format.[1]

Alternatively, they can be generated from the values detailed in the RTF, which are put together with some secondary structure information. In these cases where coordinates are not available for some or all of the atoms in the structure (such as the hydrogen atoms), they may be automatically using the web-based servers, such as H++ web server.[2]

The following paragraph illustrates a typical CHARMM program (Brooks et al. 2009) coordinates file.

Example of CHARMM Coordinates File

```
* GENERATED PSF, CRD FOR SEGMENTS
* READ CRD, ADD PATCHES (IF NEEDED) AND GENERATE TOPOLOGY FILE
* AUTHOR: HIQMET KAMBERAJ (H.KAMBERAJ.AT.GMAIL.COM)
* COPYRIGHT (C) 2021
* DATE: 3/18/22 12:25:54 CREATED BY USER: hiqmetkamber
*
21300
1 1 POPC N -0.98300 -19.78400 -9.57600 MEM1 1 0.00000
2 1 POPC C12 -1.29600 -20.61300 -8.33800 MEM1 1 0.00000
3 1 POPC H12A -1.85200 -20.07300 -7.54400 MEM1 1 0.00000
...
```

10.2 System Setup

10.2.1 Gas Phase

The Bovine Pancreatic Trypsin Inhibitor (BPTI) (Wlodawer et al. 1984) with PDB ID 5PTI is used for the setup. That protein contains 58 amino acids and its structure is investigated by joint X-ray and neutron experiments at a resolution of 1.00 Å. The protein has three disulphide bonds:

```
SSBOND 1 CYS A 5 CYS A 55
SSBOND 2 CYS A 14 CYS A 38
SSBOND 3 CYS A 30 CYS A 51
```

[1] www.rcsb.org.

[2] biophysics.cs.vt.edu.

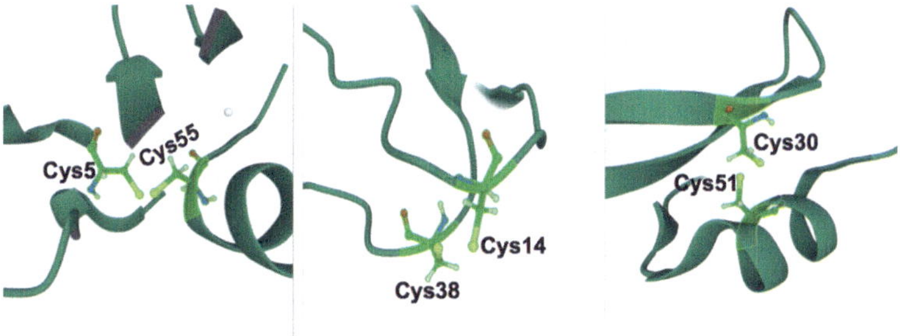

Fig. 10.1 The disulphide bonds of the BPTI structure

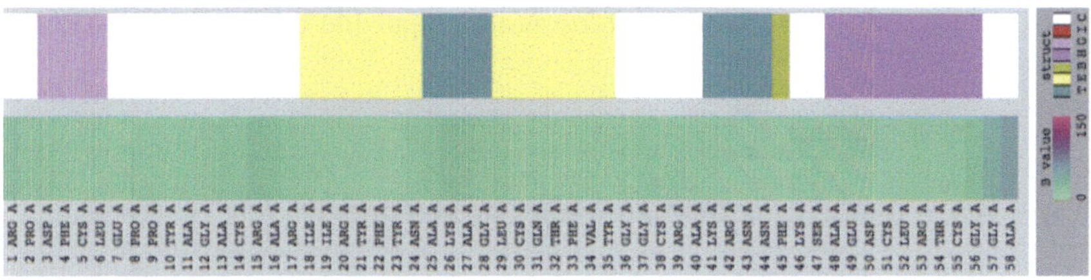

Fig. 10.2 The sequence, B-factor, and the secondary structure view of BPTI protein produced using the VMD (Humphrey et al. 1996)

Figure 10.1 shows the position of the disulphide bonds in the three-dimensional structure of the BPTI protein.

Figure 10.2 shows the sequence, secondary structure, and the so-called B-factor of BPTI protein produced using the VMD (Humphrey et al. 1996).

Figure 10.3a presents the dialog form for generating a topology (struct.psf) and coordinates files (struct.crd). Besides, the file containing the information about the simulation box side lengths and angles is generated (struct.size), used in the periodic boundary condition simulations. As indicated in Fig. 10.3a, a topology stream file is needed to read the force field topology files and parameters, which looks as the following:

```
* Created by Hiqmet Kamberaj VERSION on Nov 3, 2022.
* Stream file for topology and parameter reading
*

! protein topology and parameter
open read card unit 10 name toppar/top_all36_prot.rtf
read rtf card unit 10


open read card unit 20 name toppar/par_all36_prot.prm
read para card unit 20 flex


! nucleic acids
open read card unit 10 name toppar/top_all36_na.rtf
read rtf card unit 10 append
```

```
open read card unit 20 name toppar/par_all36_na.prm
read para card unit 20 append flex
...
```

The initial coordinates can be uploaded as PDB format, for example, as stored in the protein database server. In general, the file may contain more than one chain; therefore, more than one chain is allowed to be selected simultaneously. Note that in that case, an output file is generated for each chain, named as chain_index.crd, of structure coordinates. Current version allows two types of patches to be defined, such as the first and the last side patches (namely, NTER and CTER patches), and the disulphide patches, see also Fig. 10.3a.

After clicking on the button *Generate PSF*, four output files will be created as located in Fig. 10.4. The file genpsf.log is an output file created by CHARMM

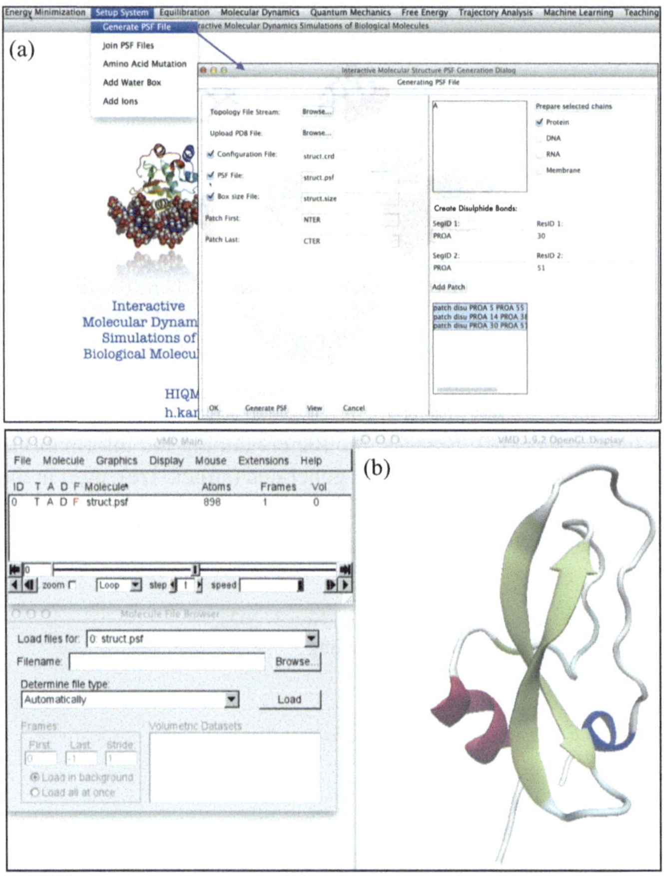

Fig. 10.3 **a** Dialog form for generating a topology file. **b** A view of BPTI protein in VMD (Humphrey et al. 1996)

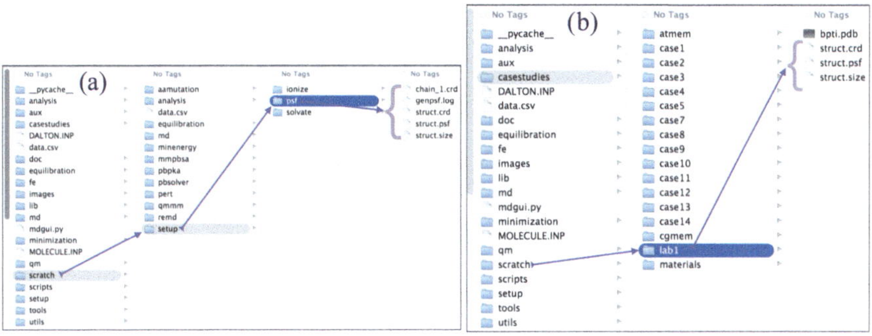

Fig. 10.4 The location of generated output files: **a** under the *scratch/setup/psf* folder and **b** under the desired *lab1* folder

program. The button *View* can be used to open VMD program for obtaining a three-dimensional view of the structure, as presented in Fig. 10.3b.

Figure 10.4a shows the created output files. The most useful files, for later use, are struct.psf, struct.crd, and struct.size, which can be copied to the desired folder; for instance, they are copied into the folder lab1, as shown in Fig. 10.4b.

10.2.2 Solvation of Molecular Biological Structure

First, the macromolecular three-dimensional structure can be immersed in a box of water molecules (the so-called *solvent*). Usually, a pre-equilibrated box of water molecules can be used at standard conditions (that is, at temperature $T = 300$ K and pressure $P = 1$ atm). Figure 10.5a shows a dialog form for solvating the structure and Fig. 10.5b presents the view of solvated BPTI protein using VMD program. The size of water simulation box is such that the distance of the solute side from the water box side is slightly larger than the long-range interactions cutoff R_c. For typically long-range interactions cutoff of $R_c = 12$ Å, that distance can be 14 Å, which allows for slight increase of the volume of solute from that of the X-ray structure and the decrease of the solvent volume during the equilibration procedure.

The choice of the simulation box length (or long-range interactions cutoff R_c) and the buffer thickness around the solute (such as protein) are illustrated in Fig. 10.6 for a system with periodic boundary conditions. In practice, the cutoff can be defined a priory (typically, in the range $10 - 12$ Å), to control the accuracy and the computational costs of the short- and long-range electrostatic and van der Waals interactions. In a periodic system, each atom must only interact with all the other atoms of system within a distance less or equal to R_c, but not with the same atoms in an image simulation box; therefore, $L/2 > R_c$, where L is the original simulation box length along one of the directions.

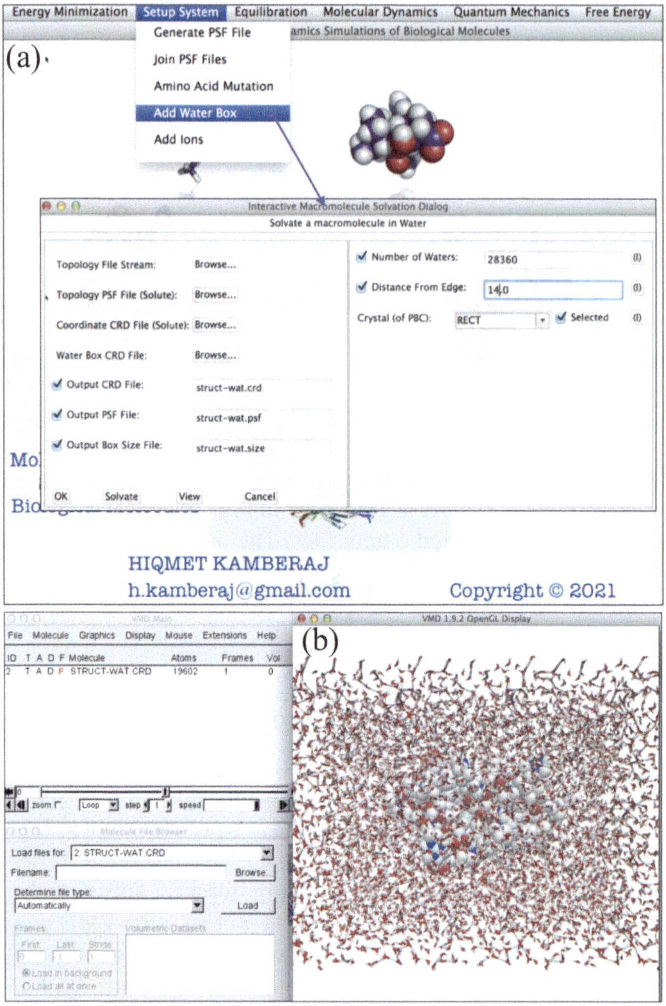

Fig. 10.5 **a** Dialog form for solvating the structure. **b** A view of solvated BPTI protein using VMD (Humphrey et al. 1996)

Furthermore, the solute in the original simulation box and in one of its images must not interact with each-other; that is, the length of the buffer around the solute in both sides along one of the directions should be greater than the cutoff, $b = b_l + b_r > R_c$, where b_l and b_r are the distances between the edges of solute and solvent on the left- and right-hand sides, respectively (see also Fig. 10.6b).

Next, counter ions are added, if the solute has a net charge different from zero (positive or negative), and salt at a selected concentration. Typically, the concentration at the standard physiological condition is 154 mM/L. Usually, to avoid clustering of ions around the surface of solute and in the bulk water, a constraint distance

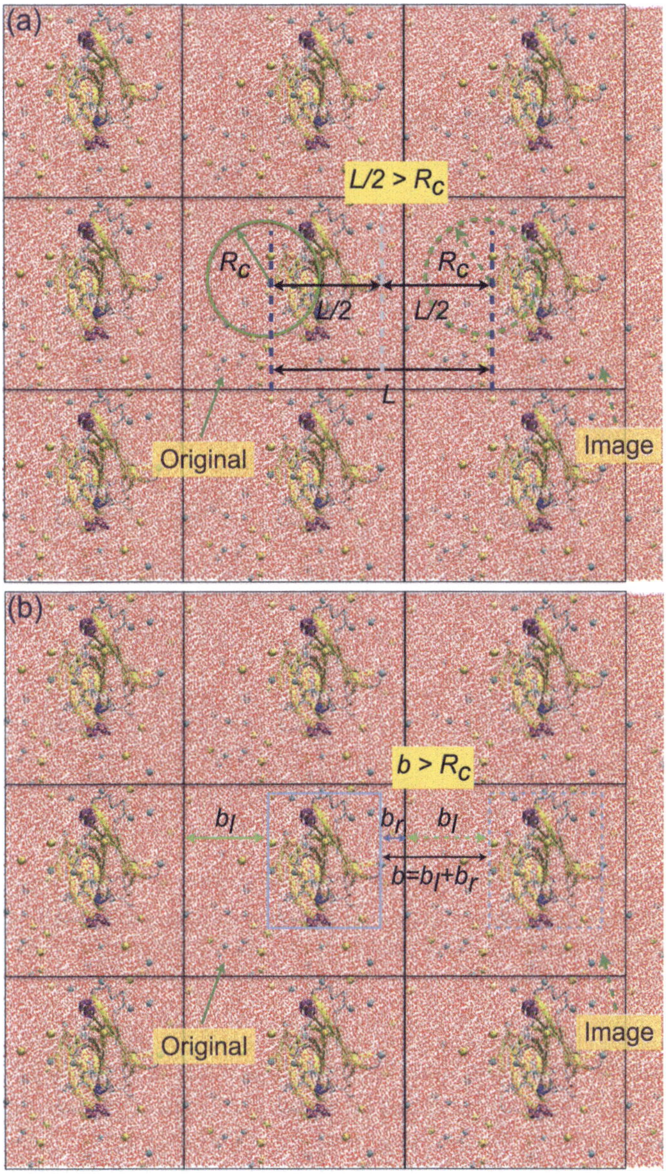

Fig. 10.6 **a** Illustration of the choice of the simulation box length (or long-range interactions cutoff R_c) and **b** the buffer thickness around the solute (such as protein)

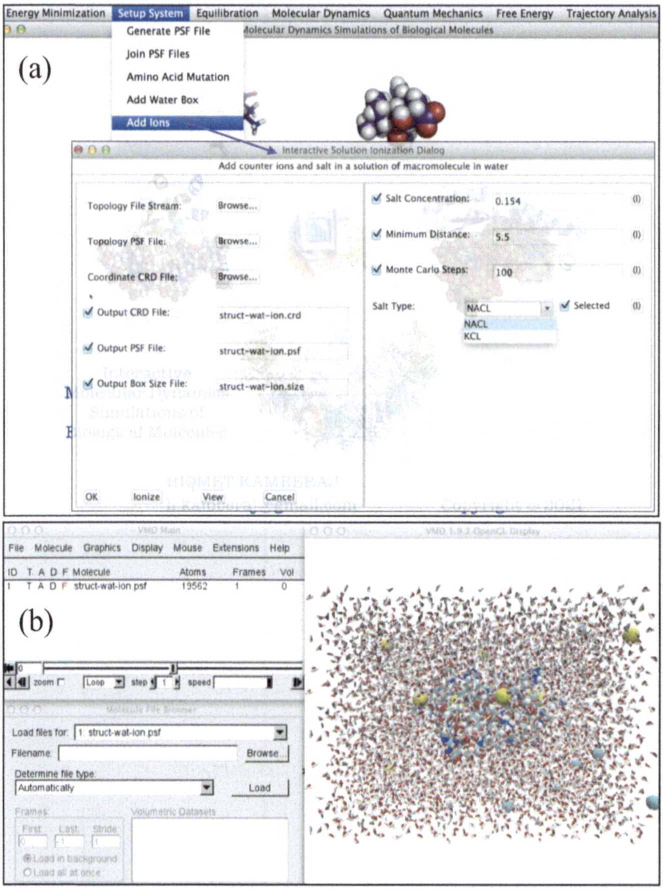

Fig. 10.7 **a** Dialog form for adding counter ions and salt concentration. **b** A view of ionised BPTI protein using VMD (Humphrey et al. 1996)

between ions is chosen, typically, 5.5 Å. Ions are distributed around the solvent using a Monte Carlo algorithms for about, typically, 100 steps, allowing obtaining a structure with minimum energy and avoiding close contacts between the ions and solvent molecules.

Figure 10.7a presents the dialog form for adding counter ions and salt concentration. In Fig. 10.7b, a view of ionised BPTI protein using VMD program is shown.

Figure 10.8a shows the created output files, which can be copied to the desired folder; for instance, they are copied into the folder lab1, as shown in Fig. 10.8b.

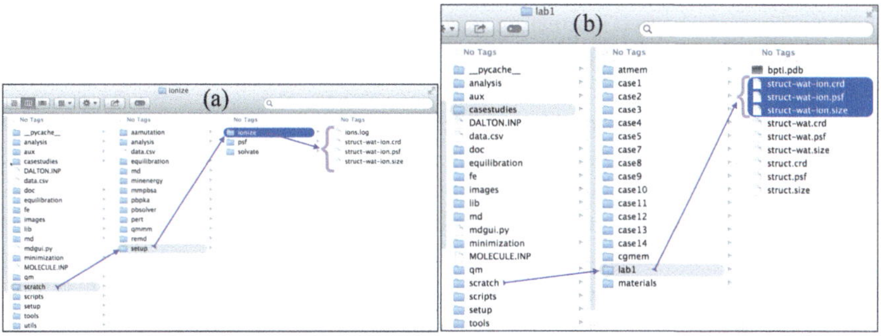

Fig. 10.8 The location of generated output files: **a** under the *scratch/setup/ionize* folder and **b** under the desired *lab1* folder

10.3 Energy Minimisation

10.3.1 Theoretical Background

In general, the potential energy of a biomolecular system is a complex multidimensional function of the atomic coordinates. The variation of the potential energy from all coordinates is called *potential energy surface*. Computation of the potential energy surface includes visualisation of the potential energy surface defined by the reaction coordinates, examination of the structure, dynamics, thermodynamics, and sampling pathways (Wales 2015). For a system of N atoms, there are $3N - 6$ coordinates, where 6 degrees of freedom are subtracted that include overall translational and rotation of system. In general, the reaction coordinate selection or projection from the full configuration space to a reduced subspace is an approximation. However, they can be optimised to preserve some properties of the observables in the reduced dimensions (Wales 2015), and in particular, for protein folding dynamics and kinetics problems (Becker and Karplus 1997; Xin et al. 2010; Doshi and Hamelberg 2011). Often, the backbone dihedral angles (ϕ, ψ) and (rmsd, rgyr) (root mean square displacement and radius of gyration, respectively) are chosen to represent the topology of multidimensional potential energy surfaces in a reduced subspace (Kamberaj and van der Vaart 2007). Construction of the potential energy surface of the system would require the entire degrees of freedom of the system. Here, a subset of the entire system's degrees of freedom is considered (namely (ϕ, ψ) backbone dihedral angles and the pair of root mean square deviations and radius of gyration), which construct the so-called potential of mean force (Wales 2015). Often, the so-called Landau free energy is used defined in terms of an occupation probability, $P(q_1, q_2)$, computed using the histogram method, as (Wales 2015):

$$F(q_1, q_2) = -k_B T \ln P(q_1, q_2), \tag{10.1}$$

where (q_1, q_2) are the two interesting reaction coordinates, and k_B and T are Boltzmann's constant and temperature, respectively. $F(q_1, q_2)$ represents the potential of mean force.

Usually, there is a large number of local minima corresponding to the potential energy surface, but only one corresponds to the so-called *global energy minimum*. The task of finding the global minimum of the potential energy function of complicated atomic structures is notoriously difficult. Since the number of locally stable structures tends to increase rapidly as the number of atoms increases, searches for the global minimum become impossible for this surface energy landscapes. Therefore, algorithms are employed to identify the potential energy minimum states and their corresponding geometrical coordinates (Kamberaj 2014; Wales 2015).

For performing energy minimisation of the potential function $U(x_1, x_2, \cdots, x_g)$, where g is the number of degrees of freedom of the system, a search is computationally performed for finding $x_1^*, x_2^*, \cdots, x_g^*$ such that for $i = 1, 2, \cdots, g$

$$\begin{aligned} \left(\frac{\partial U(x_1, x_2, \cdots, x_g)}{\partial x_i}\right)_{x_i=x_i^*} &= 0 \\ \left(\frac{\partial^2 U(x_1, x_2, \cdots, x_g)}{\partial x_i^2}\right)_{x_i=x_i^*} &> 0 \end{aligned} \tag{10.2}$$

Consider a Taylor expansion around the state point $\mathbf{x}^* = (x_1^*, x_2^*, \cdots, x_g^*)$ of the potential energy function $U(\mathbf{x})$ with $\mathbf{x} = (x_1, x_2, \cdots, x_g)$:

$$\begin{aligned} U(\mathbf{x}) = U(\mathbf{x}^*) &+ \sum_{i=1}^{g} \left(\frac{\partial U}{\partial x_i}\right)_{x_i=x_i^*} \eta_i \\ &+ \frac{1}{2}\sum_{i=1}^{g}\sum_{j=1}^{g} \left(\frac{\partial^2 U}{\partial x_i \partial x_j}\right)_{x_i=x_i^*, x_j=x_j^*} \eta_i \eta_j + \cdots \end{aligned} \tag{10.3}$$

where $\eta_i = x_i - x_i^*$.

In Eq. (10.3), $U(\mathbf{x})$ is a function of $3N$ Cartesian coordinates $\mathbf{x}$, the vector $\mathbf{x}^*$ is a $3N$ components vector for the current state point, $\mathbf{Q} \equiv \dfrac{\partial U(\mathbf{x}^*)}{\partial \mathbf{x}}$ is a $3N$ components vector, and $\mathbf{H} \equiv \dfrac{\partial^2 U}{\partial \mathbf{x}^2}$ is a $3N \times 3N$ matrix, known also as Hessian matrix and force constant matrix, with elements:

$$H_{ij} = \left(\frac{\partial^2 U}{\partial x_i \partial x_j}\right)_{x_i=x_i^*, x_j=x_j^*} \tag{10.4}$$

In a matrix form, Eq. (10.3) is written as:

$$U(\mathbf{x}) = U(\mathbf{x}^*) + \eta^T \mathbf{Q} + \frac{1}{2}\eta^T \mathbf{H}\eta + \cdots \tag{10.5}$$

where (^{T}) stands for the transpose and η and $\mathbf{g}$ are column vectors. The accuracy of approximating $U(\mathbf{x})$ depends on the order of terms included from the Taylor expansion. That is because the molecular force fields include terms of higher than quadratic terms. Therefore, any minimisation method can not perform better than the molecular mechanics potential energy function; for instance, the minimisation methods that include quadratic terms in expression of Eq. (10.5) locate the minimum state in less steps (such as the Newton-Raphson method).

Furthermore, if the system is far from the global minimum, the harmonic approximation shows to be less robust in converging; however, if the system is close to a local minimum, then the harmonic approximation is more valid and minimisation works well.

10.3.1.1 The Steepest Descents Method

The steepest descent (SD) method progresses in the direction parallel to the net force or straight downhill. For a system of g degrees of freedom, this direction is, for convenience, characterised by the unit vector $\mathbf{e}$:

$$\mathbf{e}(\mathbf{x}) = -\frac{\mathbf{Q}(\mathbf{x})}{\mid \mathbf{Q}(\mathbf{x}) \mid} \tag{10.6}$$

Then, the new Cartesian coordinates at the iteration step t is taken from the previous step using an arbitrary step size δ as:

$$\mathbf{x}_t = \mathbf{x}_{t-1} + \delta \mathbf{e}_{t-1} \tag{10.7}$$

In the SD algorithm, used in the energy minimisation of molecular systems, initial δ is arbitrary determined. If at the first iteration step the energy decreases, the step size δ increases by a factor (such as $\delta = 1.2\delta$ Leach 2001) for the following iterations. This continues until at an iteration step an increase in the energy value is observed, which indicates that the system has crossed a potential energy well. Then, when that is observed, the step size decreases by a constant factor (such as $\delta = 0.5\delta$ Leach 2001).

In general, the step size will depend on the shape of the potential energy surface; for instance, for a fairly flat surface large δ can be used; however, for more narrow potential energy landscapes, small step size would be more appropriate. The SD method using an arbitrary step size often requires few function evaluations, and hence it is computationally more efficient; however, it may require more iterations to find the potential energy function minimum. Thus, other methods are also suggested, such as line search method (Allen and Tildesley 1989; Leach 2001). In the case of molecular force fields, the direction of the gradient unit vector **e** is primary determined by the largest non-bonded forces, and thus the SD method is a suitable approach for taking over from the highest-energy state points in an initial configuration. In particular, the SD method is faster, including the cases when initial configuration is far from the minimum when the harmonic approximation in the energy expansion is not often an accurate approximation. In practice, the SD method often requires many steps with small step size as moving downhill long narrow valley (Leach 2001). Besides, the method has to do a right-angle turn at each step, which may not be the way to the minimum. Hence, the trajectory to the local minimum oscillates by continuously correcting itself.

10.3.1.2 Conjugate Gradients Minimisation

The conjugate gradients (CG) method uses a set of directions showing no oscillation-like motion behaviour as in the SD method when iterating in a narrow valley (Leach 2001). In the SD method, the directions (and so the gradients) of successive iteration steps are perpendicular (that is, $\mathbf{e}_t^T \mathbf{e}_{t-1} = 0$ and $\mathbf{Q}_t^T \mathbf{Q}_{t-1} = 0$). In contrast, in the CG method, the gradients of successive iteration steps at a state point are orthogonal; however, the directions are conjugate. The collection of conjugate directions for a quadratic function of g-dimension coordinates converges to the minimum in g iterations. In the CG method, the direction $\mathbf{e}_t$ of the movement at the t-th step is defined as (Hestenes and Stiefel 1952; Fletcher and Reeves 1964):

$$\mathbf{e}_t = -\mathbf{Q}_t + \delta_t \mathbf{e}_{t-1} \tag{10.8}$$

where δ_t is a scalar constant defined as

$$\delta_t = \frac{\mathbf{Q}_t^T \mathbf{Q}_t}{\mathbf{Q}_{t-1}^T \mathbf{Q}_{t-1}} \tag{10.9}$$

Furthermore, the CG method satisfies the following relationships between the directions of movements and gradients:

$$\begin{aligned} &\mathbf{Q}_i^T \mathbf{Q}_j = 0 \\ &\mathbf{e}_i^T \mathbf{H}(\mathbf{x}_i, \mathbf{x}_j) \mathbf{e}_j = 0 \\ &\mathbf{Q}_i^T \mathbf{e}_j = 0 \end{aligned} \tag{10.10}$$

From Eq. (10.8), in the conjugate gradients method, the first iteration step the displacement is according to the steepest descents in the direction of the gradient. Then, from the second step, the search is according to the CG algorithm.

There are other variants of the conjugate gradients method, such as the one that purposes the following scalar constant δ_t (Polak and Ribiére 1969):

$$\delta_t = \frac{(\mathbf{Q}_t - \mathbf{Q}_{t-1})^T \mathbf{Q}_t}{\mathbf{Q}_{t-1}^T \mathbf{Q}_{t-1}} \tag{10.11}$$

Note that the molecular force fields are only approximately quadratic functions, and thus these two variants of the CG method are only at best approximation identical.

10.3.1.3 The Newton-Raphson Method

The Newton-Raphson (NRAP) method is the simplest second-order method, which uses the first and second derivatives to search for the minimum. The first derivative provides information about the direction of the tangent to the curve function at a given point while the second derivative provides information about the curvature of that function.

Using Eq. (10.3), the first derivative of $U(\mathbf{x})$ is given as:

$$\frac{\partial U(\mathbf{x})}{\partial x_i} = x_i \frac{\partial U(\mathbf{x}_t)}{\partial x_i} + \frac{\partial^2 U(\mathbf{x}_t)}{\partial x_i^2} \eta_i \tag{10.12}$$

For a quadratic function (because the second derivative is the same everywhere):

$$\frac{\partial^2 U(\mathbf{x}_t)}{\partial x_i^2} = \frac{\partial^2 U(\mathbf{x})}{\partial x_i^2} \tag{10.13}$$

Furthermore, at the minimum points (that is, $\mathbf{x} = \mathbf{x}^*$), $\partial U(\mathbf{x}^*)/\partial x_i = 0$; thus,

$$\mathbf{x}^* = \mathbf{x}_t - \mathbf{Q}^T(\mathbf{x}_t)\mathbf{H}^{-1}(\mathbf{x}_t) \tag{10.14}$$

In Eq. (10.14), $\mathbf{H}^{-1}$ is the inverse of Hessian matrix, which is a $g \times g$ matrix, and hence it is computationally very demanding for large g, as it may be the case of the molecular force fields. Therefore, the NRAP method is suitable for small molecular system of atoms (Leach 2001). Note that for the NRAP method to be stable, it is required that $\mathbf{H}$ matrix is positive definite, achieved if its eigenvalues are positive (Leach 2001). Otherwise, the NRAP method becomes unstable, and the move will be toward saddle points where the energy increases. That also occurs when the minimisation starts from points far from the minimum where the harmonic approximation is not a suitable one. In such cases, it is recommended to use first a more robust minimisation method (such as the SD method) to get closer to the minimum energy configuration, characterised by a positive definite $\mathbf{H}$, then apply the NRAP method (Leach 2001).

10.3.1.4 Other Variants of the Newton-Raphson Method

There exist other variants of the NRAP method, which are primary introduced to avoid the calculation of full Hessian matrix every iteration step. Another class of the NRAP methods is the so-called quasi-Newton methods, which calculate only the first derivatives and then estimate the inverse of **H** as the iterations continue; for instance, one can use the same Hessian matrix for a few successive iteration steps of the NRAP method and only the first derivative calculated at every step.

The block-diagonal Newton-Raphson method is a variant of the Newton-Raphson method where only one atom moves at each iteration. Therefore, all the elements of the Hessian matrix $H_{ij} = 0$, except from the atom being moving. That is, if k is the atom, which is moving, then

$$\begin{aligned}
H_{xx}^{(k)} &= \frac{\partial^2 U}{\partial x_k^2} \neq 0, \quad H_{xy}^{(k)} = \frac{\partial^2 U}{\partial x_k \partial y_k} \neq 0, \quad H_{xz}^{(k)} = \frac{\partial^2 U}{\partial x_k \partial z_k} \neq 0 \\
H_{yx}^{(k)} &= \frac{\partial^2 U}{\partial y_k \partial x_k} \neq 0, \quad H_{yy}^{(k)} = \frac{\partial^2 U}{\partial y_k^2} \neq 0, \quad H_{yz}^{(k)} = \frac{\partial^2 U}{\partial y_k \partial z_k} \neq 0 \\
H_{zx}^{(k)} &= \frac{\partial^2 U}{\partial z_k \partial x_k} \neq 0, \quad H_{zy}^{(k)} = \frac{\partial^2 U}{\partial z_k \partial y_k} \neq 0, \quad H_{zz}^{(k)} = \frac{\partial^2 U}{\partial z_k^2} \neq 0
\end{aligned} \tag{10.15}$$

Here, (x_k, y_k, z_k) are the Cartesian coordinates of the atom k. Thus, the problem reduces to an 3×3 matrix that has to be inverted. However, the block-diagonal method is less efficient when the motions of atoms are coupled (Leach 2001).

10.3.1.5 Quasi-newton Methods

As mentioned above, the computation of the inverse of a matrix may be time-consuming step, and thus it is the main disadvantage of the NRAP method. Another disadvantage (Leach 2001) might be the analytical calculations of the second derivatives of the potential energy function $U(\mathbf{x})$. In the quasi-Newton method, the inverse of the Hessian matrix is built up gradually every iteration step. For that, a collection of the Hessian matrices $\mathbf{H}^{(t)}$ is calculated with the following property (Leach 2001):

$$\lim_{t \to \infty} H_{ij}^{(t)} = \left[\frac{\partial^2 U}{\partial x_i \partial x_j} \right]^{-1} \tag{10.16}$$

Then, at the t-th iteration step, the new positions $\mathbf{x}_t$ are given as

$$\mathbf{x}_t = \mathbf{x}_{t-1} - \mathbf{H}^{(t-1)} \mathbf{Q}_{t-1} \tag{10.17}$$

Equation (10.17) is exact for molecular force fields of quadratic terms for the Cartesian coordinates, which is not often the case. Thus, for these systems a line search is required along the vector $\mathbf{x}_t - \mathbf{x}_{t-1}$. After moving to the new position $\mathbf{x}_t$, the Hessian matrix $\mathbf{H}$ is updated using its value at a previous step. There exist three different approaches that give the formulas for updating the $\mathbf{H}$ matrix (Leach 2001). In the first approach, known as Davidon-Fletcher-Powell, the Hessian matrix at each iteration step is

$$\mathbf{H}_t = \frac{(\mathbf{x}_t - \mathbf{x}_{t-1}) \otimes (\mathbf{x}_t - \mathbf{x}_{t-1})}{(\mathbf{x}_t - \mathbf{x}_{t-1})^T (\mathbf{Q}_t - \mathbf{Q}_{t-1})} - \frac{\left((\mathbf{x}_t - \mathbf{x}_{t-1})^T \mathbf{H}_t\right) \otimes \left((\mathbf{x}_t - \mathbf{x}_{t-1})^T \mathbf{H}_t\right)}{(\mathbf{Q}_t - \mathbf{Q}_{t-1})^T \mathbf{H}_t (\mathbf{Q}_t - \mathbf{Q}_{t-1})} \tag{10.18}$$

In Eq. (10.18), $\otimes$ is a matrix product; that is, for any two vectors, $\mathbf{x}$ and $\mathbf{y}$ of g dimensions, a $g \times g$ matrix can be constructed, $\mathbf{A} = \mathbf{x} \otimes \mathbf{y}$, as the following:

$$A_{ij} = x_i y_j \tag{10.19}$$

For the second approach, known as Broyden-Fletcher-Goldfarb-Shanno, the update at each step is

$$\mathbf{H}_t = \frac{(\mathbf{x}_t - \mathbf{x}_{t-1}) \otimes (\mathbf{x}_t - \mathbf{x}_{t-1})}{(\mathbf{x}_t - \mathbf{x}_{t-1})^T (\mathbf{Q}_t - \mathbf{Q}_{t-1})} - \frac{\left((\mathbf{x}_t - \mathbf{x}_{t-1})^T \mathbf{H}_t\right) \otimes \left((\mathbf{x}_t - \mathbf{x}_{t-1})^T \mathbf{H}_t\right)}{(\mathbf{Q}_t - \mathbf{Q}_{t-1})^T \mathbf{H}_t (\mathbf{Q}_t - \mathbf{Q}_{t-1})} + \left((\mathbf{Q}_t - \mathbf{Q}_{t-1})^T \mathbf{H}_t (\mathbf{Q}_t - \mathbf{Q}_{t-1})\right) \mathbf{v} \otimes \mathbf{v} \tag{10.20}$$

where

$$\mathbf{v} = \frac{\mathbf{x}_t - \mathbf{x}_{t-1}}{(\mathbf{x}_t - \mathbf{x}_{t-1})^T (\mathbf{Q}_t - \mathbf{Q}_{t-1})} - \frac{(\mathbf{Q}_t - \mathbf{Q}_{t-1})^T \mathbf{H}_t}{(\mathbf{Q}_t - \mathbf{Q}_{t-1})^T \mathbf{H}_t (\mathbf{Q}_t - \mathbf{Q}_{t-1})} \tag{10.21}$$

The third approach is called Murtaugh-Sargent where the Hessian matrix is updated as the following:

$$\mathbf{H}_t = \mathbf{H}_{t-1} + \frac{\mathbf{u} \otimes \mathbf{u}}{\mathbf{u}^T (\mathbf{Q}_t - \mathbf{Q}_{t-1})} \tag{10.22}$$

where

$$\mathbf{u} = (\mathbf{x}_t - \mathbf{x}_{t-1}) - \mathbf{H}_{t-1} (\mathbf{Q}_t - \mathbf{Q}_{t-1}) \tag{10.23}$$

Note that other variants of the quasi-Newton method exist. Furthermore, for a quadratic function of g degrees of freedom, all these methods converge to the minimum state in g iteration steps. Often, the Hessian matrix is initialised to the unit matrix and faster algorithms can be obtained if better estimates of the inverse Hessian matrix exist (Leach 2001).

The practical choice of the energy minimisation method primary depends on the storage and computational demands, their relative speeds, possibility for evaluation of analytical derivatives, their computational robustness, and their availability in the common biomolecular modelling packages. Calculation of the inverse of the Hessian matrix and its storage could be quite costly for molecular force fields; therefore, for performing energy minimisation of potential function of molecular systems with large number of atoms, the SD and CG methods are preferred. On the other hand, for molecular calculations with small number of atoms systems, the Newton-Raphson method may be used; however, it may have problems if the system is initially far from the minimum state. In such cases, the steepest descents and conjugate gradients approaches can be chosen first followed by a Newton-Raphson approach.

10.3.1.6 Convergence Criteria

For a large biomolecular system immersed in solvent knowing the precise location of the minimum or saddle point is not obvious, and so the search for the so-called global minimum state may continue for ever. Therefore, in practice, we need to define a criterion for being sufficiently close to the desired minimum state of the energy function, and hence the search stops.

There exist different criteria that we may think, such as by monitoring the energy and when the energy difference between two successive iteration steps becomes less then a limit value, the search terminates. Alternatively, we can monitor the coordinates of atoms and stop when the difference between successive configurations becomes sufficiently small. Another approach is to monitor the root-mean-square of gradient as

$$\text{rms} = \left(\frac{\mathbf{Q}^T \mathbf{Q}}{g} \right)^{1/2} \tag{10.24}$$

which adds up the squares of the energy's gradients for coordinates divided by the total number of degrees of freedom, and taking the square root of the result. Note that different software packages use any of the above mentioned methods or all of them; however, any calculation is limited by the precision of representing the numbers in the computer, and so the search should terminate whenever this limit is reached.

10.3.1.7 CHARMM Program Energy Minimisation

In CHARMM software package (Brooks et al. 2009), there are several minimisation algorithms provided. They include the following methods:

1. The steepest descent;
2. The conjugate gradient;
3. Adopted basis Newton-Raphson;
4. Newton-Raphson;
5. Powell;
6. Truncated Newton method.

10.3.2 Case Study

Figure 10.9 shows different energy minimisation methods that can be chosen for optimising the input structure. A dialog form for energy minimisation of the structure using steepest descent method is also presented in Fig. 10.9.

In this case study, the BPTI in solvent is chosen as input system, prepared in the laboratory 1. Input/output configuration files can be chosen using the dialog form in Fig. 10.10, which are under the folder *casestudies/lab2/inputs*. In general, *Input Restart File* is selected if the *STATUS* option is checked (see Fig. 10.9); otherwise, it is any default file name (including an empty file). The output files are temporary stored under the folder *scratch/minenergy* for each optimisation method.

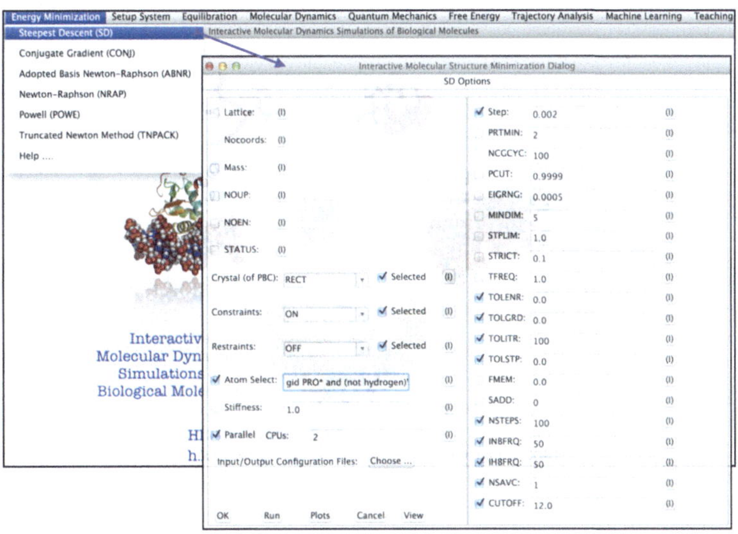

Fig. 10.9 Dialog form for energy minimisation of the structure using steepest descent method

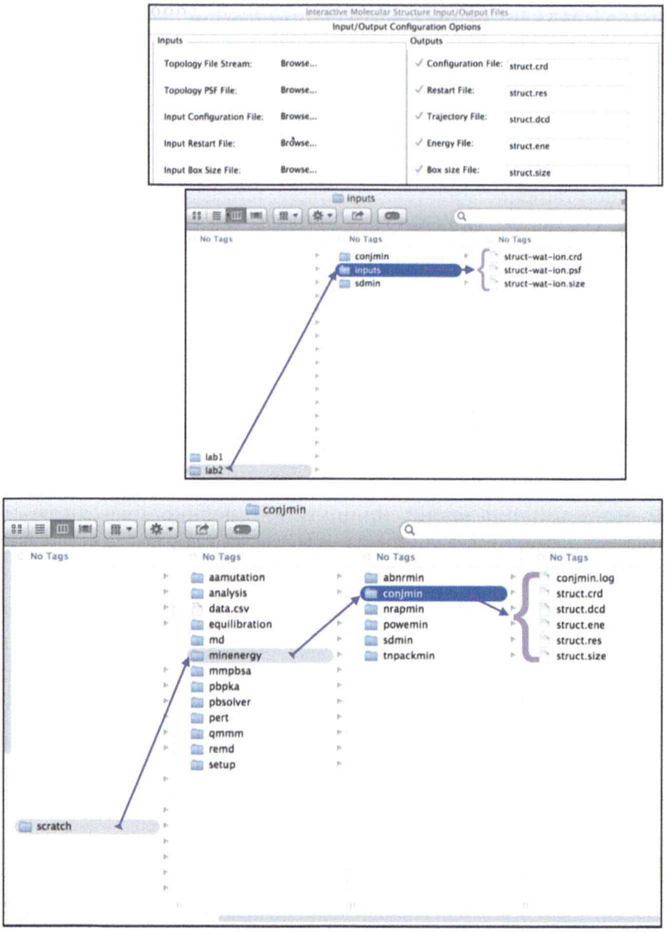

Fig. 10.10 Dialog form for the input/output files

The log file contains the information about the energy components of the system, which can be represented in a tabular or graphical format, as presented in Fig. 10.11.

Note that different constraints or restraints can be applied into the system next to the *Constraints* or *Restraints* options, as depicted in Fig. 10.9. The option *Atom Select* can be used to select the constraint/restraint atoms, such as:

'segid PRO* and (not hydrogen)'

for fixing all heavy atoms of protein at their initial X-ray structure positions; however, the hydrogens of protein and all the atoms of solvent (such as water and ions) are allowed to move freely.

Steps	Energy	Bonds	Angles	Urey-angle	Dihedrals	Impropers	Cmap	Vdw	Electrostatic	ImageVdw	ImageElec
0	-6.7e+04	55.90	156.45	11.54	477.36	6.11	-23	6743.73	-6.4e+04	-30	-2.8e+03
10	-7.4e+04	2890.30	1341.05	11.02	478.35	6.54	-25	7282.04	-7.2e+04	-64	-3.6e+03
20	-7.6e+04	2494.38	1693.17	12.95	480.76	8.93	-27	8413.97	-7.5e+04	-1.1e+02	-3.9e+03
30	-7.8e+04	2693.50	1852.06	11.35	481.94	6.44	-28	9126.24	-7.7e+04	-21	-4.2e+03
40	-7.8e+04	2599.44	1829.85	12.23	481.54	5.38	-29	9538.14	-7.8e+04	81.44	-4.4e+03
50	-7.9e+04	2832.18	1861.31	11.74	481.66	5.76	-30	9746.49	-7.9e+04	141.29	-4.6e+03
60	-8e+04	2711.30	1839.99	11.38	479.03	5.10	-30	10035.00	-8e+04	181.08	-4.7e+03
70	-8e+04	2726.58	1840.27	11.08	474.63	5.41	-31	10246.06	-8e+04	221.94	-4.9e+03
80	-8e+04	2838.99	1854.61	10.76	475.44	5.24	-32	10459.09	-8.1e+04	269.12	-5e+03
90	-8.1e+04	2919.67	1813.14	11.17	476.57	5.65	-32	10579.98	-8.1e+04	317.30	-5.1e+03
100	-8.1e+04	2976.68	1744.70	11.73	477.93	6.00	-32	10707.46	-8.1e+04	347.69	-5.2e+03
100	-8.1e+04	2976.68	1744.70	11.73	477.93	6.00	-32	10707.32	-8.1e+04	347.65	-5.2e+03
110	-8.2e+04	2901.75	1748.12	10.84	474.88	7.34	-32	10674.59	-8.2e+04	328.28	-5.2e+03
120	-8.2e+04	2952.26	1746.53	11.37	473.44	6.11	-32	10772.33	-8.2e+04	330.00	-5.3e+03
130	-8.2e+04	2959.38	1787.38	11.32	473.37	6.37	-32	10876.88	-8.2e+04	348.12	-5.3e+03
140	-8.2e+04	2954.74	1812.92	11.33	473.49	6.69	-32	10916.28	-8.3e+04	361.93	-5.4e+03
150	-8.2e+04	2962.50	1826.42	10.80	473.04	6.45	-33	10978.01	-8.3e+04	362.86	-5.4e+03
160	-8.2e+04	2967.96	1844.89	11.17	472.92	5.61	-34	11029.33	-8.3e+04	379.87	-5.4e+03
170	-8.3e+04	2984.96	1851.64	11.85	472.19	6.79	-35	11072.04	-8.3e+04	399.50	-5.5e+03
180	-8.3e+04	2971.40	1846.15	11.28	472.28	6.73	-36	11100.11	-8.3e+04	410.14	-5.5e+03
190	-8.3e+04	2997.72	1853.15	11.47	471.83	6.96	-36	11112.83	-8.3e+04	405.15	-5.5e+03
200	-8.3e+04	3005.77	1855.65	11.08	471.24	6.67	-36	11147.52	-8.3e+04	407.70	-5.5e+03
200	-8.3e+04	3005.77	1855.65	11.08	471.24	6.67	-36	11147.51	-8.3e+04	407.70	-5.5e+03
210	-8.3e+04	2986.09	1847.15	11.05	470.83	6.74	-36	11163.06	-8.3e+04	410.03	-5.6e+03
220	-8.3e+04	3025.22	1846.25	11.35	470.41	6.90	-36	11223.53	-8.4e+04	419.67	-5.6e+03
230	-8.3e+04	3017.98	1841.80	10.72	470.05	6.23	-35	11279.78	-8.4e+04	425.79	-5.6e+03

220 rows x 11 columns

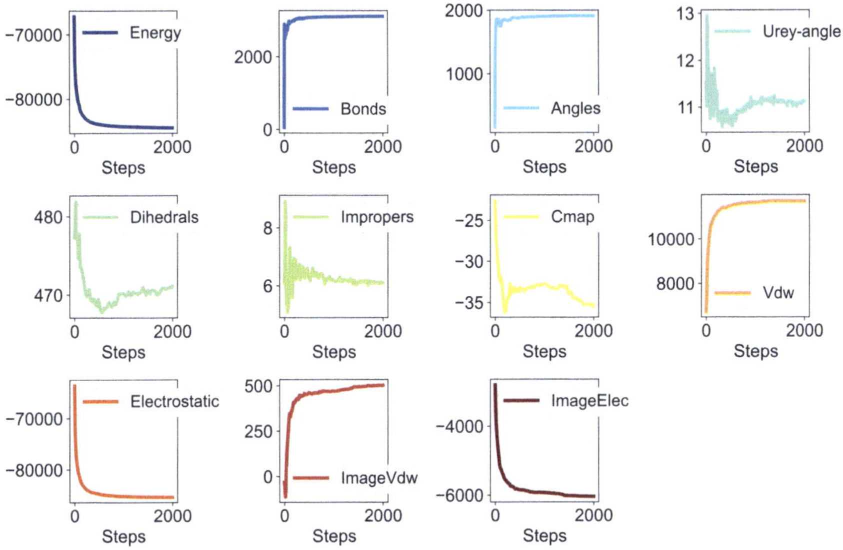

Fig. 10.11 Energy minimisation of the BPTI structure using conjugate gradient method for 2000 steps

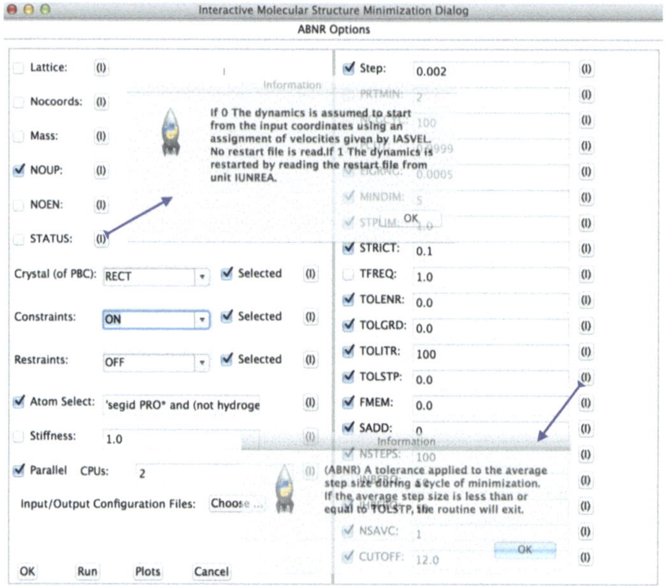

Fig. 10.12 Information about the input parameter

Furthermore, the information about the use of selected parameters in the dialog form can be obtained by clicking on the button *(I)*, as shown in Fig. 10.12. This is a useful information because it indicates when and how to use a parameter.

10.4 Equilibration of System

The equilibration process includes the adjustment of the solvent and solute volume at the standard conditions (temperature and pressure). Usually, the protein structure is refined using the X-ray structure, which is at the temperature much lower than room temperature and in a crystal structure; therefore, the volume of protein may be smaller than its volume in solution at the room temperature and standard pressure of about 1 atm. Furthermore, there may exist *bad contacts* between the water and protein, especially with the first shell of waters around protein.

The equilibration protocol, in general, includes the following procedures: constraints MD simulations, annealing MD simulations, and restraints MD simulations, as depicted in Fig. 10.13.

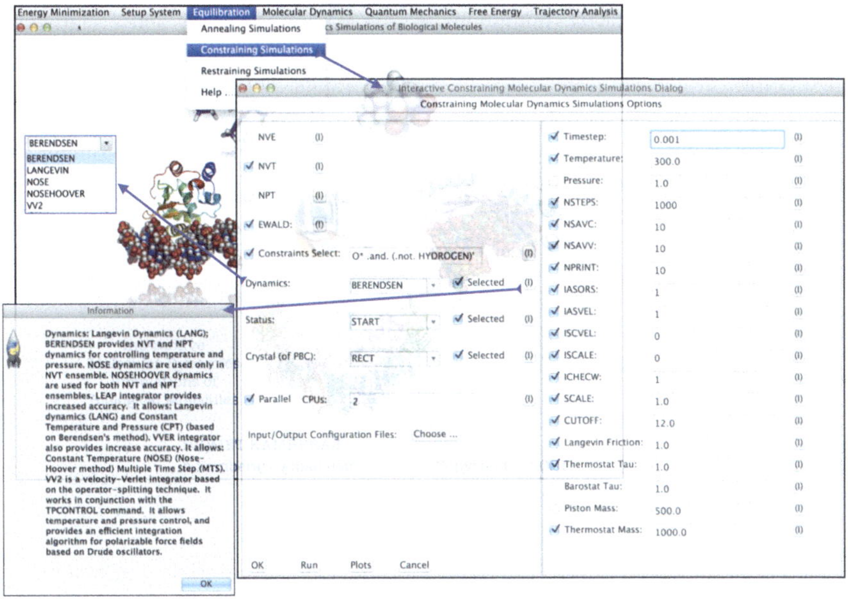

Fig. 10.13 The dialog form for constraints molecular dynamics simulations

10.4.1 Requirements

The following requirements are needed to complete the laboratory work:

- A folder named *lab4*.
- Under that folder, create a new folder and name it *inputs*.
- The output structure files of an energy minimisation run, namely *struct.crd*, *struct.res*, *struct.size*, and the topology file *struct.psf*. All these files should be copied under the folder *lab4/inputs*.

10.4.2 Constraint MD Simulations

After an energy minimisation run, which was described in the previous laboratory, the equilibration usually starts with a constraint MD simulation run. The *constraint* means that the protein atom coordinates are fixed at their initial X-ray crystal structure and hydrogen atoms may be considered fully flexible, along with solvent atoms, such as waters and ions. That is done using the constraint (as shown in Fig. 10.13):

'segid PRO* .and. (.not. HYDROGEN)'

Usually, the constraints MD simulations are run at constant N, V, and T statistical ensemble because it is expected that the solute volume should change from its initial structure, which is disabled due to the constraints. Different dynamics can be chosen (such as BERENDSEN, LANGEVIN, NOSE, NOSEHOOVER, and VV2), as explained in Fig. 10.13. It is recommended to choose BERENDSEN dynamics at this stage because it is less prone of numerical errors.

Exercise 10.1 (Activity case study) Consider solvated BPTI system from Laboratory 2. The input files are under the folder *casestudies/lab4/inputs*. Run constraint MD simulations for 20 ps with a time step between 1.000 and 2.000 fs at the temperature $T = 300$ K. Use BERENDSEN dynamics.

1. Analyse the trajectory of energy components and temperature during the simulation run using the 'Plots' toolbox.
2. Choose another algorithms for performing dynamics. Analyse convergence of the temperature at the target temperature $T = 300$ K. What you can say about the speed of that convergence?
3. Often the speed of the convergence of the temperature to the target temperature depends on the value of 'Thermostat Tau' or 'Thermostat Mass'. Check that speed for different values of, for example, 'Thermostat Tau', using the same dynamics method. What do you observe? Interpret the results based on the theoretical background from the previous chapters.

The structure output files are saved under the folder *scratch/equilibration/cons*, which should be copied into the folder *casestudies/lab4/cons*.

10.4.3 Annealing MD Simulations

In Fig. 10.14, a dialog form for annealing molecular dynamics simulations is given. Three different algorithms are given for controlling the dynamics, such as the Langevin dynamics (LANG) integrator, Leap-Frog (LEAP) integrator, and the velocity-Verlet (VVER) integrator. The step frequency for heating the molecule in increments of TEMINC degrees in the heating portion of a dynamics run. Zero means do no heating. An important parameter is IHTFRQ, which represents the step frequency for heating the molecule in increments of TEMINC degrees in the heating/cooling portion of a dynamics run. Zero means do no heating (or cooling) and a non-zero value is such that if, for example, there are needed N_{incr} from the initial temperature to the final temperature, then the total number of steps for running the dynamics should be

$$\text{NSTEPS} \geq N_{\text{incr}} \times \text{IHTFRQ} \tag{10.25}$$

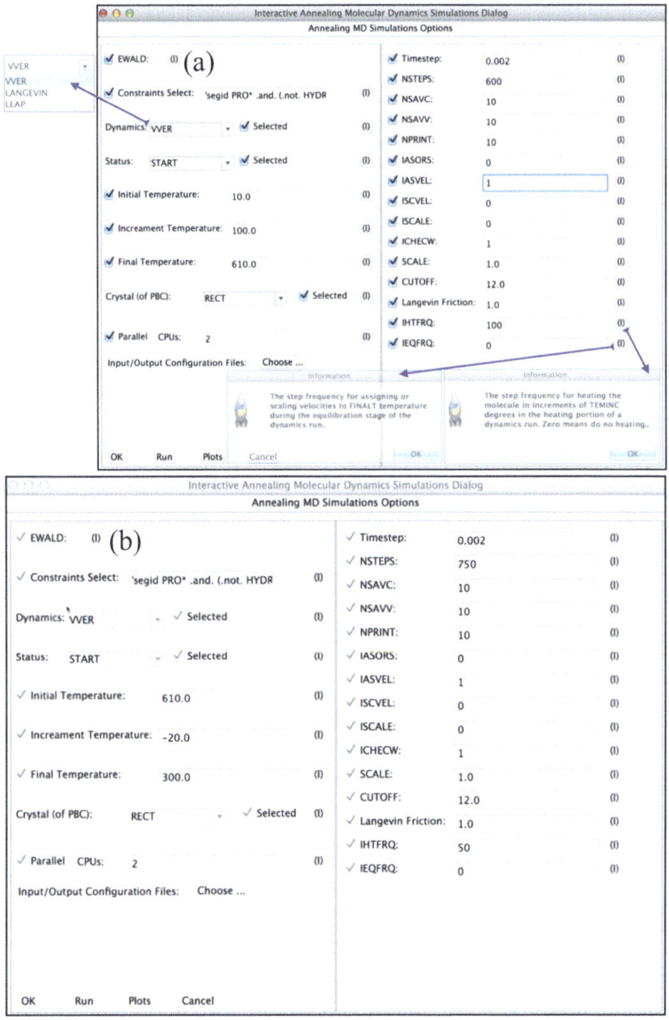

Fig. 10.14 The dialog form for annealing molecular dynamics simulations: **a** Heating protocol and **b** cooling protocol

Note that for cooling the TEMINC is negative. Another interesting parameter is IEQFRQ, which gives the step frequency for assigning or scaling velocities to FINALT temperature during the equilibration stage of the dynamics run; this parameter helps the convergence of the instantaneous temperature to the target temperature.

Figure 10.15a, b shows the convergences of properties of the system in an annealing molecular dynamics simulation run: Fig. 10.15a Cooling protocol and Fig. 10.15b heating protocol.

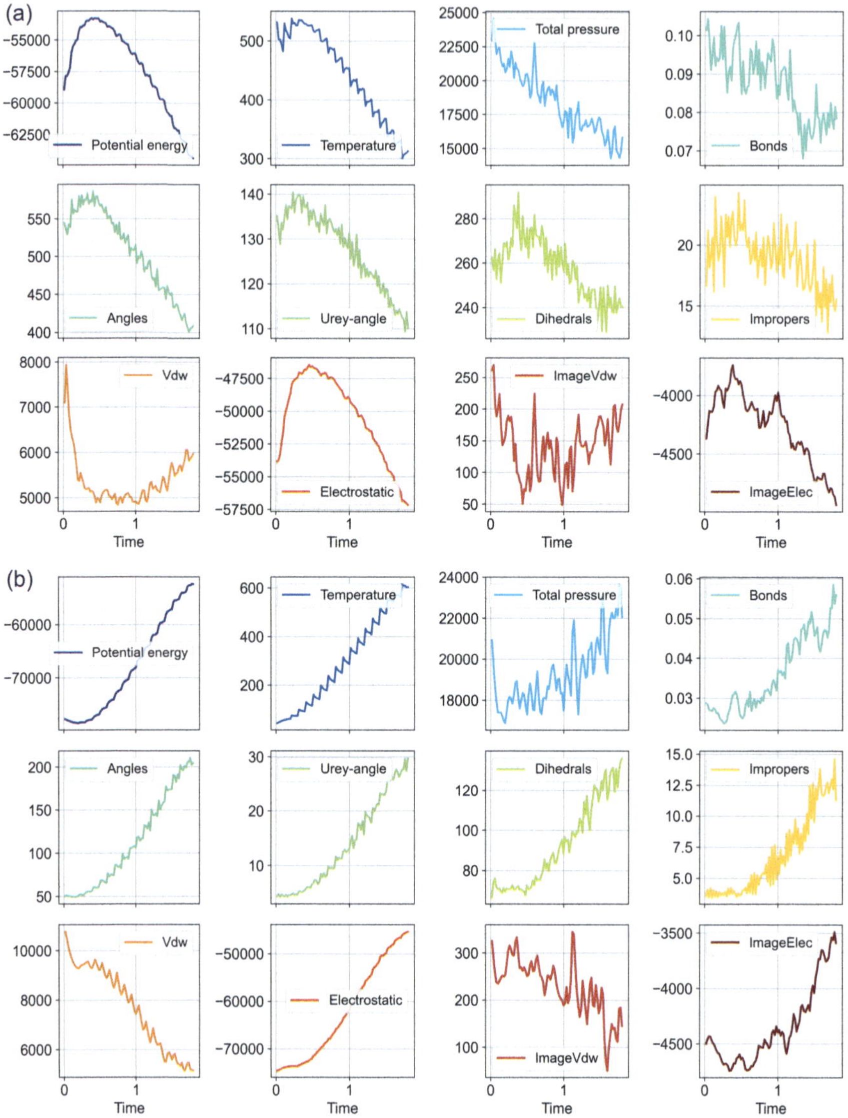

Fig. 10.15 Plots of the convergences of properties of the system in an annealing molecular dynamics simulation run: **a** Cooling protocol and **b** heating protocol. VVER algorithms was chosen to control the dynamics

The structure output files are saved under the folder *scratch/equilibration/anneal*, which should be copied into the folder *casestudies/lab4/anneal*.

Exercise 10.2 (Activity case study) Consider solvated BPTI system. The input files are under the folder *casestudies/lab4/inputs*. Run annealing MD simulations for 50 ps with a time step of 2.000 fs. For heating the temperature varies from $T = 10$ K to $T = 800$ K with a TEMINC of 50 K and for cooling you can vary the temperature from $T = 800$ K to $T = 300$ K with a TEMINC of -25 K. Use VVER dynamics.

1. Analyse the trajectory of energy components and temperature during the simulation run using the 'Plots' toolbox.
2. Choose another algorithms for performing dynamics. Analyse convergence of the temperature at the target temperature. What you can say about the speed of that convergence?
3. Visualise the trajectories during the heating and the cooling using VMD program. What can you say about the mobility of the water molecules and ions in solution?

10.4.4 Restraints MD Simulations

The aim is to learn how to add restraints in a certain type of atoms during the MD simulation runs. The restraints dynamics of the solute to their initial structure by applying a restraint potential of the form:

$$U_{\text{restraint}} = \sum_{i=1}^{N_r} K_i \left(\mathbf{r}_i(t) - \mathbf{r}_i(0)\right)^2 \tag{10.26}$$

where N_r is the number of atoms that are restraint, K_i is an harmonic force constant, $\mathbf{r}_i(t)$ is the position of atom i at instant t, and $\mathbf{r}_i(0)$ its initial position. Usually, K_i can be chosen the same constant value for all atom types; however, K_i can be chosen to be mass dependent constants. Then, K_i can be decreased slowly until the solvent density reaches the value of about 1 g/cm^3, which is its value at the standard conditions; that is, at the temperature of $T = 300$ K and pressure of $p = 1$ atm. Note that the restraint force constant K_i (1 unit = $2 \cdot m_i \cdot$ kcal/mol/Å^2, where m_i is mass of atom i) can be set to any positive value. The force constants may also be obtained from the weight array, in which case the force constant keyword is not read. When using this option, a negative values may be used for some atoms, however, the total weight must be positive. The statistical ensemble is NPT (constant number of atoms N, pressure p and temperature T), and hence the volume of the simulation box is

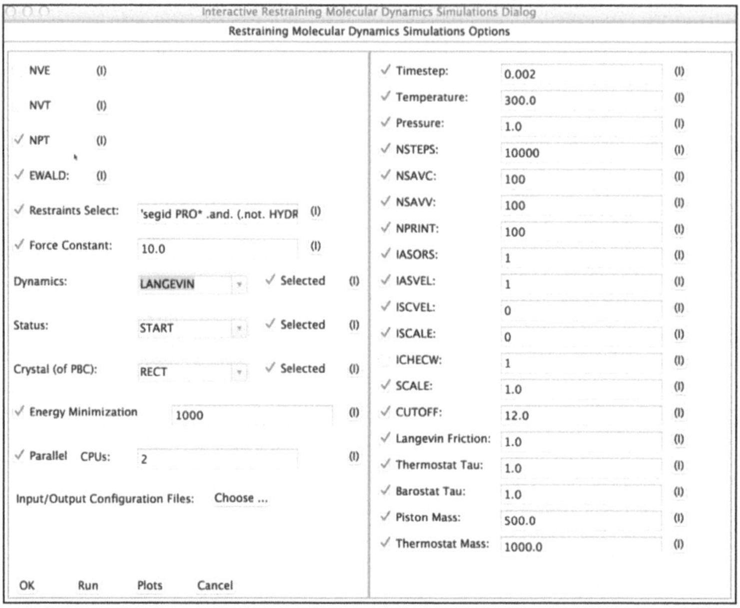

Fig. 10.16 A dialog form for running restraint MD simulations

gently allowed to fluctuate to accommodate for any changes in the volume of the solute (such as protein) from its initial crystal structure. The restraint MD simulation for each run (that is, for a new force constant) should restart from the configuration structure of the previous run (that is, $\mathbf{r}_{\text{new}}(0) = \mathbf{r}_{\text{old}}(t_{\text{last}})$).

Often, some energy minimisation is required before starting the standard dynamics of the system; therefore, *energy minimisation* option, along with the number of steps, can be updated for running restraints energy minimisation using the conjugated gradient method, as shown in Fig. 10.16. The force constant is gradually decreased until the value becomes zero. Typically, the total simulation time of restraint MD simulations is between 100 ps and 1 ns, depending on the size of system.

The structure output files are saved under the folder *scratch/equilibration/restraints*, which should be copied into the folder *casestudies/lab4/restraints* for later use.

Usually, to finish the equilibration stage of the protocol, the standard MD simulations of free constraints/restraints are continued for about 1 ns at constant temperature of $T = 300$ K and pressure $p = 1$ atm using NPT statistical ensemble.

10.5 Mini Project

Exercise 10.3 (Activity case study) Consider solvated BPTI system. The input topology file is under the folder *casestudies/lab4/inputs* and other files are under the folder *casestudies/lab4/anneal*. Run restraint MD simulations using 20 ps with a time step of 2.000 fs for each run. Use NPT statistical ensemble with temperature $T = 300$ K and pressure $p = 1$ atm. Try to reduce force constant from a value of 10 kcal/mol/Å^2 to zero in steps of 0.5-1 kcal/mol/Å^2. Choose your own algorithm for dynamics.

1. Analyse the trajectory of energy components, pressure and temperature during the simulation run using the 'Plots' toolbox.
2. Visualise the trajectories during the restraint MD simulations using VMD program. What can you say about the mobility of the protein?
3. Analyse the volume of the simulation box. What do you observe?
4. Continue the equilibration stage using standard MD simulations (free constraints/restraints) for about 1 ns at constant temperature of $T = 300$ K and pressure $p = 1$ atm using NPT statistical ensemble. Calculate the root mean square deviations from the X-ray crystal structure of protein (C_α and backbone atoms). What do you observe?

10.6 Summary

We introduced the setup protocols for preparing a molecular biological system simulation in explicit solvent. The energy minimisation methods are discussed in details.

Furthermore, we presented a complete protocol of the equilibration of solvent for a protein system in explicit solvent. Moreover, the case studies are shown for each step of equilibration and a mini-project is written for testing the knowledge gained.

References

M.P. Allen, D.J. Tildesley, *Computer Simulation of Liquids*. (Oxford University Press, 1989)

M. Becker, M. Karplus, The topology of multidimensional potential energy surfaces: theory and application to peptide structure and kinetics. J. Chem. Phys. **106**, 1495 (1997)

B.R. Brooks, C.L. Brooks, A.D. MacKerell, L. Nilsson, R.J. Petrella, B. Roux, Y. Won, G. Archontis, C. Bartels, S. Boresch, A. Caflisch, L. Caves, Q. Cui, A.R. Dinner, M. Feig, S. Fischer, J. Gao, M. Hodoscek, W. Im, K. Kuczera, T. Lazaridis, J. Ma, V. Ovchinnikov, E. Paci, R.W. Pastor, C.B. Post, J.Z. Pu, M. Schaefer, B. Tidor, R.M. Venable, H.L. Woodcock, X. Wu, W. Yang, D.M. York, M. Karplus, Charmm: the bimolecular simulation program. J. Comput. Chem. **30**(10), 1545–1614 (2009)

U. Doshi, D. Hamelberg, Extracting realistic kinetics of rare activated processes from accelerated molecular dynamics using Kramer's theory. J. Chem. Theory Comput. **7**, 575–581 (2011)

R. Fletcher, C.M. Reeves, Function minimisation by conjugate gradients. Comput. J. **7**(2), 149–154 (1964)

R.M. Hestenes, E. Stiefel, Methods of conjugate gradients for solving linear systems. J. Res. Natl. Bureau Stand. **49**(6), 409 (1952)

W. Humphrey, A. Dalke, K. Schulten, VMD-visual molecular dynamics. J. Mol. Graph. **14**, 33–38 (1996)

H. Kamberaj, Q-gaussian swarm quantum particle intelligence on predicting global minimum of potential energy function. Appl. Math. Comput. **229**, 94–106 (2014)

H. Kamberaj, A. van der Vaart, Multiple scaling replica exchange for the conformational sampling of biomolecules in explicit water. J. Chem. Phys. **127**, 234102–234109 (2007)

A.R. Leach, *Molecular Modelling: Principles and Applications*, 2nd edn. (Prentice Hall, Pearson Education Limited, Edingburgh Gate, 2001)

E. Polak, G. Ribiére, Note sur la convergence de méthodes de directions conjuguées. Revue Française d'Automatique, Informatique, Recherche Opérationnelle **3**(1), 35–43 (1969)

D.J. Wales, Perspective: Insight into reaction coordinates and dynamics from the potential energy landscape. J. Chem. Phys. **142**, 130901–12 (2015)

A. Wlodawer, J. Walter, R. Huber, L. Sjolin, Structure of bovine pancreatic trypsin inhibitor. Results of joint neutron and X-ray refinement of crystal form II. J. Mol. Biol. **180**, 301–329 (1984)

Y. Xin, U. Doshi, D. Hamelberg, Examining the limits of time reweighing and Kramer's rate theory to obtain correct kinetics from accelerated molecular dynamics. J. Chem. Phys. **132**, 224101 (2010)

Chapter 11
Normal Mode Analysis

The chapter aims to introduce the normal mode analysis of a protein.

This chapter introduces the normal mode analysis of a protein, which is associated with the energy minimisation.

11.1 Theoretical Background

The basis for determining the force constants, inter-atomic distances, and bond dissociation energies of molecular structures are molecular vibrational spectra of small molecules (Lifson and Warshel 1968). The vibrational bond motions represent the small deviations from the equilibrium states. In particular, all possible vibrational motions of a molecule can be characterised as a superposition of different basic types of vibration (called *normal modes*). For a molecule of N atoms, there are $3N$ degrees of freedom to describe the motion of atoms in a three-dimensional space related to atomic coordinates. Omitting the three translational and three rotational degrees of freedom of the molecule as a whole, then there are $3N - 6$ normal modes necessary to describe the internal motion of molecules.

Experimental methods used to detect the vibrational energy levels of a molecule include the spectroscopic techniques, such as the vibrational absorption of infrared radiation (IR) and Raman scattering. IR spectroscopy is a powerful technique that captures information on the transitions between vibrational quantum states since these transitions lead to absorption and emission of infrared radiation. IR wavelength is in the range 1–100 μm. IR transitions can occur if there exists a change in the dipole moment of the molecule during the transition. On the other hand, Raman spectroscopy, which also captures the transitions between vibrational levels, occurs when the polarisability of the molecular system changes during the transition.

H. Kamberaj, *Computer Simulations in Molecular Biology*, Scientific Computation,
https://doi.org/10.1007/978-3-031-34839-6_11

These two techniques are considered complementary methods, since some transitions that have a change in the dipole moment absorb light, and others have a change in polarisability and scatter light. For small symmetric molecules, the observed transitions are complementary, whereas, for large and asymmetric molecules, both, Raman and IR spectra are mostly the same.

The units of normal modes are *hertz* (Hz), that is, the inverse of seconds. However, often they are also reported in *wave-numbers*, that is, the number of waves per centimetre. Higher the frequency, more energetically expensive is the deformation along that particular mode. For example, bond-stretching modes have, in general, higher frequencies than angle-bending modes, which in turn have higher frequencies than torsion angle modes.

In the following, the general theory of normal modes is described based on the literature (Kamberaj 2020). The general theory of the normal modes has been described at a great detail in Goldstein (2002).

11.1.1 General Theory of Normal Modes

In the molecular mechanics or quantum mechanics energy functions, the global minimum energy corresponds to a stable state that the system can not escape, or hypothetically a *frozen* state at temperature of 0 K (Leach 2001). On the other hand, the experimental results correspond with molecules being at some finite temperature T in which molecules participate some translational, rotational, and vibrational motion, and hence the total internal energy can be decomposed as follows:

$$U(T) = U_{\text{trans}}(T) + U_{\text{rot}}(T) + U_{\text{vib}}(T) + U_{\text{vib}}(0) \tag{11.1}$$

In Table 11.1, we show the thermodynamic properties of ten compounds calculated using quantum mechanics at the Hartree-Fock level. The units of energy are Hartree. The pressure was 1 atm and the temperature was 298.15 K. Assuming that all the translational and rotational modes are fully accessible, based on the equipartition theorem:

$$U_{\text{trans}}(T) = U_{\text{rot}}(T) = \frac{d}{2}k_B T = 1.42 \times 10^{-3}\ \text{Eh} \tag{11.2}$$

where k_B is Boltzmann's constant and d is the dimensionality (that is, for a typical molecule $d = 3$ and for a linear molecule $d = 2$). The vibrational term at the room temperature (in which vibrational energy levels are partially excited) is calculated as difference in the vibrational enthalpy at $T > 0$ K and $T = 0$ K as:

$$U_{\mathrm{vib}}(T) = \sum_{\text{all modes}} \left(\frac{\hbar\omega_i}{2} + \frac{\hbar\omega_i}{\exp(\hbar\omega_i/k_B T) + 1} \right) \tag{11.3}$$
$$= U_{\mathrm{vib}}(0) + \sum_{\text{all modes}} \frac{\hbar\omega_i}{\exp(\hbar\omega_i/k_B T) + 1}$$

Here, the summation index i runs over all normal modes of vibrational motion, ω_i is the angular frequency of the mode i, and $\hbar = h/2\pi$ where $h = 6.63 \times 10^{-34}\ J \cdot s$ is the Planck's constant. Thus, even at $T = 0$ K (the so-called zero-point energy), $U_{\mathrm{vib}}(0)$ can be significantly different from zero, as indicated in Table 11.1. The other thermodynamic properties (such as entropy and free energy) are calculated using vibrational frequencies using the statistical mechanics. For example, the vibrational entropy is given as:

$$S_{\mathrm{vib}}(T) = k_B \sum_{\text{all modes}} \left(\frac{\hbar\omega_i/k_B T}{\exp(\hbar\omega_i/k_B T) - 1} - \ln\left(1 - \exp(-\hbar\omega_i/k_B T)\right) \right) \tag{11.4}$$

Note that normal modes are useful because they enable analysis of collective motions of the atoms in a macromolecular system.

Based on the description in Ref. Goldstein (2002) (and a more recently work Kamberaj 2020), consider a conservative system with a potential energy a function of coordinates only, the *equilibrium* state of the system is characterised by

$$Q_i = -\left(\frac{\partial U}{\partial x_i}\right)_{x_i = x_i^*}, \quad i = 1, 2, \ldots, g \tag{11.5}$$

where $x_1, x_2, \ldots, x_g$ are the generalised coordinates of the system, which do not involve the time explicitly. Eq. (11.5) indicates that potential energy has an extremum at the equilibrium configuration of the system, $x_1^*, x_2^*, \ldots, x_g^*$. If the system is initially at the equilibrium position, with zero initial velocities ($\dot{x}_i$), then the system will continue in equilibrium indefinitely. An equilibrium position is classified as *stable* if a small disturbance of the system from equilibrium produces only a small bounded motion about the equilibrium position. The equilibrium is *unstable* if a small disturbance will eventually result in an unbounded motion. For example, a pendulum at rest is in stable equilibrium, but the egg that is standing on an end is an unstable equilibrium (Goldstein 2002).

Following the discussion, when the extremum of U is a minimum, the equilibrium is a stable one. That is, if the system is at equilibrium and it has deviated from this position, then this disturbance will produce an increase in energy by an amount of dE above the equilibrium value U. Since the system is conservative, based on the conservation law of energy, the velocities must then decrease and with time become zero, which is a bound motion. On the other hand, if U decreases due to a deviation from equilibrium, the velocities (and so the kinetic energy) increase indefinitely, resulting in an unbounded motion.

Table 11.1 The thermodynamic properties of ten compounds calculated using quantum mechanics Hartree-Fock method at $T = 298.15$ K. The units of energy are Hartree (Eh) and for entropy the units are Hartree/Kelvin (Eh/K). H denotes enthalpy, U denotes internal energy, S denotes the entropy, and G denotes the Gibbs free energy. $U_{\text{vib}}(0)$ is vibrational energy at $T = 0K$

Compound	1	2	3	4	5	6	7	8	9	10
Name	methyl propanoate	1,3-dichloro-benzene	methanol	2-chloro-phenol	but-1-ene	2-chloro-propane	1,1,1,2,2,2-hexa-chloro-ethane	1,1,2-trichloro-ethane	1,2,3,4,5-penta-chloro-6-nitro-benzene	4-methyl-pyridine
$U_{\text{vib}}(0)$	–305.864	–1148.495	–115.032	–764.448	–156.102	–577.166	–2832.537	–1455.910	–2728.544	–285.730
U_{elec}	–305.864	–1148.495	–115.032	–764.448	–156.102	–577.166	–2832.537	–1455.910	–2728.544	–285.730
H_{elec}	–305.864	–1148.495	–115.032	–764.448	–156.102	–577.166	–2832.537	–1455.910	–2728.544	–285.730
S_{trans}	6.27E-05	6.51E-05	5.79E-05	6.45E-05	6.05E-05	6.21E-05	6.74E-05	6.47E-05	6.84E-05	6.30E-05
U_{trans}	1.42E-03	1.42E-03	1.42E-03	1.42E-03	1.42E-03	1.42E-03	1.42E-03	1.42E-03	1.42E-03	1.42E-03
H_{trans}	2.36E-03	2.36E-03	2.36E-03	2.36E-03	2.36E-03	2.36E-03	2.36E-03	2.36E-03	2.36E-03	2.36E-03
S_{rot}	4.24E-05	4.73E-05	3.02E-05	4.57E-05	3.87E-05	4.05E-05	4.96E-05	4.51E-05	5.33E-05	4.28E-05
U_{rot}	1.42E-03	1.42E-03	1.42E-03	1.42E-03	1.42E-03	1.42E-03	1.42E-03	1.42E-03	1.42E-03	1.42E-03
H_{rot}	1.42E-03	1.42E-03	1.42E-03	1.42E-03	1.42E-03	1.42E-03	1.42E-03	1.42E-03	1.42E-03	1.42E-03
S_{vib}	1.84E-05	1.73E-05	7.60E-07	1.54E-05	9.79E-06	8.72E-06	3.45E-05	1.44E-05	4.71E-05	1.13E-05
U_{vib}	1.30E-01	9.01E-02	5.52E-02	1.05E-01	1.18E-01	1.03E-01	2.56E-02	5.49E-02	6.65E-02	1.26E-01
H_{vib}	1.30E-01	9.01E-02	5.52E-02	1.05E-01	1.18E-01	1.03E-01	2.56E-02	5.49E-02	6.65E-02	1.26E-01
G_{elec}	–305.864	–1148.495	–115.032	–764.448	–156.102	–577.166	–2832.537	–1455.910	–2728.544	–285.730
G_{trans}	–1.63E-02	–1.71E-02	–1.49E-02	–1.69E-02	–1.57E-02	–1.62E-02	–1.77E-02	–1.69E-02	–1.80E-02	–1.64E-02
G_{rot}	–1.12E-02	–1.27E-02	–7.60E-03	–1.22E-02	–1.01E-02	–1.07E-02	–1.34E-02	–1.20E-02	–1.45E-02	–1.14E-02
G_{vib}	1.25E-01	8.50E-02	5.50E-02	1.00E-01	1.15E-01	1.00E-01	1.53E-02	5.05E-02	5.25E-02	1.23E-01
S_{tot}	1.24E-04	1.30E-04	8.89E-05	1.26E-04	1.09E-04	1.11E-04	1.51E-04	1.24E-04	1.69E-04	1.17E-04
$U_{\text{vib}}(0)$	–305.737	–1148.409	–114.977	–764.346	–155.986	–577.064	–2832.517	–1455.858	–2728.486	–285.606
U_{tot}	–305.731	–1148.402	–114.974	–764.340	–155.981	–577.060	–2832.508	–1455.852	–2728.475	–285.601
H_{tot}	–305.730	–1148.401	–114.973	–764.340	–155.981	–577.059	–2832.507	–1455.851	–2728.474	–285.600
G_{tot}	–305.767	–1148.440	–115.000	–764.377	–156.013	–577.092	–2832.552	–1455.888	–2728.524	–285.635

Now, we consider the motion of the system close to the configuration of stable equilibrium. Because the deviations from the equilibrium are too small, we can expand the potential energy function in the Taylor series about the equilibrium and keeping only the lowest-order energy terms as given by Eq. (11.3). Note that terms linear in η_i vanish automatically due to equilibrium conditions. The first term in the series is the potential energy of the equilibrium position, and by shifting the arbitrary zero of potential to be the equilibrium minimum potential energy value, then

$$U(x_1, x_2, \ldots, x_g) = \frac{1}{2}\left(\frac{\partial^2 U}{\partial x_i \partial x_j}\right)_{x_i=x_i^*, x_j=x_j^*} \eta_i \eta_j = \frac{1}{2} H_{ij} \eta_i \eta_j \tag{11.6}$$

where the second derivatives of U have been denoted by H_{ij} (the so-called Hessian matrix) depend only on the equilibrium values of the x_i. Based on the definition, it can be seen that H_{ij} are symmetric, $H_{ij} = H_{ji}$.

The kinetic energy is written as the following summation:

$$T = \frac{1}{2}\sum_{i,j} m_{ij}\dot{x}_i\dot{x}_j = \frac{1}{2}\sum_{i,j} m_{ij}\dot{\eta}_i\dot{\eta}_j \tag{11.7}$$

where the mass coefficients m_{ij} are in general functions of the coordinates x_k, but they may be expanded in a Taylor series about the equilibrium configuration:

$$\begin{aligned} m_{ij}(x_1, x_2, \ldots, x_g) &= m_{ij}(x_1^*, x_2^* \ldots, x_g^*) \\ &+ \sum_{k=1}^{g}\left(\frac{\partial m_{ij}}{\partial x_k}\right)_{x_i=x_i^*} \eta_k + \cdots \end{aligned} \tag{11.8}$$

Denoting the values of m_{ij} at equilibrium as T_{ij}, we get an alternative expression as:

$$T = \frac{1}{2}\sum_{i,j} T_{ij}\dot{\eta}_i\dot{\eta}_j \tag{11.9}$$

where T_{ij} is symmetric, $T_{ij} = T_{ji}$.

The Lagrangian of the system is:

$$L = \frac{1}{2}\sum_{i,j} T_{ij}\dot{\eta}_i\dot{\eta}_j - \frac{1}{2}\sum_{i,j} H_{ij}\eta_i\eta_j \tag{11.10}$$

Applying the Lagrangian's equations of motion, we get the following equations of motion:

$$\sum_{j=1}^{g}\left(T_{ij}\ddot{\eta}_j + H_{ij}\eta_j\right) = 0 \tag{11.11}$$

for $i = 1, \ldots, g$.

Equation (11.11) represents a linear differential equation of motion with constant coefficients. The real part of the general form of the solution is given by

$$\eta_j = \sum_{k=1}^{g} A_{jk}\alpha_k \cos(\omega_k t + \delta_k) \tag{11.12}$$

where $A_{jk}\alpha_k$ is the real amplitude of the oscillation for each coordinate η_j. Then, the second derivative of Eq. (11.12) with respect to time is

$$\ddot{\eta}_j = -\sum_{k=1}^{g} A_{jk}\alpha_k\omega_k^2 \cos(\omega_k t + \delta_k) \tag{11.13}$$

Substituting this expression and the expression given by Eq. (11.12) into Eq. (11.11), gives (Kamberaj 2020)

$$\sum_{k=1}^{g}\left[\sum_{j=1}^{g} -T_{ij}A_{jk}\omega_k^2 + \sum_{j=1}^{g} H_{ij}A_{jk}\right]\alpha_k \cos(\omega_k t + \delta_k) = 0 \tag{11.14}$$

This equation holds for all time t if

$$\left(\sum_{j=1}^{g} T_{ij}A_{jk}\right)\omega_k^2 = \sum_{j=1}^{g} H_{ij}A_{jk} \tag{11.15}$$

for $k = 1, 2, \ldots, g$. By introducing the matrices, $\mathbf{T}$ with elements T_{ij}, $\mathbf{H}$ with elements H_{ij}, $\mathbf{A}$ with elements A_{jk} and the diagonal matrix $\boldsymbol{\Lambda}$ with diagonal elements $\Lambda_{kk} = \omega_k^2$, we can write Eq. (11.15) in a matrix form as

$$\mathbf{TA\Lambda} = \mathbf{HA} \tag{11.16}$$

Following Levitt et al. (1985), before we determine $\boldsymbol{\Lambda}$ and $\mathbf{A}$ by solving Eq. (11.16), a normalisation condition is established. For that Q_k is defined as

$$Q_k = \alpha_k \cos(\omega_k t + \delta_k) \tag{11.17}$$

Then, Eq. (11.11) is written as

$$\eta_j = \sum_{k=1}^{g} A_{jk}Q_k \equiv \mathbf{AQ} \tag{11.18}$$

where $\mathbf{Q}$ is a vector with elements Q_k. Eq. (11.18) relates the generalised coordinates η and the coordinates $\mathbf{Q}$. Then, the potential energy and the kinetic energy can be written as

$$U = \frac{1}{2}\mathbf{Q}^T\mathbf{A}^T\mathbf{HAQ} \tag{11.19}$$

$$T = \frac{1}{2}\dot{\mathbf{Q}}^T\mathbf{A}^T\mathbf{TA}\dot{\mathbf{Q}}$$

By choosing the normalisation condition as (Levitt et al. 1985)

$$\mathbf{A}^T\mathbf{TA} = \mathbf{1} \tag{11.20}$$

then, the kinetic energy can be written as

$$T = \sum_{k=1}^{g} \frac{1}{2}\dot{Q}_k^2 \equiv \frac{1}{2}\dot{\mathbf{Q}}^T\dot{\mathbf{Q}} \tag{11.21}$$

If we multiply both sides of Eq. (11.16) by $\mathbf{A}^T$ and using the normalisation condition, Eq. (11.20), we get

$$\boldsymbol{\Lambda} = \mathbf{A}^T\mathbf{HA} = \mathbf{A}^T\mathbf{TA}\boldsymbol{\Lambda} \tag{11.22}$$

Thus, the potential energy can be written as

$$U = \frac{1}{2}\mathbf{Q}^T\boldsymbol{\Lambda}\mathbf{Q} = \frac{1}{2}\sum_{k=1}^{g} \Lambda_{kk}Q_k^2 = \frac{1}{2}\sum_{k=1}^{g} \omega_k^2 Q_k^2 \tag{11.23}$$

It can be seen that both potential and kinetic energy are simple sums of squares of coordinates Q_k and its time derivative $\dot{x}_k$, hence the coordinates Q_k are normal coordinates and mass scaled one, which is indicated by the form of the kinetic energy.

11.1.2 Dynamical Behaviour of System

We can search for solution of Eq. (11.16) using the normalisation condition, Eq. (11.20), using standard methods. Then, the dynamics of the system are characterised as (Kamberaj 2020)

$$x_j = x_j^* + \sum_{k=1}^{g} A_{jk}\alpha_k \cos(\omega_k t + \delta_k) \tag{11.24}$$

where δ_k is the phase of the kth normal mode of motion, and ω_k is an angular frequency which is determined as: $\omega_k = \sqrt{\Lambda_{kk}}$. In general, ω_k, α_k and δ_k will depend on the initial positions and velocities at $t = 0$.

First, we determine $\mathbf{x}^*$, which represents the coordinates at the thermodynamic equilibrium state, by applying Eq. (11.25). The thermodynamic equilibrium state is the configuration for which the potential energy of the system has a minimum value, which can be determined using numerical methods for minimising general functions (Andricioaei and Straub 1996, 1998).

As it has been suggested elsewhere Levitt et al. (1985), the derivative of the potential energy with respect to generalised coordinates, $\partial U/\partial x_j$, should be calculated analytically, then the elements H_{ij} of the matrix **H** are calculated numerically as

$$H_{ij} = \frac{1}{\epsilon}\left[\left(\frac{\partial U}{\partial x_i}\right)_{x_j=x_j^*+\epsilon} - \left(\frac{\partial U}{\partial x_i}\right)_{x_j=x_j^*}\right] \tag{11.25}$$

For calculation of the matrix **T**, one can start with expression of the kinetic energy in terms of the time derivative of the Cartesian coordinates: $T = (1/2)\sum_{i=1}^{N} m_i \dot{\mathbf{r}}_i^2$. For small changes of the Cartesian coordinates, $\delta\mathbf{r}_i$ can be expressed in terms of the small changes on the generalised coordinates δx_i as

$$\delta\mathbf{r}_i = \sum_{k=1}^{g} \frac{\partial \mathbf{r}_i}{\partial x_k}\delta x_k \tag{11.26}$$

Assuming that these small changes occur in time interval δt, then

$$\dot{\mathbf{r}}_i = \sum_{k=1}^{g} \frac{\partial \mathbf{r}_i}{\partial x_k}\frac{\delta x_k}{\delta t} = \sum_{k=1}^{g} \frac{\partial \mathbf{r}_i}{\partial x_k}\dot{x}_k \tag{11.27}$$

Thus, the kinetic energy can be written as

$$\begin{aligned} T &= \frac{1}{2}\sum_{i=1}^{N} m_i \sum_{k=1}^{g} \frac{\partial \mathbf{r}_i}{\partial x_k}\dot{x}_k \sum_{j=1}^{g} \frac{\partial \mathbf{r}_i}{\partial x_j}\dot{x}_j \\ &= \frac{1}{2}\sum_{k=1}^{g}\sum_{j=1}^{g}\left(\sum_{i=1}^{N} m_i \frac{\partial \mathbf{r}_i}{\partial x_k}\cdot\frac{\partial \mathbf{r}_i}{\partial x_j}\right)\dot{x}_k\dot{x}_j \\ &= \frac{1}{2}\sum_{k=1}^{g}\sum_{j=1}^{g} T_{jk}\dot{x}_k\dot{x}_j \end{aligned} \tag{11.28}$$

where T_{jk} are given as:

$$T_{jk} = \sum_{i=1}^{N} m_i \frac{\partial \mathbf{r}_i}{\partial x_k}\cdot\frac{\partial \mathbf{r}_i}{\partial x_j} \tag{11.29}$$

Since the kinetic energy should include the overall translational and rotational kinetic energy, when calculating $\partial\mathbf{r}_i/\partial x_k$, the coordinate $\mathbf{r}_i$ must be calculated before the overall translational and rotational degrees of freedom are removed from the system.

The algorithm for calculation of normal modes is described in Kamberaj (2020). Here, the implementation in CHARMM program (Brooks et al. 2009) is used.

11.1.3 Time Averages of Properties

The trajectory of every atomic motion depends on the initial conditions, positions, and velocities, which determine the amplitude α_k and phase δ_k of each normal mode. Time averaged properties along the trajectory depend only on the amplitudes α_k and, thus, are of more general importance. For example, let Δp_i be some property of the system along the ith degree of freedom that is a linear function of the change in the generalised coordinate Δx_k:

$$\Delta p_i = \sum_{j=1}^{g} P_{ij} \Delta x_j \tag{11.30}$$

Then, the correlation coefficients of that property between two different degrees of freedom of the system i and j is

$$\begin{aligned} \langle \Delta p_i(\tau) \Delta p_j(\tau) \rangle &= \sum_{k=1}^{g} \sum_{m=1}^{g} P_{ik} P_{jm} \langle \Delta x_k(\tau) \Delta x_m(\tau) \rangle \\ &= \frac{1}{2} \sum_{n=1}^{g} \sum_{k=1}^{g} \sum_{m=1}^{g} P_{ik} P_{jm} A_{kn} A_{mn} \alpha_n^2 \\ &= \frac{1}{2} \sum_{n=1}^{g} P'_{in} P'_{jn} \alpha_n^2 \end{aligned} \tag{11.31}$$

where

$$\begin{aligned} P'_{in} &= \sum_{k=1}^{g} P_{ik} A_{kn} \\ P'_{jn} &= \sum_{m=1}^{g} P_{jm} A_{mn} \end{aligned} \tag{11.32}$$

11.1.4 Thermal Fluctuations

The amplitude α_k for every normal mode k depends on the temperature T (Levitt et al. 1985). The average potential energy of every normal mode is

$$
\begin{aligned}
\langle U \rangle_k &= \frac{1}{2}\omega_k^2 \langle Q_k^2(\tau) \rangle \\
&= \frac{1}{2}\omega_k^2 \alpha_k^2 \langle \cos^2(\omega_k \tau + \delta_k) \rangle \\
&= \frac{1}{4}\omega_k^2 \alpha_k^2 \\
&= \frac{1}{2} k_B T
\end{aligned}
$$

for classical dynamics. Thus,

$$
\langle Q_k^2(\tau) \rangle = \frac{k_B T}{\omega_k^2} \tag{11.33}
$$

where $\langle Q_k^2(\tau) \rangle$ is the classical mean square fluctuation of Q_k. Moreover,

$$
\alpha_k = \sqrt{\frac{2k_B T}{\omega_k^2}} \tag{11.34}
$$

Hence, since all the time averaged properties depend on α, they also depend on the temperature T.

11.1.5 Calculation of Normal Modes

The frequencies of the normal modes, along the displacement of atoms are calculated using the molecular mechanics force fields using the Hessian matrix **H**. That is done after an appropriate energy minimisation of the protein, which determines a local state point of the system, and hence **H** can be calculated. The Hessian matrix is converted into an equivalent force-constant matrix of the mass-weighted-coordinates (Leach 2001):

$$
\mathbf{F} = \mathbf{M}^{-1/2}\mathbf{H}\mathbf{M}^{1/2} \tag{11.35}
$$

where **M** is a $g \times g$ diagonal matrix of atomic masses. Next, the secular equation is solved to determine the eigenvalues and eigenvectors of matrix **F**:

$$
\mid \mathbf{F} - \mathbf{I} \mid = 0 \tag{11.36}
$$

For that, a standard matrix diagonalisation procedure is taken. If the Hessian matrix is expressed in terms of the Cartesian coordinates, then six of its eigenvalues must be zero since they correspond to overall translational and rotational motion of the entire system. The frequency ω_i for each mode i is calculated from the corresponding eigenvalue λ_i as

$$
\omega_i = \sqrt{\lambda_i}. \tag{11.37}
$$

11.2 Case Study

The normal modes are calculated for a single protein in a vacuum. The starting configuration of the protein is obtained using an energy minimisation of the X-ray structure with respect to Cartesian coordinates, including the hydrogen atoms. Two protein X-ray structures were studied, the Fc fragment of Protein IgG and C2 fragment of streptococcal protein G (C2) (Sauer-Eriksson et al. 1995). A graph of the three-dimensional structures of C2 (left) and Fc (right) fragments, respectively, of protein G and IgG, is shown in Fig. 11.1.

For the minimisation of each structure, 5000 steps of Conjugated Gradient, followed by 10000 steps of the Adopted Basis Newton-Raphson method, were used, as implemented in the CHARMM program (Brooks et al. 2009). The force field is CHARMM22 (MacKerell et al. 1998). No cutoffs were used during the calculations of non-bounded interactions. All the degrees of freedom are included in calculations.

In Fig. 11.2, we show the spectra of the distribution of the wave-numbers (k) for the dihedral angles, bending angles, and stretching bonds based on the force field constants in CHARMM22 (MacKerell et al. 1998):

$$k = \frac{\omega}{c} \tag{11.38}$$

where c is the speed of light in vacuum ($c = 3 \times 10^8$ m/s) and ω is the angular frequency:

$$\omega = \sqrt{\frac{K}{\mu}} \tag{11.39}$$

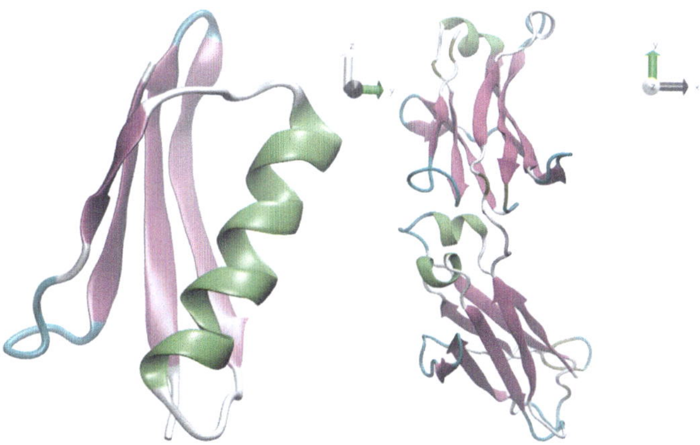

Fig. 11.1 Three dimensional structure of C2 (left) and Fc (right) fragments of protein G and IgG, respectively. The plot was created using VMD program (Humphrey et al. 1996). Alpha helices are coloured in green, 3–10 Helix structure in blue, beta sheets in mauve, turn structures in cyan and coil structures in white

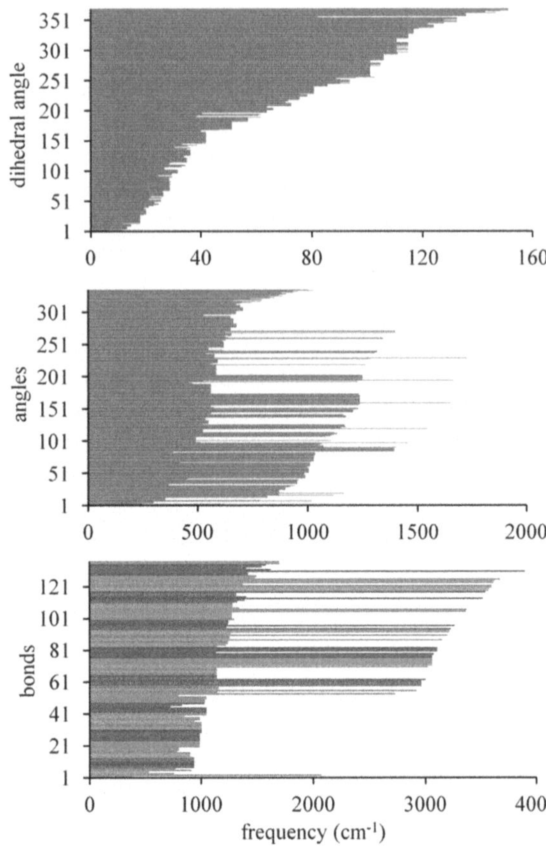

Fig. 11.2 The spectra of the frequency (in cm^{-1}) for the dihedral angles, bending angles and bond stretching based on the force field constants in CHARMM22 (MacKerell et al. 1998)

where μ is the reduced mass, and K is the force field constant. Here, k is simply the frequency given in cm^{-1}. Then, the relation with the period of motion (or the time-scale of the dynamics of the motion) is:

$$T = \frac{2\pi}{\omega} \tag{11.40}$$

In Table 11.2, we have summarised the results of histogram analysis for the frequency mode distributions based on the force field constant values of the CHARMM22. We created six histogram bins, as shown in Table 11.2. It can be seen that bond stretching modes are dominant at the frequency range greater than 1500 cm^{-1}; then, the angle bending modes are dominant at the frequencies in the range 300 − 1000 cm^{-1}; while the dihedral angle modes characterise the frequencies smaller than 300 cm^{-1}. The frequencies in the range between 1000–1500 cm^{-1} are dominated by the bond length stretching and the angle bending modes.

Figure 11.3 shows a dialog form for performing the NMA.

Table 11.2 The histogram frequency (in %) of the frequency mode distributions for the dihedral angles, bending angles and covalent bond based on the force field constants in CHARMM22 (MacKerell et al. 1998)

	$k \leq 20$ cm^{-1}	(20, 300] cm^{-1}	(300, 500] cm^{-1}	(500, 1000] cm^{-1}	(1000, 1500] cm^{-1}	$k > 1500$ cm^{-1}
Bonds	0	0	1	24	43	32
Angles	0	1	7	58	33	1
Dihedrals	8	92	0	0	0	0

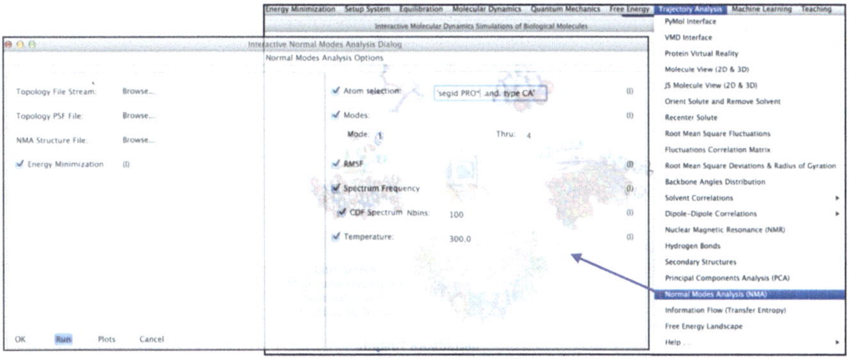

Fig. 11.3 Dialog form for normal mode analysis

Following the normal mode analysis presented in Ref. Brooks and Karplus (1983), Figs. 11.4 and 11.5 show the number of normal modes versus frequency in a 5 cm^{-1} interval, giving the density of states distribution. We present the data up to the frequencies of 2000 cm^{-1}, as depicted in Figs. 11.4a and 11.5a. The hydrogen stretching mode frequencies are omitted. In Figs. 11.4b and 11.5b, we show the results of the cumulative distribution for the number of modes below 200 cm^{-1}. Furthermore, in Figs. 11.4c and 11.5c, we present the cumulative distributions in the low-frequency region (up to 400 modes of main interest).

The results are presented for the C2 fragment of protein G and Fc fragment of protein IgG. It is interesting to note that for both proteins, there exists a reasonably continuous frequency distribution between 6 and 1300 cm^{-1} (with a drop off in the frequency range from 500 cm^{-1} to 600 cm^{-1}). While between 1300 cm^{-1} and 1800 cm^{-1}, the frequencies are clustered; besides, these frequencies are dominated by the bond-stretching vibration modes, as indicated by Table 11.2. These high-frequency modes represent the fastest motion in our systems, and as a result, they characterise the short-time vibration motions in proteins, such as bond-stretching.

From Figs. 11.4a and 11.5a, we notice another clustered region of high-frequency modes in the range 900–1300 cm^{-1}. As shown in Table 11.2, this range of frequencies is dominated by both the bond-stretching and the angle-bending vibration modes. A peak appears at the frequency of about 950 cm^{-1}, which is characteristic of the C-C

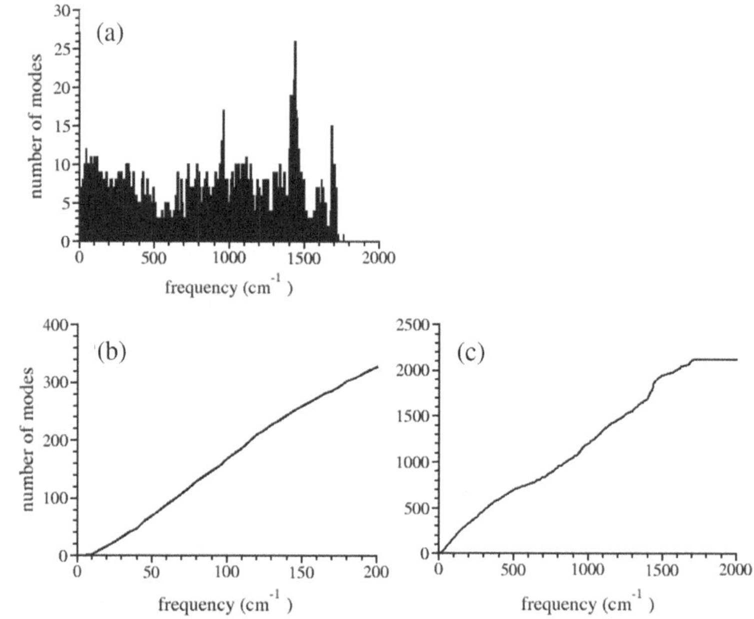

Fig. 11.4 Normal mode frequencies: **a** The number of normal modes with frequencies in a 5 cm^{-1} interval versus frequency for C2 fragment of G; **b** The number of modes up to a given frequency in the low-frequency region for C2 fragment of G; **c** The number of modes up to a given frequency for C2 fragment of G

and C-N bond stretching and H-C-C and H-N-C angle bending vibration modes. This range includes vibration modes with a relaxation time of less than 1 ps, and thus, it is a fast vibration mode.

Note that the range of frequencies higher than about 500 cm^{-1} is present in the computer simulations spectra corresponding to the molecular dynamics of proteins in solution (Karplus and McCammon 2002).

Of interest is the region of the frequency range less than about 500 cm^{-1}, which is characteristic of the dihedral angle vibration modes, as indicated from Table 11.2. The spectrum distributions exhibit a peak of around 50 cm^{-1} for C2 fragment of G protein, whereas, for Fc fragment, that peak is around the frequency 70 cm^{-1}. The range of the spectrum in 20–300 cm^{-1} is characteristic of the dihedral angle vibration modes, with a typical time scale of up to approximately 50 ps. From Fig. 11.4b, it can be seen that the total number of modes up to the frequency 20 cm^{-1} is around 20, which represents the low-frequency modes characteristic of the slow conformation motions in both systems; therefore, they correspond to the long-timescale movements in the system. In the view of the spectrum of the interactions, they represent long-range interactions, such as van der Waals and electrostatic interactions in the system. The low-frequency modes describe the rigid-body motions of the domains (which are blocks of amino-acids) in proteins. The low-frequency dynamical modes are not

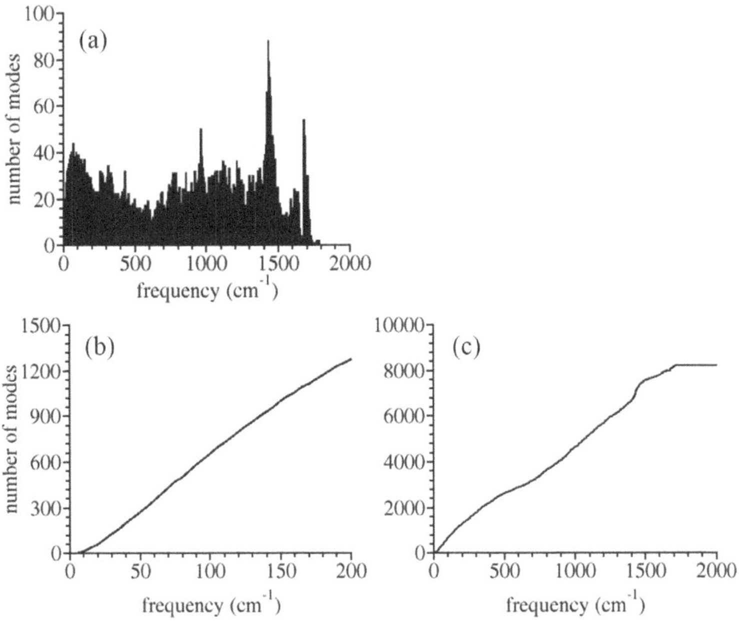

Fig. 11.5 Normal mode frequencies: **a** The number of normal modes with frequencies in a 5 cm^{-1} interval versus frequency for Fc fragment of IgG; **b** The number of modes up to a given frequency in the low-frequency region for Fc fragment of IgG; **c** The number of modes up to a given frequency for Fc fragment of IgG

always evident in computer simulations, such as the molecular dynamics simulations, since some of these modes exceed the time scales covered by the simulations.

In Fig. 11.6 are shown the root mean square fluctuations (rmsf) calculated at $T = 1000$ K using the formula (Brooks and Karplus 1983):

$$\Delta r_k^2 = k_B T \sum_i \frac{| \mathbf{a}_{ik} |^2}{\omega_i^2} \tag{11.41}$$

where $\mathbf{a}_{ik}$ is the vector of the projections of the ith normal mode with frequency ω_i on the Cartesian components of the displacement vector for the kth atom, k_B the Boltzmann's constant, and T is the temperature.

Besides, we have shown the rmsf calculated from molecular dynamics (MD) simulations at $T = 300$ K for both the C-α and backbone atoms. The data are reported for C2 fragment of protein G (see Fig. 11.7). As expected, the values of the rmsf are smaller in a magnitude comparable to those obtained from the normal modes because the results of the MD runs include contributions from the solvent environment and different configurations. Also, the normal modes sample a single local minimum,

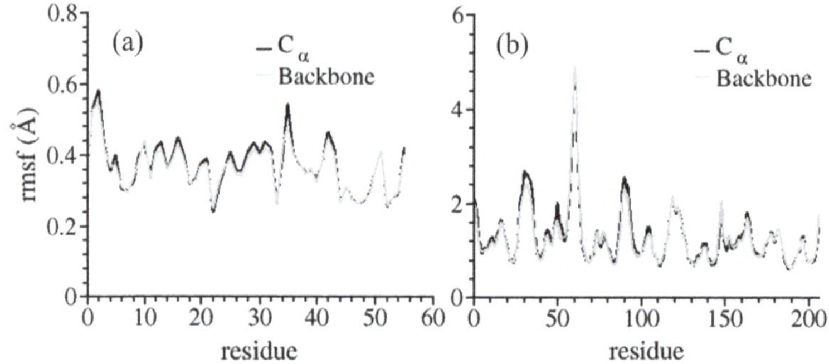

Fig. 11.6 **a** The root mean square fluctuations of the C2 fragment of protein G for C−α and backbone atoms of each residue averaged over all normal modes at at 1000 K. **b** For Fc fragment of protein G

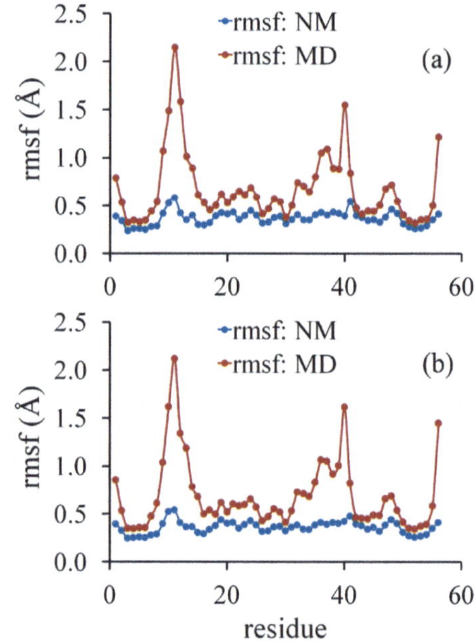

Fig. 11.7 The root mean square fluctuations of the C2 fragment of protein G for **a** C−α and **b** backbone atoms of each residue averaged over all normal modes at at 1000 K and from MD simulations

while MD simulations consider contributions from multiple local minimum configurations. Moreover, the normal mode results do not include the anharmonicity effects of the potential energy surface around the local minimum. However, the results are interesting because they both have similarities in the distribution spectrum along the chain of the amino acids, as shown in Fig. 11.7.

11.3 Mini Project

Exercise 11.1 (Activity case study) Consider solvated BPTI system. The input files are under the folder *casestudies/lab3/inputs*. Run normal mode analysis module at the temperature $T = 300$ K.

1. Analyse using the VMD program the protein dynamics of the modes 1 through 4. What you can say about the these modes? What are the dominant global motions?
2. Plot the root mean square fluctuations for each residue. What you can say about the most flexible regions?
3. Analyse the normal mode frequency spectrum and normal mode eigenvalues. Determine the number of normal modes that can include about 65–85% of the motion.

11.4 Summary

In this chapter, we introduced the normal mode analysis of a protein. In general, the NMA is associated with the energy minimisation; thus, first an energy minimisation is run to minimise the energy state of the system.

Furthermore, a detailed discussion of the theory of the normal mode analysis is presented from the mechanics viewpoint. Moreover, the case studies are shown for calculation of normal modes of a system and a mini-project is written for testing the knowledge gained.

References

I. Andricioaei, J.E. Straub, Finding the needle in the haystack: algorithms for conformational optimisation. Comput. Phys. **10**, 449–454 (1996)

I. Andricioaei, J.E. Straub, Global optimisation using bad derivatives: derivative-free method for molecular energy minimisation. J. Comput. Chem. **19**(13), 1445–1455 (1998)

B. Brooks, M. Karplus, Harmonic dynamics of protein: normal modes and fluctuations in bovine pancreatic trypsin inhibitor. Proc. Natl. Acad. Sci. U.S.A. **80**, 6571–6575 (1983)

B.R. Brooks, C.L. Brooks, A.D. MacKerell, L. Nilsson, R.J. Petrella, B. Roux, Y. Won, G. Archontis, C. Bartels, S. Boresch, A. Caflisch, L. Caves, Q. Cui, A.R. Dinner, M. Feig, S. Fischer, J. Gao, M. Hodoscek, W. Im, K. Kuczera, T. Lazaridis, J. Ma, V. Ovchinnikov, E. Paci, R.W. Pastor, C.B. Post, J.Z. Pu, M. Schaefer, B. Tidor, R.M. Venable, H.L. Woodcock, X. Wu, W. Yang, D.M. York, M. Karplus, Charmm: the bimolecular simulation program. J. Comput. Chem. **30**(10), 1545–1614 (2009)

H. Goldstein, *Classical Mechanics*, 3rd edn. (Addison-Wesley, United State of America, 2002)

W. Humphrey, A. Dalke, K. Schulten, VMD-visual molecular dynamics. J. Mol. Graph. **14**, 33–38 (1996)

H. Kamberaj, *Molecular Dynamics Simulations in Statistical Physics: Theory and Applications* (Computational Series. Springer Nature, Switzerland, 2020)

M. Karplus, J.A. McCammon, Molecular dynamics simulations of biomolecules. Nat. Struct. Biol. **9**(9), 646–652. With corrigenda in Nat. Struct. Biol. **9**(10), 788 (2002)

A.R. Leach, *Molecular Modelling. Principles and Applications*, 2nd edn. (Prentice Hall, Pearson Education Limited, Edingburgh Gate, 2001)

M. Levitt, C. Sander, P.S. Stern, Protein normal-mode dynamics: trypsin inhibitor, cram bin, ribonuclease and lysozyme. J. Mol. Biol. **181**, 423–447 (1985)

S. Lifson, A. Warshel, Consistent force field for calculations of conformations, vibrational spectra, and enthalpies of cycloalkane and n-alkane molecules. J. Chem. Phys. **49**, 5116–5129 (1968)

A.D. MacKerell, Jr, D. Bashford, M. Bellot, R.L. Dunbrack Jr., J.D. Evanseck, M.J. Field, S. Fisher, J. Gao, H. Guo, S. Ha, D. Joseph-McCarthy, L. Kuchnir, K. Kuczera, F.T.K. Lau, C. Mattos, S. Michnick, T. Ngo, D.T. Nguyen, B. Prodhom, W.E. Reiher, B. Roux, M. Schlenkrich, J.C. Smith, R. Stote, J. Straub, M. Watanabe, J. Wiorkiewicz-Kuczera, D. Yin, M. Karplus, All-atom empirical potential for molecular modelling and dynamics studies of proteins. J. Phys. Chem. B **102**(18), 3586–3616

A.E. Sauer-Eriksson, G.J. Kleywegt, M. Uhlen, T.A. Jones, Crystal structure of the C2 fragment of streptococcal protein G in complex with the Fc domain of human IgG. Structure **3**, 265–278 (1995)

Chapter 12
Standard Molecular Dynamics Simulations of Biological Systems

The chapter aims to show running standard molecular dynamics simulations of biological systems in explicit solvent.

This chapter describes how to run standard MD simulations of a biological system in explicit solvent using different statistical ensembles and numerical integrators.

12.1 Setup Molecular Dynamics Simulations

In Fig. 12.1 is shown a dialog form for running standard MD simulations. Different dynamics and statistical ensembles can be chosen, as shown in Fig. 12.1. Typical standard MD simulations run in constant N, P, and T statistical ensemble; usually, BERENDSEN dynamics are chosen at the equilibration stage and NOSE for the NVT (canonical) ensemble simulations. The other dynamics, such as LANGEVIN, NOSEHOOVER, and VV2, are chosen for running the data production standard MD simulations. Periodic boundary conditions were applied in this study, and all bonds, including the hydrogen atom, were constraints around the force field equilibrium lengths using the SHAKE algorithm (Ryckaert et al. 1977) with a tolerance of 1.0×10^{-8}.

The input files of the final step of the equilibration stage are saved in the folder *casestudies/lab5/inputs* and are used to start the simulations. The numerical time step was $\Delta t = 2$ fs.

Here, the BPTI protein in explicit solvent is considered as a case study system. Total number of atoms was $N = 19562$ atoms. Figure 12.2 shows the trajectories of energy components (in kcal/mol), temperature (in Kelvin), and pressure (in atm) from an output of standard MD simulations of about 200 ps. LANGEVIN dynamics were used with a piston mass of $250 - 500$ amu and the thermostat mass of 500–1000

H. Kamberaj, *Computer Simulations in Molecular Biology*, Scientific Computation,
https://doi.org/10.1007/978-3-031-34839-6_12

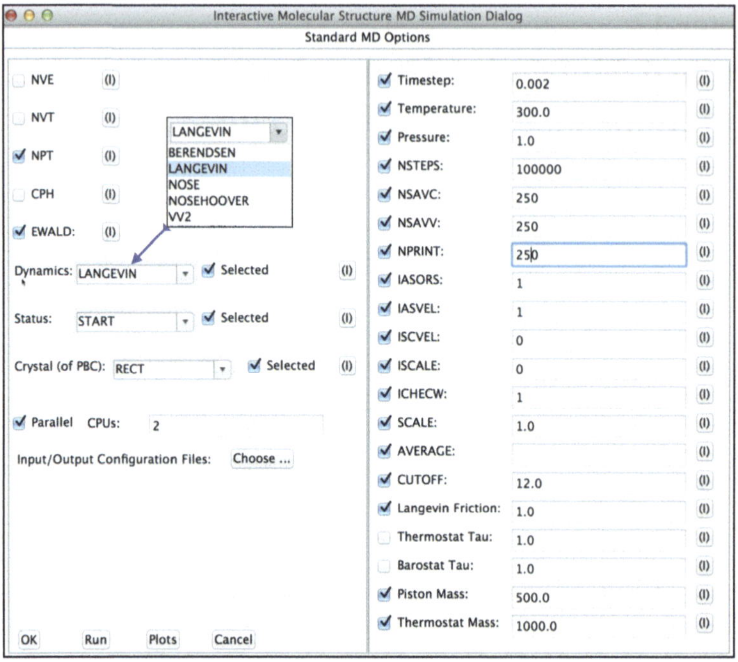

Fig. 12.1 A dialog form for running standard MD simulations

kcal·mol^{-1}·ps^2, and a cutoff of the non-bonded interactions of 12 Å. The constant N, P, and T statistical ensemble was used for MD simulations at $P = 1$ atm and $T = 300$ K. The snapshots of the MD simulation were collected every 0.2 ps (or equivalently 100 steps).

12.2 Data Analysis of MD Trajectory

12.2.1 Root Mean Square Displacement and Radius of Gyration

The root mean square displacement (rmsd) of a group of n atoms (such as C$-\alpha$ and backbone) can be calculated according to:

$$\text{rmsd} = \sqrt{\frac{1}{n}\sum_{i=1}^{n} \mid \mathbf{r}_i(t) - \mathbf{r}_i(0) \mid^2} \tag{12.1}$$

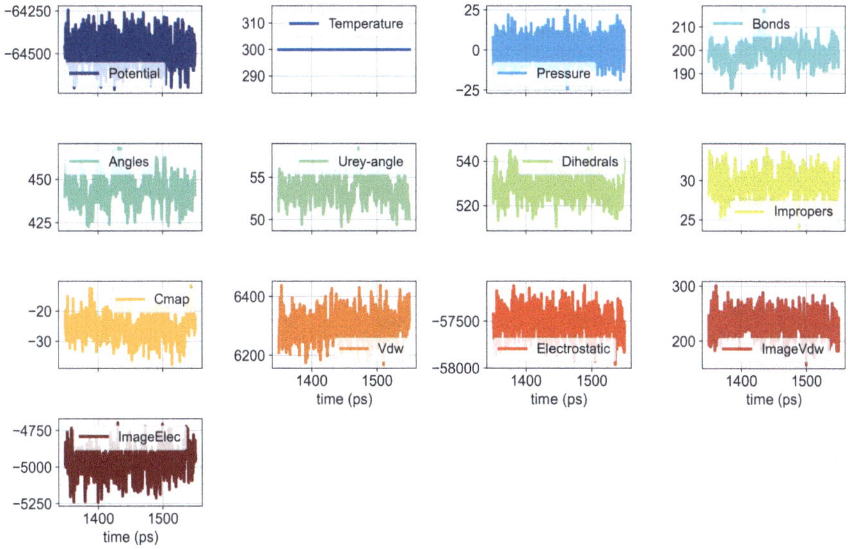

Fig. 12.2 The trajectories of energy components (in kcal/mol), temperature (in Kelvin), and pressure (in atm) from an output of standard MD simulations of about 200 ps

where $\mathbf{r}_i(0)$ is the position vector of the i-th atom of the group at the reference structure. The radius of gyration (rgyr) of the same group of n atoms (that is, C$-\alpha$ or backbone atoms) is calculated according to

$$\text{rgyr} = \sqrt{\frac{\sum_{i=1}^{n} m_i \mid \mathbf{r}_i(t) - \mathbf{r}_{\text{com}}(t) \mid^2}{\sum_{i=1}^{n} m_i}} \tag{12.2}$$

where $\mathbf{r}_{\text{com}}(t)$ is the centre of mass of that group of atoms at time t and m_i is the mass of atom i. Both rmsd and rgyr are measured in units of Å. Figure 12.3 presents a dialog form for calculation of rmsd and rgyr from MD simulations. In these calculations, the formulas implemented in CHARMM program (Brooks et al. 2009) are used as in module CORMAN and CORREL. Besides, the clustering based on the rmsd quantity can be calculated using a predefined cluster radius, as depicted in Fig. 12.3.

Figure 12.4 presents the trajectories of the rmsd (in Å) and rgyr (in Å) of the C$-\alpha$ atoms from an output of standard MD simulations of about 2.5 ns. The protein structure snapshots were aligned with X-ray structure of BPTI protein to remove overall rotational and translational motion of protein (Ichiye and Karplus 1991; Kamberaj 2011). The rmsd values fluctuate between 0.34 Å and 1.49Å and they equilibrate after about 1 ns.

Besides, in Fig. 12.5, a scatter plot is shown of the rmsd (in Å) and rgyr (in Å) of the backbone atoms from the output of standard MD simulations of about 2.5 ns. The scatter points are coloured according to the cluster index they belong. For clustering,

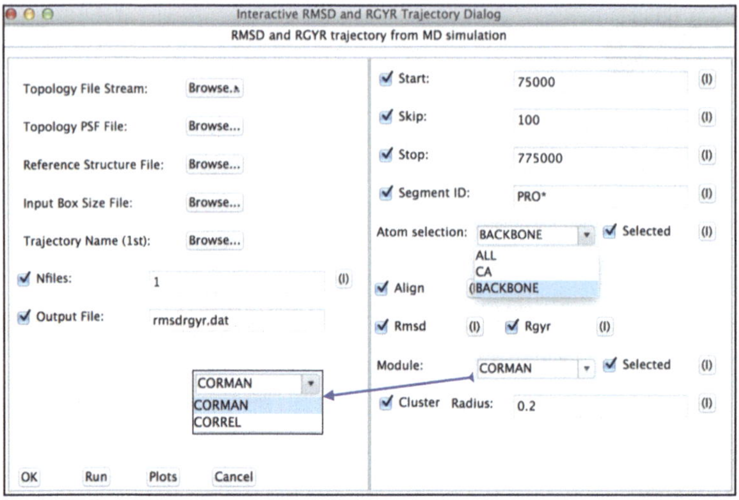

Fig. 12.3 A dialog form for calculation of rmsd and rgyr from MD simulations

Fig. 12.4 **a** The rmsd (in Å) and **b** rgyr (in Å) of the C$-\alpha$ atoms from an output of standard MD simulations of about 2.5 ns

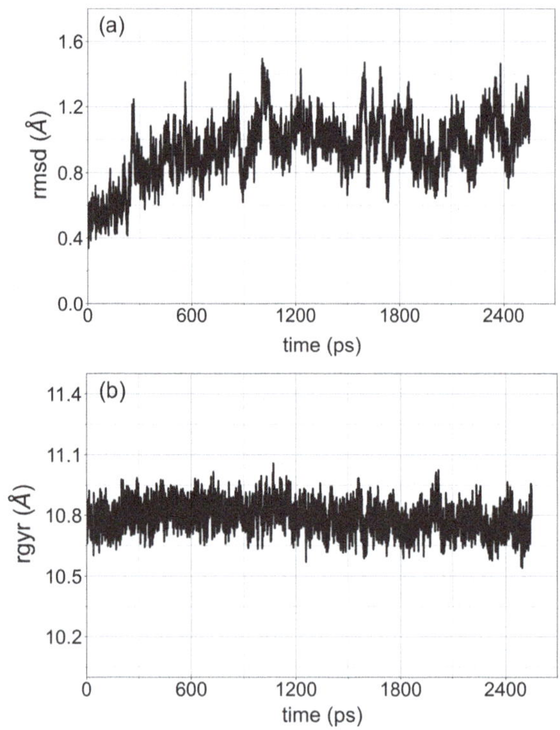

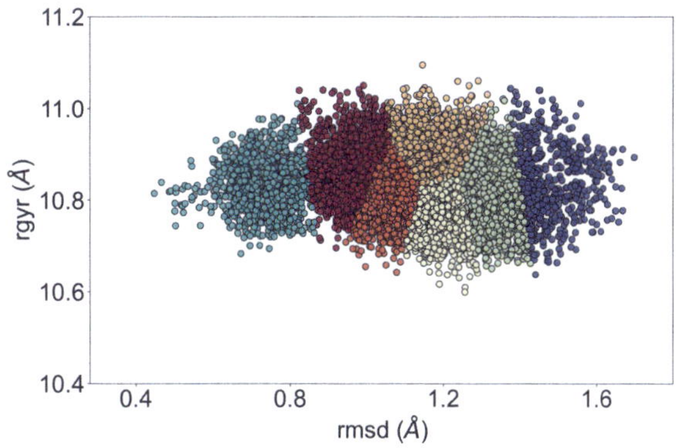

Fig. 12.5 A scatter plot of the rmsd (in Å) and rgyr (in Å) of the backbone atoms from an output of standard MD simulations of about 2.5 ns

a radius of 0.2Å was used, for which seven different clusters were identified, as shown in different colours in Fig. 12.5.

12.2.2 *Solvent Properties*

The mean square displacement (msd) is calculated as

$$\text{msd} = \frac{1}{n}\sum_{i=1}^{n} \mid \mathbf{r}_i(t) - \mathbf{r}_i(0) \mid^2, \tag{12.3}$$

where n is the number of water molecules in the subset and $\mathbf{r}_i(0)$ is position of oxygen atom at $t = 0$. Usually, msd is presented in Å^2.

Furthermore, the rotation diffusion order parameter measured from the correlation function of second order Legendre polynomials

$$P_2 = \frac{3}{2}\cos^2\theta - \frac{1}{2}, \tag{12.4}$$

where θ is the angle between a fixed z-axis and the direction perpendicular to the plane of water molecule TIP3 geometry. For P_2 long time correlation is zero and normalised to one.

Figure 12.6a shows the msd (in Å^2) for the first water shell (0–5 Å), second water shell (5–10 Å), and all shells of solvent around the protein. Figure 12.6b presents the rotational diffusion order parameter P_2 for the first and second water shells. The standard MD simulations were for about 2.5 ns. For the first water shell around the protein, the estimated value of diffusion coefficient from the last 80 data points of msd(t) was $D = (0.1342 \pm 0.001327)$ Å^2/ps. For the first shell, the estimated value of diffusion coefficient from the last 80 data points of msd(t) was $D = (0.3635 \pm 0.001082)$ Å^2/ps, and for all shells around the protein, the estimated value

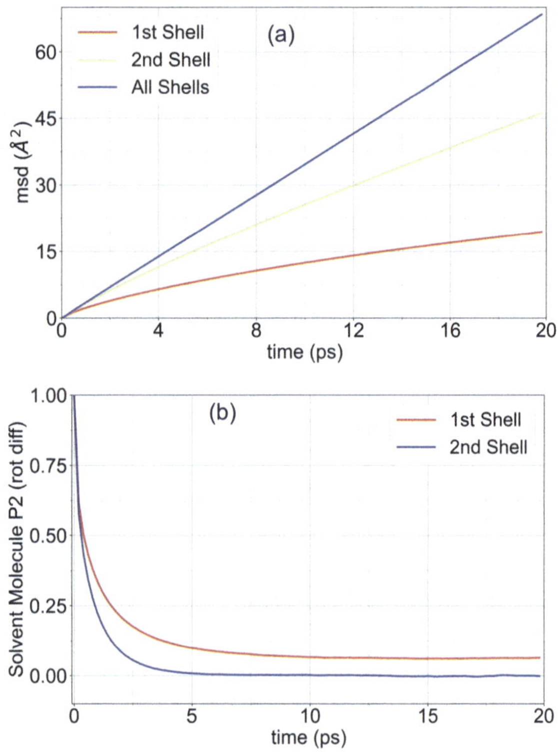

Fig. 12.6 **a** The msd (in $Å^2$) for the first (0–5 Å), second (5–10 Å), and all shells of solvent around the protein. **b** The rotational diffusion order parameter P_2 for the first and second shells. The standard MD simulations were for about 2.5 ns

of diffusion coefficient from the last 80 data points of msd(t) was $D = (0.5757 \pm 0.0001044)$ $Å^2$/ps.

The rotation diffusion relaxation time is defined by fitting $P_2(t)$ to the following function:

$$P_2^{\text{fit}} = (C_0 - C_\infty)\, e^{-t/\tau} + C_\infty, \tag{12.5}$$

where τ is the rotation diffusion relaxation time (in picoseconds) and C_∞ and C_0 are, respectively, the long time and short time (i.e., $t = 0$) correlation coefficients. The estimated value for the first shell is $\tau = 2.2027$ ps (assuming normalised $C(t)$ decays to zero; that is $C_\infty = 0$), and for the second shell $\tau = 0.69458$ ps.

In Fig. 12.7, we show the radial pair distribution functions of waters from the standard MD simulations using NPT ensemble for about 5 ns. In Fig. 12.7a, $g_{HH}(r)$; Fig. 12.7b, $g_{OH}(r)$; and Fig. 12.7c, $g_{OO}(r)$. The step width along r was $\Delta r = 0.1$ Å. The radial distribution function of a pair of atoms X and Y is defined as

$$g_{XY}(r) = \frac{V}{N^2} \langle \sum_{\substack{i,j=1 \\ i \neq j}}^{N} \delta(r - r_{ij}) \rangle \ , \tag{12.6}$$

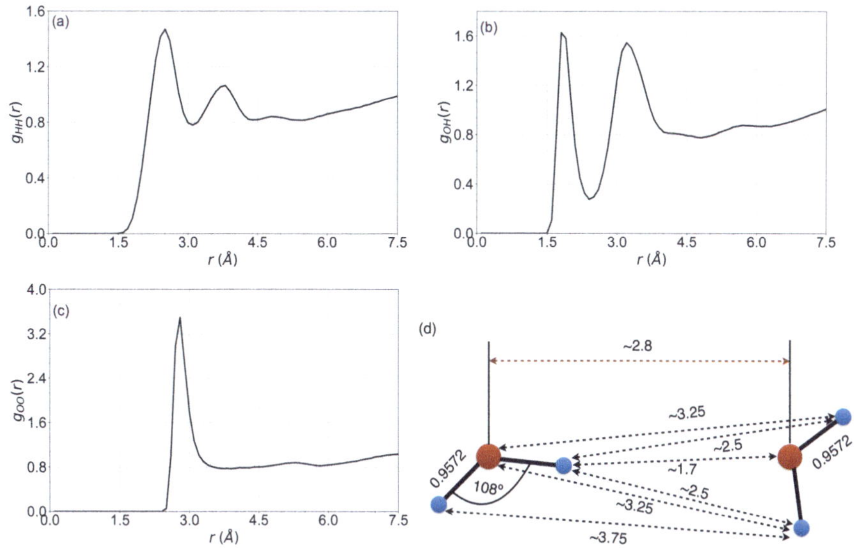

Fig. 12.7 The pair radial distribution functions of waters from the standard MD simulations using NPT ensemble for about 5 ns. **a** $g_{HH}(r)$; **b** $g_{OH}(r)$; **c** $g_{OO}(r)$; and **d** The relative most probable orientation of two water molecules and the approximate distances between O and H atoms in Å corresponding to the first two peaks of the pair radial distribution functions **a**–**c**. The step width along r was $\Delta r = 0.1$ Å

where $\delta(\cdots)$ is the Dirac delta function, normalised such that $\langle\delta(\mathbf{r})\rangle = V^{-1}$ where V is the volume of the simulation cell and N number of atoms X or Y. This function examines the structure of the phase by finding the probability of finding a pair of particles at a distance r. In practice, a certain maximum distance $r_{\max}$, typically $r_{\max} < R_c$ (long-range non-bonded interactions cutoff distance), is split in small intervals with width Δr, and r represents the centre of each of those intervals. For $r > R_c$, the radial distributions functions converge to one, which is the limit of the bulk water behaviour. Then,

$$\delta(r_k - r_{ij}) = \begin{cases} 1 \text{ if } r_k - \Delta r/2 \leq r_{ij} \leq r_k + \Delta r/2, \\ 0 \text{ otherwise} \end{cases} \tag{12.7}$$

where r_k is the centre of the k-th interval.

The results presented in Fig. 12.7a for $g_{HH}(r)$; Fig. 12.7b for $g_{OH}(r)$; and Fig. 12.7c for $g_{OO}(r)$ indicated a relative most probable orientation of two water molecules and the approximate distances between O and H atoms in Å corresponding to the first two peaks of the pair radial distribution functions, as in Fig. 12.7d.

Figure 12.8 presents a dialog form for calculation of solvent properties (such as solvent density, MSD, and the radial pair distributions of waters) from MD simulations.

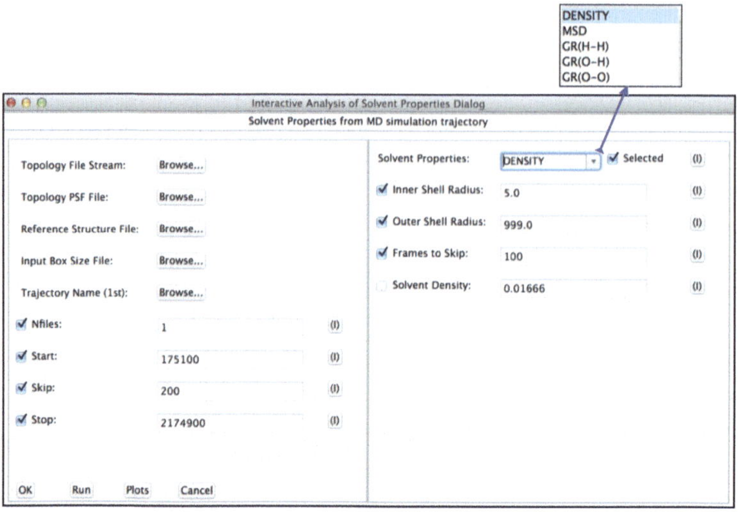

Fig. 12.8 A dialog form for calculation of solvent properties from MD simulations

Other solvent properties can also be analysed, as shown in Fig. 12.9a. For example, in Fig. 12.9b, it is shown water-water reorientation correlation function and the exponential fit. The calculated relaxation time was about 2.345 ± 0.02161 ps.

In Fig. 12.10a, it is shown the mean residence time and in Fig. 12.10b the mean residence histogram time of the first and second shells of waters around the protein from 5 ns MD simulations at an NPT ensemble for the BPTI protein.

Figure 12.11 presents a dialog form for calculation of the mean residence histogram time of the first and second shells of waters from MD simulations.

12.2.3 Solute Properties

Next, we show how to calculate some properties of the protein (solute) from MD simulations. Here, the data from MD simulations of BPTI protein are used at constant N, P, T ensemble for about 5 ns. The configuration snapshots were saved every picoseconds. Initially, the protein configurations were aligned with X-ray structure configuration using the C-α selection to remove overall rotational and translational motion.

The root mean square fluctuations (rmsf) are computed according to (for $i = 1, \ldots, N_g$):

$$\mathrm{rmsf}_i = \sqrt{\langle |\ \mathbf{r}_i(t) - \langle \mathbf{r}_i \rangle_t\ |^2 \rangle_t} \tag{12.8}$$

where $\langle \cdots \rangle_t$ denotes a time average and N_g is the number of atoms of selected group (such as C-α or backbone atoms).

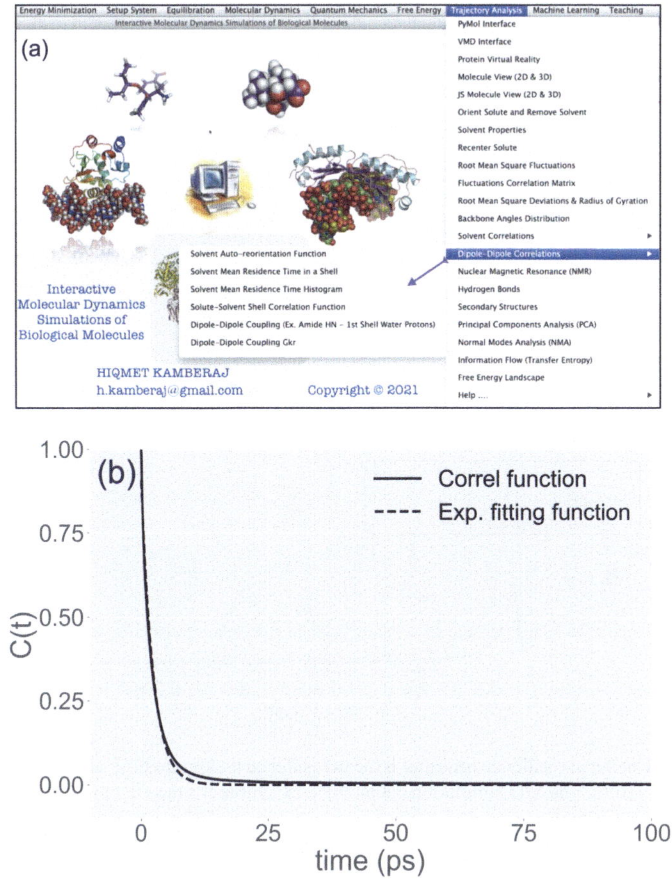

Fig. 12.9 **a** A dialog form for calculation of solvent-solvent reorientation correlation function from MD simulations. **b** Water-water reorientation correlation function from 5 ns MD simulations at an NPT ensemble for the BPTI protein

The normalised covariance matrix is computed as the following:

$$C_{ij} = \frac{\langle (\mathbf{r}_i(t) - \langle \mathbf{r}_i \rangle_t) \cdot (\mathbf{r}_j(t) - \langle \mathbf{r}_j \rangle_t) \rangle_t}{\sqrt{\langle | \mathbf{r}_i(t) - \langle \mathbf{r}_i \rangle_t |^2 \rangle_t} \cdot \sqrt{\langle | \mathbf{r}_j(t) - \langle \mathbf{r}_j \rangle_t |^2 \rangle_t}} \tag{12.9}$$

In Fig. 12.12a, b are shown the dialog forms for calculation of the root mean square fluctuations of the solute and normalised covariance matrix from MD simulations.

In Fig. 12.13a, it is presented the normalised covariance matrix of the C-α atoms for the BPTI protein as a colormap matrix; the colour bar indicates the C_{ij} scale from minimum value -1 (corresponding to blue colour) to a maximum value of $+1$ (corresponding to a red colour). As it can be seen, the correlated and anti-correlated

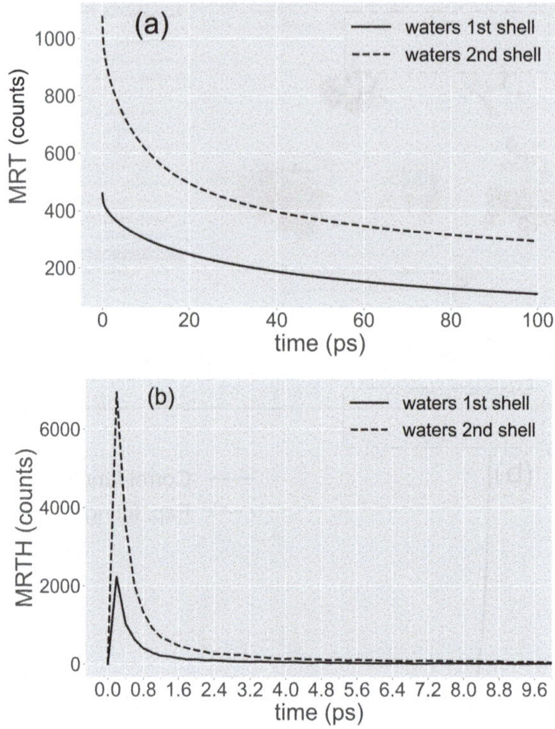

Fig. 12.10 **a** The mean residence time and **b** mean residence histogram time of the first and second shells of waters from 5 ns MD simulations at an NPT ensemble for the BPTI protein

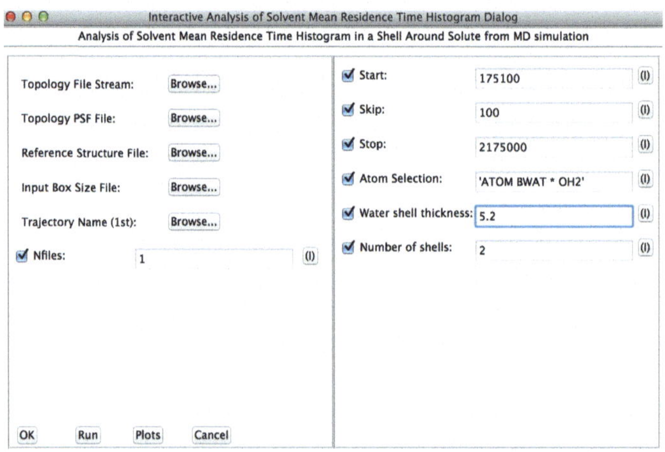

Fig. 12.11 A dialog form for calculation of the mean residence histogram time of the first and second shells of waters from MD simulations

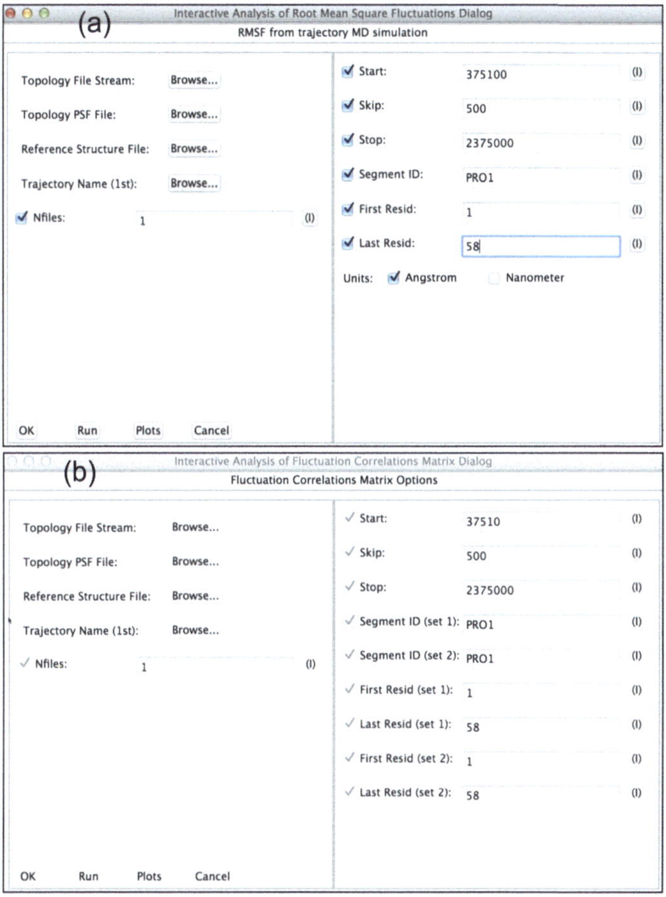

Fig. 12.12 **a** A dialog form for calculation of the root mean square fluctuations of the solute and **b** normalised covariance matrix from MD simulations

fluctuations are clearly identified. Root mean square fluctuations plot of C-α atoms of the BPTI protein is also shown in Fig. 12.13b, which indicates the high flexible regions (high values of rmsf) and the low flexible regions (small values of rmsf).

12.2.4 Nuclear Magnetic Resonance Analysis

The nuclear magnetic resonance (NMR) analysis can be used to obtain a set of time series for a number of NMR properties from a trajectory of MD simulations. For example, the relaxation rates due to dipole-dipole fluctuations (T_1, T_2, NOE, ROE),

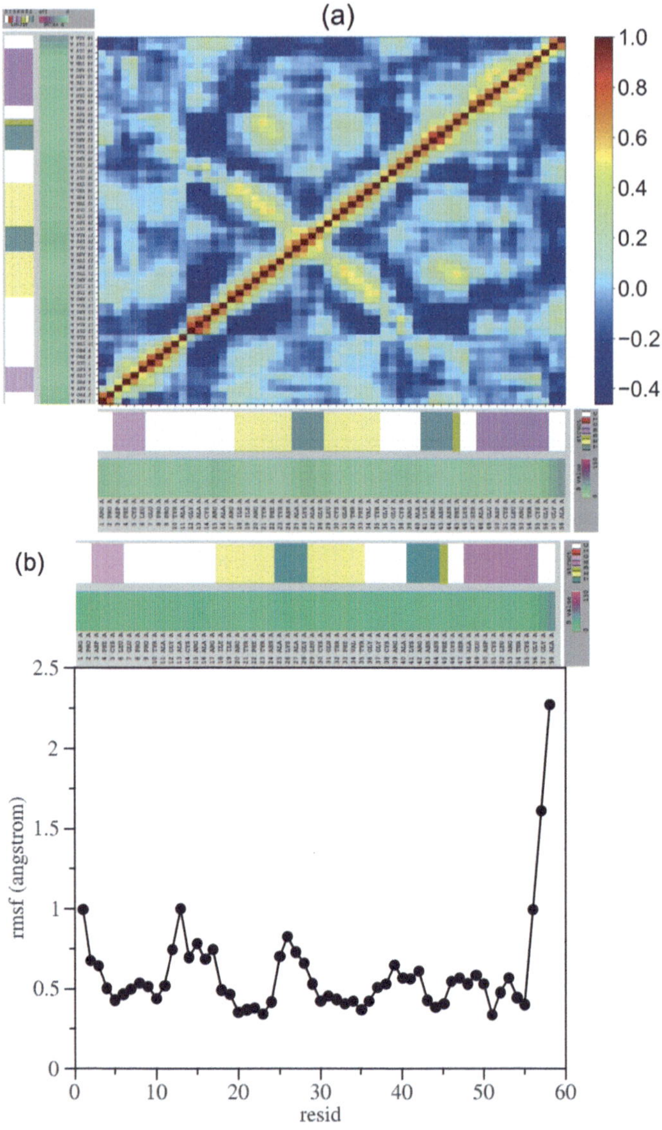

Fig. 12.13 **a** Normalised covariance matrix of the C-α atoms of the BPTI protein; **b** Root mean square fluctuations of C-α atoms of the BPTI protein

chemical shift anisotropy and Deuterium order parameters for oriented samples can be computed. The relaxation times in proteins are given as

$$\frac{1}{T_1} = \frac{N\hbar^2\gamma_N^2\gamma_H^2}{r^6}\frac{4\pi}{10}\left[J_0(\omega_N-\omega_H)+3J_1(\omega_N)\right.$$
$$\left.+6J_2(\omega_N+\omega_H)\right]$$
$$\frac{1}{T_2} = \frac{N\hbar^2\gamma_N^2\gamma_H^2}{r^6}\frac{4\pi}{20}\left[4J_0(0)+3J_1(\omega_N)\right.$$
$$+J_0(\omega_N-\omega_H)+3J_1(\omega_N)$$
$$\left.+6J_1(\omega_H)+6J_2(\omega_N+\omega_H)\right]$$
$$\eta = \frac{\gamma_H}{\gamma_N}\left[\frac{6J_2(\omega_N+\omega_H)-J_0(\omega_N-\omega_H)}{J_0(\omega_N-\omega_H)+3J_1(\omega_N)+6J_2(\omega_N+\omega_H)}\right]$$

where γ_N and γ_H are the gyromagnetic ratios of the N and H nuclei, ω_N and ω_H are the Larmor frequencies of the N and H nuclei, r is the distance of the N-H bond, and N is the number of protons directly bounded to the relaxing N nucleus. For the dipole-dipole relaxation rate properties the rotational tumbling time are taken in picoseconds, the magnetic field in TESLA, where 11.74 T yields 500 MHz for a proton H. The relaxation times are in 1/s.

Figure 12.14a presents the long relaxation time rate R_1 (s^{-1}) and short relaxation time rate R_2 (s^{-1}) of N-H bond for the BPTI protein. Figure 12.14b shows the average value of the order parameter of N-H bond for the BPTI protein.

The results clearly indicate the flexible regions of protein for the long (slow part of protein) and short (fast part of protein) relaxation time-scales (corresponding to large values of R_1 and R_2 or small values of the average order parameter $\langle S_2\rangle$) and less flexible regions of protein (corresponding to small values of R_1 and R_3 or large values of the average order parameter $\langle S_2\rangle$).

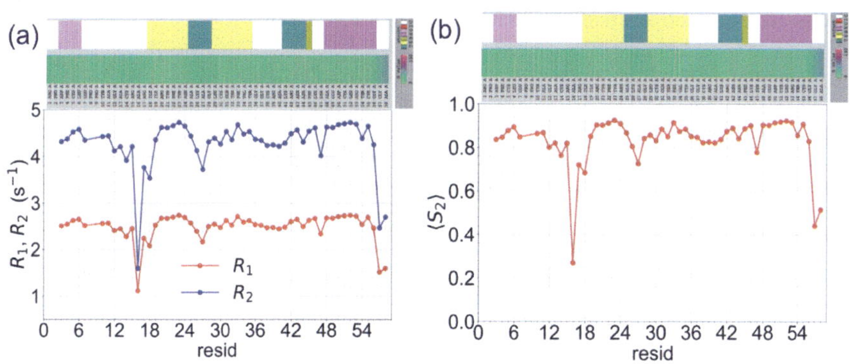

Fig. 12.14 **a** Long relaxation time rate R_1 (s^{-1}) and short relaxation time rate R_2 (s^{-1}) of N-H bond for the BPTI protein; **b** Average value of the order parameter of N-H bond for the BPTI protein

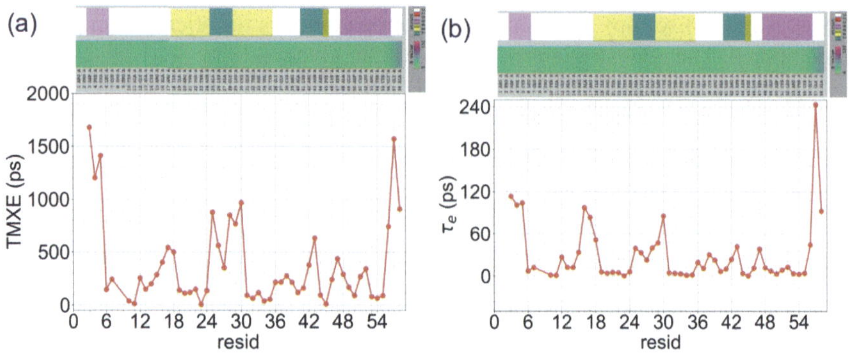

Fig. 12.15 **a** TMXE (in ps) of N-H bond for the BPTI protein; **b** τ_e (in ps) of N-H bond for the BPTI protein. τ_e (in ps) is computed from the integral of the dipole-dipole relaxation correlation function $C(t)$ out to TMXE (in ps), the first time when $C(t) \leq \langle S_2 \rangle$

In Fig. 12.15, it is shown τ_e (in ps), which is the effective correlation time for this motion computed from the integral of the dipole-dipole relaxation correlation function $C(t)$ out to TMXE (in ps), the first time when $C(t) \leq \langle S_2 \rangle$.

12.2.5 *Backbone Dihedral Angles* (ϕ, ψ)

Figure 12.16 shows a scatter plot of the backbone dihedral angles (ϕ, ψ) in degrees for the BPTI protein. The results are obtained from 5 ns of standard MD simulations at constant N, P, and T.

Besides, backbone ϕ angle (in degrees) distributions for each amino acid of the BPTI protein are plotted in Fig. 12.17a and the backbone ψ angle (in degrees) distributions for each amino acid of the BPTI protein are presented in Fig. 12.17b. The results are obtained from standard MD simulations of 5 ns.

Fig. 12.16 Scatter plot of the backbone dihedral angles (ϕ, ψ) for the BPTI protein, obtained from 5 ns of MD simulations

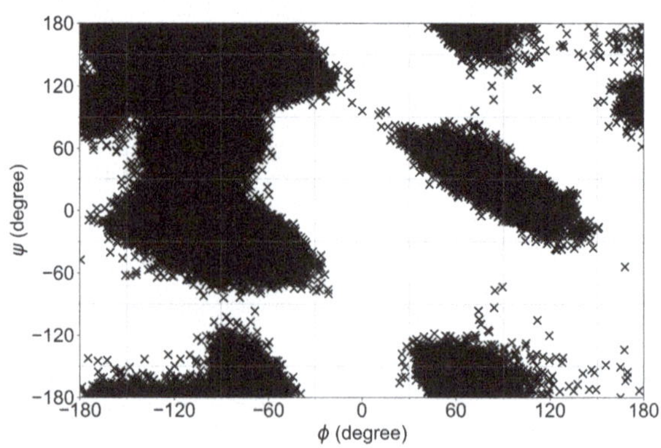

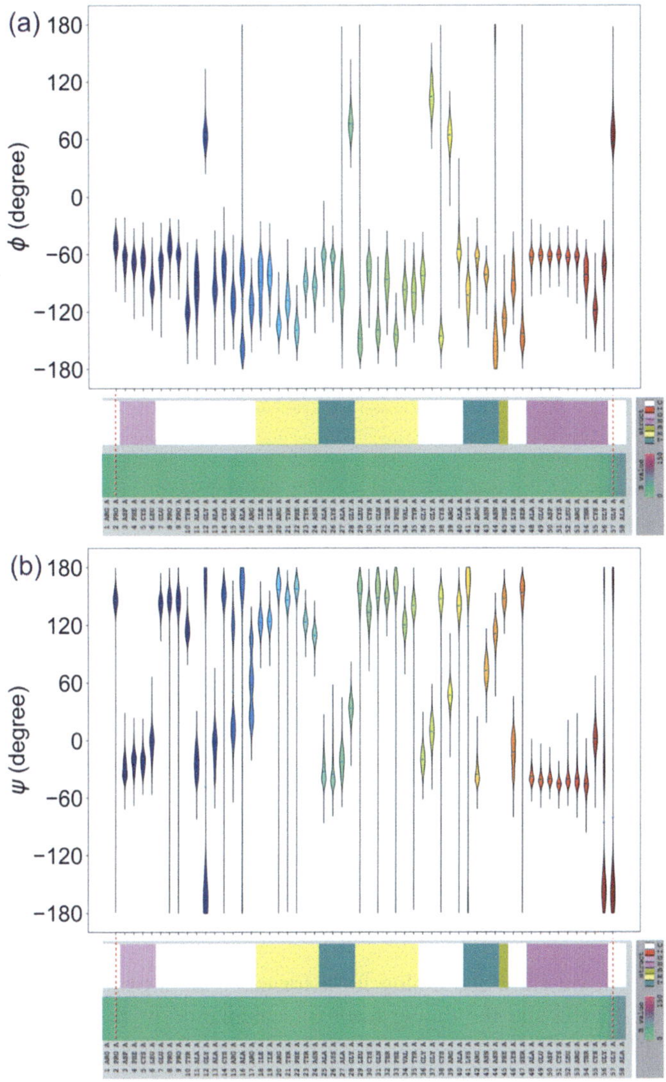

Fig. 12.17 **a** Backbone ϕ angle (in degrees) distributions for each amino acid of the BPTI protein; **b** Backbone ψ angle (in degrees) distributions for each amino acid of the BPTI protein. The results are obtained from standard MD simulations of 5 ns

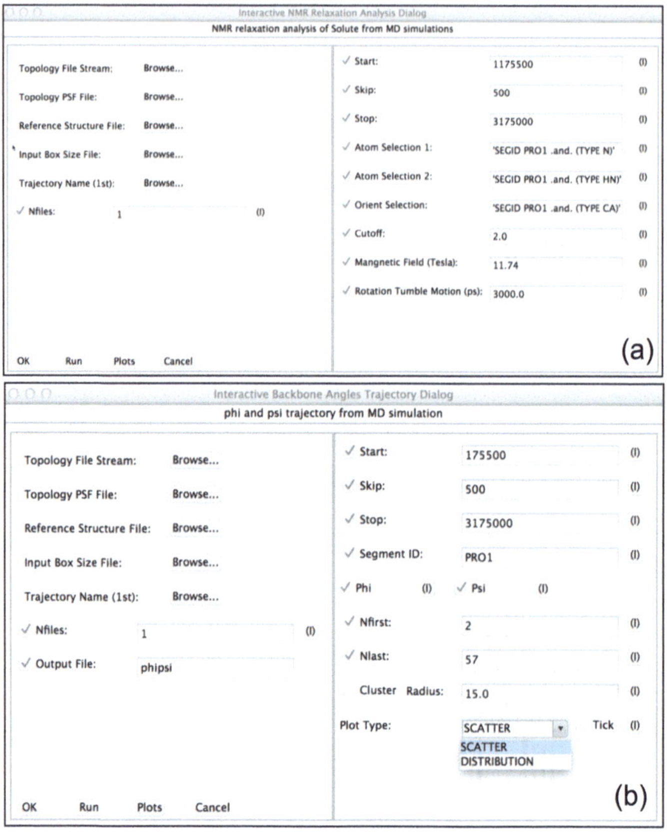

Fig. 12.18 **a** The dialog form for calculation of the NMR time correlations; **b** The dialog form for calculation of the backbone angles (ϕ, ψ)

In Fig. 12.18, we show the dialog form for calculation of the NMR time correlations Fig. 12.18a and the dialog form for calculation of the backbone angles (ϕ, ψ) Fig. 12.18b.

12.2.6 Information Flow

A trajectory in the configuration space from the molecular dynamics simulations represents a Markovian process of a biomolecule structure, where $p(\mathbf{x}_{t+1}|\mathbf{x}_t)$ gives the propagation probability at each time step (Kamberaj 2020). It produces a Boltzmann distribution in the phase space of variables, which for Newton dynamics is:

$$p(\mathbf{x}_{t+1}|\mathbf{x}_t) = \delta\left(\mathbf{x}_{t+1} - \Phi_{\Delta t}(\mathbf{x}_t)\right) \tag{12.10}$$

In Eq. (12.10), δ is the delta function and $\Phi_{\Delta t}(\mathbf{x}_t)$ denotes a discrete flow map for single time step Δt of a propagation operator. In this way, a trajectory of configuration snapshots is produced with an initial state sampled from the canonical distribution and then propagating in time using the Hamiltonian dynamics. Then, the probability of obtaining a trajectory, $\mathbf{X}_N$, of N snapshots in the configuration space is (Kamberaj 2020)

$$P(\mathbf{X}_N) = P(\mathbf{x}_0) \prod_{t=0}^{N-1} p(\mathbf{x}_{t+1}|\mathbf{x}_t) \tag{12.11}$$

where $P(\mathbf{x}_0)$ denotes probability of obtaining the initial configuration from an equilibrium Boltzmann's probability distribution with an initial energy of system $E(\mathbf{x}_0)$ (Kamberaj 2020).

A biological molecule has a large number of degrees of freedom. However, often, to describe the fluctuations of internal degrees of freedom, only a few of them are selected; for instance, C_α or backbone atoms are saved at regular time-intervals to represent the fluctuations of the amino acid of a protein, which is often called coarse-graining. Furthermore, a term of information coding can be introduced to code the fluctuations of amino acids into a single vector for each amino acid:

$$\mathbf{X}_N = \{x_0, x_1, \ldots, x_{N-1}\} \tag{12.12}$$

Here, x_k encodes the fluctuations from the equilibrium structure of Cartesian coordinates for an amino acid at time t_k. In general, x_k represents a vector of dimension m, which is the dimension of the encoded sub-space of the amino acid. Besides, $m \ll M$, where $M = 3n$ is the number of degrees of freedom of a residue of n atoms in Cartesian coordinates space.

Time is considered discrete with $t_k = k\Delta t$, where k is an integer. Then $\mathbf{X} = \{x(t_k)\}_{k=0,\ldots,N-1}$ is a time-discrete process, with N being the total number of time snapshots. For simplicity, the following notation is used: $x(t_k) \equiv x_k$. The covariance matrix's principal components are calculated to determine the essential degrees of freedom for each amino acid or nucleotide (Kamberaj 2011). Next, the natural degrees of freedom of each amino acid and nucleotide encode using the first three principal components with the largest eigenvalues of the covariance matrix.

The time delayed embedding method (Packard et al. 1980; Takens 1981; Grassberger and Procaccia 1983; Sauer et al. 1991) is often used to reconstruct the phase space of the original dynamical biomolecular system from the time series encoding the trajectories of the Cartesian coordinates of the amino acids of the system (Kamberaj and van der Vaart 2009b). A state vector in a m-dimensional space of a discrete process $x(t)$ is obtained as:

$$\mathbf{x}_k^{\mu} = \left(x_{k-(m-1)\tau}, x_{k-(m-1)\tau+\tau}, \ldots, x_k\right) \tag{12.13}$$

where τ is the time lag, which, in general, is a multiple of Δt, and $k = (m-1)\tau, \ldots, N-1$. Here, the subscript μ represent the pair of embedded parameters

(m, τ). The vector $\mathbf{x}_k^{\mu}$ reconstructs the phase space of the original component of the dynamical system in Cartesian coordinates $\mathbf{r}$.

With proper values of m and τ embedding parameters, a smooth discrete-time process is defined, which reconstructs the underlying dynamics. The correct choice of these two parameters is crucial for the proper characterisation of the time series structure (Noakes 1991; Sauer et al. 1991; Kennel et al. 1992; Abarbanel and Kennel 1993; Cellucci et al. 2003). The optimal embedding parameters are calculated as in Ref. Kamberaj and van der Vaart (2009b). First, the time delay τ is computed as the first minimum of the auto-mutual information functions (Gencaga et al. 2015; Chodera et al. 2007). Note that the global minimum of the auto-mutual information function provided large values of τ, which could be useful for very long trajectories. In contrast, the auto-correlation function is accurate to only first order, and it does not capture correlations of two mutually perpendicular vectors. m is calculated by minimising the number of false nearest neighbours.

The vector $\mathbf{X}^{\mu_x} = \{\mathbf{x}_k^{\mu_x}\}_{k=k_0}^{N-1}$ is used for the first process and the vector $\mathbf{Y}^{\mu_y} = \{\mathbf{y}_k^{\mu_y}\}_{k=k_0}^{N-1}$ for the second process to characterise the dynamics (Kamberaj and van der Vaart 2009b):

$$\mathbf{y}_k^{\mu_y} = \left\{y_{k-(m_y-1)\tau_y}, y_{k-(m_y-1)\tau_y+\tau_y}, \ldots, y_{k-\tau_y}, y_k\right\} \tag{12.14}$$

where $k_0 = \max\left((m_x - 1)\tau_x, (m_y - 1)\tau_y\right)$.

In terms of Shannon entropy (Schreiber 2000), the transfer entropy, $T_{X\rightarrow Y}$, is written as:

$$T_{X\rightarrow Y} = H(\mathbf{Y}^{\mu_y+1}) - H(\mathbf{Y}^{\mu_y}) - H(\mathbf{X}^{\mu_x}, \mathbf{Y}^{\mu_y+1}) + H(\mathbf{X}^{\mu_x}, \mathbf{Y}^{\mu_y}) \tag{12.15}$$

where $\mathbf{Y}^{\mu_y+1} = \{\mathbf{y}_k^{\mu_y+1}\}_{k=k_0}^{N-1}$ with

$$\mathbf{y}_k^{\mu_y+1} = \left\{y_{k-(m_y-1)\tau_y}, y_{k-(m_y-1)\tau_y+\tau_y}, \ldots, y_{k-\tau_y}, y_k, y_{k+\delta}\right\} \tag{12.16}$$

The local transfer entropy is given in terms of the local conditional entropy as (Nebiu and Kamberaj 2020):

$$t_{X\rightarrow Y}(k+\delta) = h(y_{k+\delta} \mid \mathbf{y}_k^{\mu_x}) - h(y_{k+\delta} \mid \mathbf{x}_k^{\mu_x}, \mathbf{y}_k^{\mu_y}) \tag{12.17}$$

where

$$h(y_{k+\delta} \mid \mathbf{x}_k^{\mu_x}, \mathbf{y}_k^{\mu_y}) = -\left(\log_2 p(\mathbf{x}_k^{\mu_x}, \mathbf{y}_k^{\mu_y+1}) - \log_2 p(\mathbf{x}_k^{\mu_x}, \mathbf{y}_k^{\mu_y})\right) \tag{12.18}$$

$$h(y_{k+\delta} \mid \mathbf{y}_k^{\mu_x}) = -\left(\log_2 p(\mathbf{y}_k^{\mu_y+1}) - \log_2 p(\mathbf{y}_k^{\mu_y})\right) \tag{12.19}$$

Here, $p(\cdots)$ is the probability mass function of a discrete random process.

For the above Markovian discrete random processes, the probability distributions are not known a priory. Different recipes are discussed (Gencaga et al. 2015). The symbolic approach can be used to estimate the transfer entropy and the information flow (Staniek and Lehnertz 2008; Kamberaj and van der Vaart 2009b; Kamberaj 2020) in which the state vectors are converted into the vectors of symbols as the following:

$$\hat{\mathbf{x}}_k^{\mu} = \left(\hat{x}_{k-(m-1)\tau}, \hat{x}_{k-(m-1)\tau+\tau}, \ldots, \hat{x}_k\right) \tag{12.20}$$

where any of the symbols $\hat{x}_t$ is drawn from some finite alphabetic $\hat{x}_t \in \{0, 1, \ldots, D-1\}$. If $D = 2$, a binary symbolic state vector is obtained; otherwise, for $D > 2$, a D-ary state vector is obtained. Besides, $D + 1$ critical points, $x_c^{(0)} < x_c^{(1)} < \cdots < x_c^{(D)}$, are chosen to encode the real-valued state vectors into symbols as:

$$\hat{x}_t = i, \quad \text{if}\, x_c^{(i)} < x_t \leq x_c^{(i+1)}, \; \forall i = 0, 1, 2, \ldots, D-1 \tag{12.21}$$

One will be interested in to what degree the state vector can be encoded to a sequence of symbols without losing any information, which in turn depends on the choice of the critical points. In general, they can be distributed at equal spacing between the minimum and maximum value of the time-series. A Monte Carlo approach can be used to choose the critical points, which reduces the loses of any information (Kamberaj and van der Vaart 2009b). The entropy of the encoded sequence of the symbols trajectory of state vectors is computed as (Grassberger 1988):

$$H(\hat{\mathbf{X}}^{\mu}) = \ln N - \frac{1}{N}\sum_{k}^{N_k} f_k \psi(f_k), \quad \text{(in nats)} \tag{12.22}$$

where N_k is the total number of state vectors of symbols for which $f_k > 0$, f_k is the frequency of obtaining a state vector of symbols of length m, and $\Psi(f_k)$ is the logarithmic derivative of the Gamma function. The joint entropy of two symbolic trajectories $\hat{X}^{\mu_x}$ and $\hat{Y}^{\mu_y}$ is calculated by first concatenating the symbolic state vectors as:

$$\hat{\mathbf{x}\mathbf{y}}_k^{\mu} = \left(\hat{x}_{k-(m_x-1)\tau_x}, \ldots, \hat{x}_k, \hat{y}_{k-(m_y-1)\tau_y}, \ldots, \hat{y}_k\right) \tag{12.23}$$

Then, Eq. (12.22) computes the joint entropy. The entropy is expressed in the units of bits by dividing with ln 2. Here, D is determined using the Kraft's inequality (Thomas and Joy 2006) as:

$$\sum_{k=1}^{N} D^{-m} \leq 1 \tag{12.24}$$

In Eq. (12.24), m is the length of sequence of symbols in a state vector. D is determined as the minimum value of the solution of Eq. (12.24) for D.

12.2.7 Relationship Between Transfer Entropy and Thermodynamics

The formalism introduced in Refs. Schneider (1991), Prokopenko et al. (2013), Prokopenko and Lizier (2014) can be used to derive a relationship between the transfer entropy and the laws of thermodynamics. For a physical system characterised by f macroscopic parameters (such as temperature, pressure, the number density of each component), considered in a quasi-equilibrium thermodynamic state, the state vector $\mathbf{y}_t \in R^f$ represents the macroscopic state of the system at any time t. The set of these vectors forms the phase space, and every macrostate can be created by many microstate vectors, each having the same probability of being at this macrostate. Furthermore, the physical system interacts with its surrounding, which is characterised by macrostate vector $\mathbf{x}_t$ at time t, due to its coupling with this surrounding. However, the nature of this coupling is not known. Here, $p(y_{t+\delta} \mid \mathbf{x}_t, \mathbf{y}_t)$ denotes the transition probability in the presence of the environment, which is obtained by sampling both $\mathbf{X}$ and $\mathbf{Y}$, as described in this study.

Then, following Refs. Schneider (1991), Prokopenko et al. (2013), Prokopenko and Lizier (2014), it can be shown that for processes close to the equilibrium, both reversible or irreversible, at constant temperature (Kamberaj 2020):

$$t_{X\rightarrow Y}(k+\delta) = -\frac{\Delta q_{\text{ext}}}{k_B T \ln 2} \tag{12.25}$$

Here, Δq_{ext} is the heat flow between the system Y and the surrounding in the context of the source X. The sign minus indicates the heat flow into the physical system Y attributed to the surrounding X and the local transfer entropy is the opposite. Equation (12.25) gives the minimum amount of energy that a physical system (characterised by the dynamical component Y) must dissipate into the environment (characterised by the dynamical component X) to gain one bit of information. If $t_{X\rightarrow Y}(k+\delta) > 0$, and hence the physical system Y gains information, then the heat flows to the surrounding X from the system Y, $\Delta q_{\text{ext}} < 0$. In that case, a measure of the source X improves our predictions about the macroscopic transition state of Y. For thermally isolated transitions, that is $\Delta q_{\text{ext}} = 0$, then $t_{X\rightarrow Y}(k+\delta) = 0$, indicating that Y and X are independent, in other words, X and Y do not interact with each other.

Therefore, a functional relationship exists between the local transfer entropy and the second law of thermodynamics such that $t_{X\rightarrow Y}(k+\delta)$ is proportional to $-\Delta q_{\text{ext}}$, which gives the external heat flow during the irreversible transition in the context of the source X. Besides, the interactions between the system Y and the surrounding X cause the heat to continuously flow between the system and surrounding. Since there is an average non-negative transfer entropy, $T_{X\rightarrow Y} \geq 0$ determines the minimum average energy, as $Q_{\text{ext}} = T_{X\rightarrow Y} k_B T \ln 2$, that dissipates into the dynamical component X by the dynamical component Y in order that Y gains $T_{X\rightarrow Y}$ bits of information. Thus, the net average transfer entropy

$$D_{X\to Y} = T_{X\to Y} - T_{Y\to X} \tag{12.26}$$

gives the net heat flow (or minimum net average energy dissipation) between any two Markovian random processes X and Y, such that, if $D_{X\to Y} > 0$, then there is net heat flow from Y to X, and vice-versa. Knowing that transfer entropy determines the information flow, if $D_{X\to Y} > 0$, then the information flows from X to Y, and vice-versa. Thus, the heat flow and information flow have opposite directions. For any two Markovian random processes X and Y, if the information flows from X to Y, then X *influences* Y (or Y is *influenced by* X).

In the underlying dynamics of MD simulations, in which the system is coupled to a thermostat/barostat, there is a continuous exchange of energy (and heat) between the system and thermostat due to the interactions between the system and thermostat/barostat to adjust the temperature and pressure. That is also shown in the equations of motion (Martyna et al. 1996; Kamberaj 2020). Therefore, Eq. (12.25) can also be used to calculate the information flow between the components of the system and the surrounding in the context of the thermostat/barostat as a source of the transfer entropy. However, when using Eq. (12.25), for calculating the local transfer entropy $t_{X\to Y}(k+\delta)$, that is the local transfer entropy from the surrounding in the context of the source X to the component of system Y, one must consider both the fluctuation vector of a component of the system (that is, Y vector) and fluctuation vector of surrounding in the context of the source X and hence the fluctuation vectors of all other of the variables, including those characterising either thermostat or barostat, do not (directly) interplay in the calculations using Eq.12.25.

The fluctuation vectors of both system component Y and surrounding in the context of the source X are dynamically coupled to the thermostat/barostat. Hence, the equilibrium structure fluctuations are related to the thermostat and barostat degrees of freedom. Therefore, the exchange of the energy (and heat) between the system and thermostat/barostat will interplay in the equilibrium structure fluctuations. Nevertheless, the probability distribution of equilibrium structure fluctuations should be equivalent for all statistical ensembles in the thermodynamic limit. Furthermore, this equivalence is more evident on the sub-nanoseconds time scale that this study is addressing for the calculation of the information and heat flow because the biomolecular system and its components and surrounding have a short time to sample more than a minimum local state or at most a few minimum states of the energy landscape.

Moreover, the relationship between the heat flow in thermodynamics and the heat flow in the information theory is defined in the context of the relationship that exists between the Shannon entropy and Boltzmann entropy (Thomas and Joy 2006) or that between the heat dissipated and information gain as given above (Schneider 1991; Prokopenko and Lizier 2014; Prokopenko et al. 2013).

Figure 12.19a shows the local transfer entropy, $t_{X\to Y}$, from water to protein (in bits) at every instant for the entire trajectory sampling the folded states of the BPTI protein to support the findings that in the folded state, the reaction occurs with a phase transition iceberg $\to$ solvent. It can be seen that when the protein samples the folded state, the local transfer entropy is positive, and hence the heat flows from the protein to the surrounding water (that is, $\Delta q_{\text{ext}} < 0$). In these computations, X is

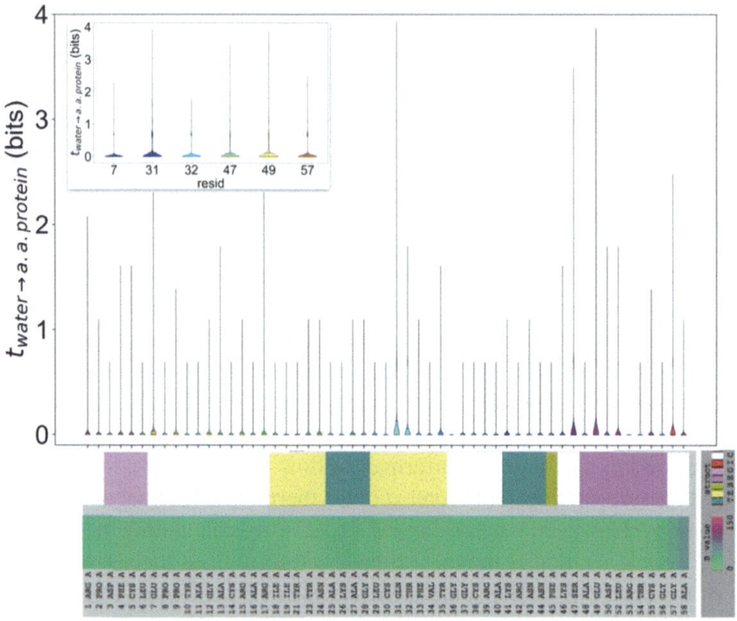

Fig. 12.19 Local transfer entropy $t_{X \to Y}$ from water to amino acids of protein (in bits) at every instant for the full trajectory sampling the folded state of BPTI protein

the fluctuations of the dipole momentum vector of the first shell water of thickness 4.5 Å around each amino acid of protein and Y is the fluctuation dipole momentum vector of the amino acid backbone atoms.

In Fig. 12.20, it is shown a dialog form for calculation of the information flow. The values of symbolic local transfer entropy are subject to statistical errors, and thus, we applied a statistical significance of 90% for each value. A normal distribution is assumed for the averaged values. Only if the value of $t_{X \to Y}$ is statistically significantly different from zero equals the average value; otherwise, it is assigned to zero. The shuffling method, proposed in Refs. Kamberaj and van der Vaart (2009b), Kamberaj (2020), is used to remove any bias on the calculation of $t_{X \to Y}$; 100 permutations of the original trajectories are used for the shuffling. The cutoff value of the statistical test for zero value was 0.002. The embedded parameters (m, τ), respectively, the embedded dimension and time lag, are needed as described in Refs. Kamberaj and van der Vaart (2009b), Kamberaj (2020).

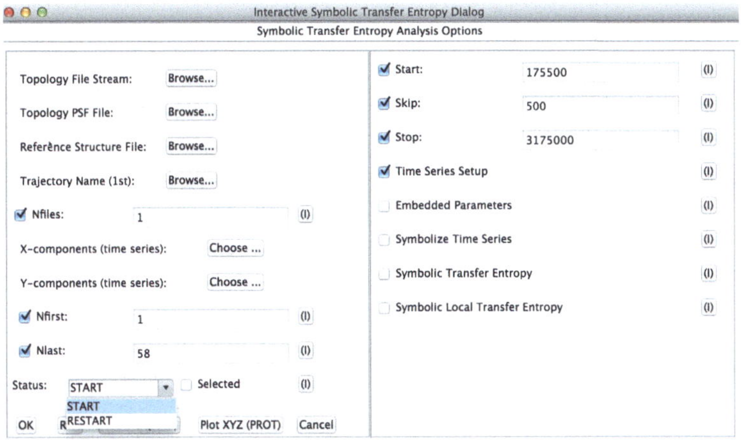

Fig. 12.20 The dialog form for calculation of the information flow

12.3 Mini Project

Exercise 12.1 (Activity case study) Consider again the BPTI system in explicit solvent. The input files are under the folder *casestudies/lab5/inputs*. Run standard MD simulations for up to 15 ns at the temperature $T = 300$ K, $P = 1$ atm.

1. Analyse the rmsd of the C-α and backbone atoms. What can you conclude about the equilibration of protein?
2. Analyse the trajectories of temperature T and pressure P. Make plots of running averages using different block sizes. What can you conclude by comparing the running averages and their target values?
3. Analyse using the VMD program the dynamics of the waters in the first shell around the protein. What can you say about the these waters? Look at the hydrogen bonds water-water and water-protein.
4. Plot the root mean square fluctuations for each residue. What can you say about the most flexible regions?
5. Analyse the properties of the solvent in different shells around the protein.
6. Calculate the radial distribution functions water-water, ion-ion, and water-ions.
7. Calculate the msd of ions in the solution. What are the values of diffusion coefficients?

12.4 Summary

This chapter described the standard MD simulations of a biological system in explicit solvent. Besides, the use of different statistical ensembles and numerical integrators is shown.

Furthermore, we showed how to analyse various properties of interest of the solvent and solute from MD trajectory using a case study system. A mini-project is written for testing the knowledge gained.

References

H.D.I. Abarbanel, M.B. Kennel, Local false nearest neighbors and dynamical dimensions from observed chaotic data. Phys. Rev. E **47**(5), 3057–3068 (1993)

B.R. Brooks, C.L. Brooks, A.D. MacKerell, L. Nilsson, R.J. Petrella, B. Roux, Y. Won, G. Archontis, C. Bartels, S. Boresch, A. Caflisch, L. Caves, Q. Cui, A.R. Dinner, M. Feig, S. Fischer, J. Gao, M. Hodoscek, W. Im, K. Kuczera, T. Lazaridis, J. Ma, V. Ovchinnikov, E. Paci, R.W. Pastor, C.B. Post, J.Z. Pu, M. Schaefer, B. Tidor, R.M. Venable, H.L. Woodcock, X. Wu, W. Yang, D.M. York, M. Karplus, Charmm: the bimolecular simulation program. J. Comput. Chem. **30**(10), 1545–1614 (2009)

C.J. Cellucci, A.M. Albano, P.E. Rapp, Comparative study of embedding methods. Phys. Rev. E **67**, 066210–13 (2003)

J.D. Chodera, W.C. Swope, J.W. Pitera, C. Seok, K.A. Dill, Use of the weighted histogram analysis method for the analysis of simulated and parallel tempering simulations. J. Chem. Theory Comput. **3**, 26–41 (2007)

D. Gencaga, K.H. Knuth, W.B. Rossow, A recipe for the estimation of information flow in a dynamical system. Entropy **17**, 438–470 (2015)

P. Grassberger, Finite sample corrections to entropy and dimension estimates. Phys. Lett. A **128**, 369–373 (1988)

P. Grassberger, I. Procaccia, Measuring the strangeness of strange attractors. Phys. D **189**, 9 (1983)

T. Ichiye, M. Karplus, Collective motions in proteins: a covariance analysis of atomic fluctuations in molecular dynamics and normal modes simulations. Proteins **11**(3), 205–217 (1991)

H. Kamberaj, A theoretical model for the collective motion of proteins by means of principal component analysis. Cent. Eur. J. Phys. **9**(1), 96–109 (2011)

H. Kamberaj, *Molecular Dynamics Simulations in Statistical Physics: Theory and Applications* Computational Series. (Springer Nature, Switzerland, 2020)

H. Kamberaj, A. van der Vaart, Extracting the causality of correlated motions from molecular dynamic simulations. Biophys. J . **97**, 1747–1755 (2009)

M.B. Kennel, R. Brown, H.D.I. Abarbanel, Determining embedding dimension for phase-space reconstruction using a geometrical construction. Phys. Rev. A **45**, 3403–3411 (1992)

G.J. Martyna, M.E. Tuckerman, D.J. Tobias, M.L. Klein, Explicit reversible integrators for extended systems dynamics. Mol. Phys. **87**(5), 1117–1157 (1996)

D. Nebiu, H. Kamberaj, Symbolic information flow measurement (SIFM): a software for measurement of information flow using symbolic analysis. SoftwareX **11**, 100470 (2020)

L. Noakes, The Takens embedding theorem. Int. J. Bifurc. Chaos Appl. Sci. Eng. **1**, 867–872 (1991)

N.H. Packard, J.P. Crutchfield, J.D. Farmer, R.S. Shaw, Geometry from a time series. Phys. Rev. Lett. **45**(9), 712–716 (1980)

M. Prokopenko, J.T. Lizier, Transfer entropy and transient limits of computation. Sci. Rep. **4**, 5394 (2014)

M. Prokopenko, J.T. Lizier, D.C. Price, On thermodynamic interpretation of transfer entropy. Entropy **15**, 524–543 (2013)

J.P. Ryckaert, G. Ciccotti, H.J.C. Berendsen, Numerical integration of the Cartesian equations of motion of a system with constraints: molecular dynamics of n-alkanes. J. Comput. Phys. **23**, 327–341 (1977)

T. Sauer, J.A. Yorke, M. Casdagli, Embedology. J. Stat. Phys. **65**, 579–616 (1991)

T.D. Schneider, Theory of molecular machines. II. Energy dissipation from molecular machines. J. Theor. Biol. **148**, 125–137 (1991)

T. Schreiber, Measuring information transfer. Phys. Rev. Lett. **85**, 461–464 (2000)

M. Staniek, K. Lehnertz, Symbolic transfer entropy. Phys. Rev. Lett. **100**, 158101 (2008)

F. Takens, *Detecting Strange Attractors in Fluid Turbulence* Dynamical Systems and Turbulence. (Springer, Berlin, 1981)

M.C. Thomas, A.T. Joy, *Elements of Information Theory* (Wiley Inc, Hoboken, 2006)

Chapter 13
Principal Components Analysis of Biological Systems

The chapter aims to describe principal components analysis of biological systems in explicit solvent.

This chapter describes the principal components analysis of biological systems in explicit solvent and the stability of the principal components analysis by comparing the quasi-harmonic modes of different simulation trajectories of the same system.

13.1 Theoretical Background

The principal component analysis (PCA) (Karhunen 1947) enables decreasing the system's number of degrees of freedom of computer simulations in molecular biology (Brooks et al. 1995). This is also known as quasi-harmonic analysis (Karplus and Jushick 1981; Ichiye and Karplus 1991). Besides, the method has been used to represent a molecular dynamics trajectory in terms of a smaller number of variables, responsible for slow conformational transitions (Go 1990; Kitao et al. 1991; Garcia 1992; Amadei et al. 1993; Aalten et al. 1993; Grubmüller 1995). These are also called essential degrees of freedom obtained from the eigenvectors associated with significantly non-zero eigenvalues calculated from the covariance matrix of the atomic position fluctuations of a MD trajectory (Amadei et al. 1993). Using the PCA, we can determine a small number of essential modes and then project equations of motion on the resulting low-dimensional configurational subspace to obtain a new smaller set of the differential equations on the reduced phase space (Lange et al. 2006; Stepanova 2007; Kamberaj 2011), which have the form of well-known generalised Langevin equation (Albers et al. 1971).

There exist various software packages for obtaining the PCA (Janezic and Brooks 1995; Brooks et al. 1995) from a MD trajectory. In particular, in the following, we

H. Kamberaj, *Computer Simulations in Molecular Biology*, Scientific Computation,
https://doi.org/10.1007/978-3-031-34839-6_13

describe the algorithm introduced in Refs. Kamberaj and van der Vaart (2009a), Kamberaj (2011). After removing the overall translational and rotational motions of the system, its configurations of N selected group of atoms (often the C-α or backbone atoms), stored at regular time steps from an MD trajectory, are represented by single vector

$$\mathbf{q}(t) = \left(q_1(t),\, q_2(t),\, \ldots,\, q_g(t)\right)^T \tag{13.1}$$

where

$$q_i(t) \equiv q_i(t) - \langle q_i(t)\rangle, \quad (i = 1,\, 2,\, \ldots,\, g) \tag{13.2}$$

denotes the fluctuations of atomic positions and g is the number of degrees of freedom of the selected group of atoms ($g = 3N$). These represent configurational points distributed in the space obtained by computer simulations of a biological molecule. The so-called quasi-harmonic approach (Karplus and Jushick 1981) describes the distribution of the coordinates' fluctuations by a multivariate Gaussian distribution with covariance matrix $\mathbf{C}$ with elements C_{ij}

$$C_{ij} = \langle q_i(t) q_j(t)\rangle_t \tag{13.3}$$

where $\langle \cdots \rangle_t$ denotes a time average. The covariance matrix can also be expressed in terms of the mass weighted coordinates,

$$q_i'(t) = \sqrt{m_i} q_i(t) \tag{13.4}$$

Hence, the mass weighted covariance matrix is obtained as

$$C_{ij}' = \sqrt{m_i m_j} \langle q_i(t) q_j(t)\rangle_t \tag{13.5}$$

Here, $\mathbf{C}'$ and $\mathbf{C}$ represents $3N \times 3N$ symmetric matrices. Using Jacobi algorithm from Numerical Recipes book (Press et al. 1996), either $\mathbf{C}'$ or $\mathbf{C}$ is diagonalised to obtain the vector of eigenvalues $\mathbf{\Lambda}$ and matrix of eigenvectors $\mathbf{E}$:

$$\mathbf{\Lambda} = \left(\lambda_1, \lambda_2, \ldots, \lambda_i, \ldots, \lambda_g\right)^T \tag{13.6}$$

$$\mathbf{E} = \begin{bmatrix} E_{11} & E_{12} & \cdots & E_{1i} & \cdots & E_{1g} \\ E_{21} & E_{22} & \cdots & E_{2i} & \cdots & E_{2g} \\ \cdots & \cdots & \cdots & \cdots & \cdots & \cdots \\ E_{g1} & E_{g2} & \cdots & E_{gi} & \cdots & E_{gg} \end{bmatrix}$$

Here, $\mathbf{E}$ is an orthogonal matrix whose i-th column is the i-th eigenvector of the covariance matrix $\mathbf{C}$ (which represents either standard $\mathbf{C}$ or $\mathbf{C}'$ in the following discussion), called *principal components* (PC):

$$\mathbf{e}_i \equiv \left(E_{1i}, E_{2i}, \ldots, E_{gi}\right)^T \tag{13.7}$$

Then, a new set of coordinates is obtained

$$\mathbf{X} = (X_1, X_2, \ldots, X_j, \ldots, X_g)^T \tag{13.8}$$

where

$$X_j = \sum_{i=1}^{g} E_{ij} q_i \tag{13.9}$$

The eigenvalues of the matrix $\mathbf{C}$, λ_i (for $i = 1, \ldots, g$), represent the mean-square fluctuations along each principal axis:

$$\lambda_i = \langle X_i^2(t) \rangle = \mathbf{e}_i^T \mathbf{C} \mathbf{e}_i \tag{13.10}$$

which is a matrix form representation.

Often, in a coarse-grained model, M coordinates are proposed (Grubmüller 1995) with the largest eigenvalues to represent the slow motions in a biological molecule:

$$\mathbf{c}(t) = (X_1(t), \ldots, X_M(t))^T \tag{13.11}$$

These coordinates are also called *collective coordinates*. This choice is based on the observation from a molecular dynamics simulation that the remaining X_i $(i = M + 1, \ldots, g)$ with small eigenvalues describe localised, high frequency (or fast motions), nearly harmonic vibration modes with small amplitudes, which are expected not to reflect the conformational transitions. On the other hand, the dynamics of M conformational coordinates is essentially slow and non-harmonic, dominating the collective motions in a protein (Go 1990; Kitao et al. 1991; Garcia 1992; Amadei et al. 1993; Aalten et al. 1993; Grubmüller 1995). Therefore, it is assumed that with this choice we capture the relevant degrees of freedom of the protein dynamics. The number M of the conformational degrees of freedom, which are explicitly considered, will determine the level of coarse-graining.

13.2 Diffusive Motion in a Protein

Often, Green-Kubo equation describes the diffusive motion from computer simulations of the biological molecules (such as proteins). It relates the correlation function integrals with the transport coefficients (Balucani and Zoppi 1994; Kamberaj 2011):

$$D_m(t) = \frac{1}{d} \int_0^{\mathcal{T}} d\tau \langle v_m(t_0) v_m(t_0 + \tau) \rangle_{t_0} \tag{13.12}$$

In Eq. (13.12), $\mathbf{v}$ is the velocity vector of the collective coordinates

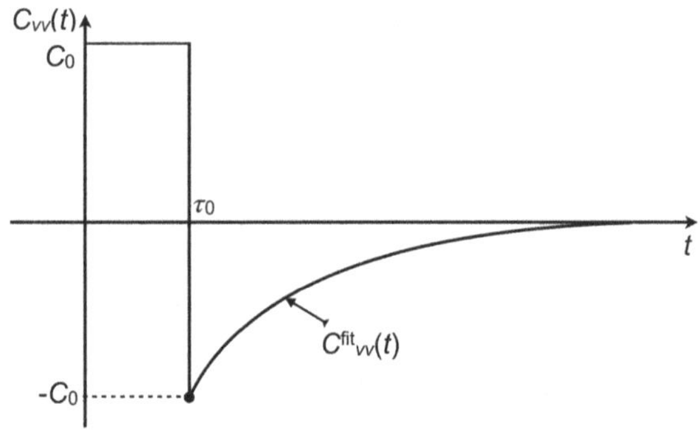

Fig. 13.1 Graphical illustration of $C_{vv}^{\text{fit}}(t)$ function

$$\mathbf{v} = (\dot{c}_1, \dot{c}_2, \ldots, \dot{c}_M)^T \tag{13.13}$$

In Eq. (13.12), d gives the dimension (here, $d = 1$), $\langle \cdots \rangle_{t_0}$ denotes an ensemble average over many starting times t_0, and t is the simulation time. In Eq. (13.12), the term

$$C_m^{vv}(t) = \langle v_m(t_0) v_m(t_0 + \tau) \rangle_{t_0} \tag{13.14}$$

represents the velocity auto-correlation function and τ is a time shift.

A double diffusion model (Amadei et al. 1999), is proposed to represent the diffusive motion of collective coordinates (Kamberaj 2011). For that, a short time diffusion D_0, within a localised configuration space region and approximated by a single harmonic well, follows a slow diffusion motion between harmonic wells and approximated by a long time diffusion constant D_∞. In this model (Kamberaj 2011), the following fitting curve models the short time decay of the negative tail of the auto-correlation function

$$C_{vv}^{\text{fit}}(t) = -C_0 \exp\left(-\frac{t - \tau_0}{\tau_c} \right) \tag{13.15}$$

In Eq. (13.15), τ_0 measures the fast relaxation time (which is the time when $C_{vv}(t)$ becomes first negative), τ_c is the slow relaxation time, and C_0 is a constant denoting $C_0 = -C_{vv}^{\text{fit}}(t = \tau_0)$. Here, Eq. (13.15) is valid only for $t \geq \tau_0$. Figure 13.1 illustrates graphically $C_{vv}^{\text{fit}}(t)$ function.

Integrating the expression given by Eq. (13.15) and adding the constant term of $C_0\tau_0$ for the integration for $0 \leq t \leq \tau_0$ (see also Fig. 13.1), a general expression for the double diffusion model is obtained:

$$
\begin{aligned}
D^{\text{fit}}(t) &= C_0\tau_0 - \int_{\tau_0}^{t} C_0 \exp\left(-\frac{t'-\tau_0}{\tau_c}\right) dt' \\
&= C_0\tau_0 + C_0\tau_c \int_{\tau_0}^{t} \exp\left(-\frac{t'-\tau_0}{\tau_c}\right) d\left(-\frac{t'-\tau_0}{\tau_c}\right) \\
&= C_0\tau_0 + C_0\tau_c \left[\exp\left(-\frac{t-\tau_0}{\tau_c}\right) - 1\right] \\
&= -(C_0\tau_0 - C_0\tau_c)\exp\left(-\frac{t-\tau_0}{\tau_c}\right) + (C_0\tau_0 - C_0\tau_c) \\
&\quad + C_0\tau_0 \exp\left(-\frac{t-\tau_0}{\tau_c}\right) \\
&\equiv -D_\infty \exp\left(-\frac{t-\tau_0}{\tau_c}\right) + D_\infty + D_0 \exp\left(-\frac{t-\tau_0}{\tau_c}\right) \\
&= (D_0 - D_\infty)\exp\left(-\frac{t-\tau_0}{\tau_c}\right) + D_\infty
\end{aligned}
\tag{13.16}
$$

where $D_0 = C_0\tau_0$ and $D_\infty = C_0\tau_0 - C_0\tau_c$. Note that Eq. (13.16) is valid for $t \geq \tau_0$ only. In Eq. (13.16), the first term describes the short time diffusion, D_∞ is the long-time diffusion constant, and $D_0 > D_\infty$. First τ_0 can be defined from $C_{vv}(t)$ as the time when it first becomes negative, then using a nonlinear least squares fitting procedure of $D(t)$, τ_c and D_∞ can be calculated (Kamberaj 2011. The short time diffusion coefficient is calculated as $D_0 = D(\tau_0)$.

13.3 Convergence of the PCA Calculations

The entire trajectory obtained from computer simulation of a biological molecule is divided into time windows with a length of T_w. Then, the subspaces from these time windows are computed to obtain measures of the similarities between the subspaces (Lange and Grubmüller 2006). In general, the similarity measures can be defined as an average over the time windows, N_w:

$$
t = T_w,\ 2T_w,\ \ldots,\ T \tag{13.17}
$$

with T being the length of full trajectory.

The convergence establishes for sufficiently large PCA subspaces. Besides, this convergence is obtained only if a relatively long MD simulation is obtained, which is not always available. However, in the context of our discussion, the quality of the chosen PCA subspace has to be established based on the short run of MD simulations available (Lange and Grubmüller 2006).

In particular, the inner product matrix **P** (Amadei et al. 1993) is a good measure suggested for comparing the basis vectors of two subspaces. The inner product matrix **P** is calculated as:

$$\mathbf{P} = \frac{2}{N_w(N_w - 1)} \sum_{i=1}^{N_w-1} \sum_{j=i+1}^{N_w} \mathbf{E}_i^T \mathbf{E}_j \tag{13.18}$$

where $\mathbf{E}_i$ and $\mathbf{E}_j$ are eigenvector matrices obtained for the i-th and j-th trajectory windows.

Furthermore, the overlap or similarity between two subspaces is calculated by introducing two quantities:

$$s_i^{(k)} = \frac{1}{M(N_w - 1)} \sum_{k'=1 \neq k}^{N_w} \sum_{j=1}^{M} \mid \mathbf{e}_{k,i}^T \mathbf{e}_{k',j} \mid^2 \tag{13.19}$$

$$S^{(k)} = \sum_{i=1}^{g} s_i^{(k)}$$

Note that $s_i^{(k)}$ is a measure of the degree of the average overlap of each eigenvector of the subspace defined by the time window k of the entire trajectory with the subspaces of all the other time windows of the essential first M eigenvectors. $S^{(k)}$ is the total degree of the overlap of the subspace defined in the k-th time window with the essential subspaces of all other time windows.

Moreover, the following quantities can be calculated to measure the differences in PCA modes between the various time windows of the same trajectory (Kamberaj and van der Vaart 2009a) (and the reference therein Tournier and Smith 2003). In particular, the fluctuations of the k-th window trajectory are projected onto the subspace determined by the other windows of the entire trajectory as:

$$\lambda_{k,i}^{\text{proj}} = \frac{1}{N_w - 1} \sum_{k'=1 \neq k}^{N_w} \sum_{j=1}^{g} P_{ij}^{(kk')} \lambda_{k,j} \tag{13.20}$$

where $\lambda_{k,j}$ is the j-the eigenvalue of the k time window. This measure accounts that the eigenvalues may also not be ordered accordingly for the two different time windows, which should be considered as an advantage. Then, $\lambda_{k,i}^{\text{proj}}$ determines the differences in amplitudes of the fluctuations extracted by the PCA on two trajectories associated with the eigenvector $\mathbf{e}_{k,i}$ as.

$$\rho_{k,i} = \left(\frac{\lambda_{k,i}^{\text{proj}}}{\lambda_{k,i}} \right)^{1/2} \tag{13.21}$$

In general, if $\rho_{k,i} < 1$, the amplitude of fluctuations along the mode i is smaller than those in all the other time windows while if $\rho_{k,i} > 1$ the amplitude of the fluctuations along the mode i is larger in all the other time windows.

Usually, PCA is carried out on subsets of the protein atoms, such as C_α or backbone atoms.

13.4 Case Study

We considered again the BPTI protein in explicit solvent. The NPT ensemble MD simulations were performed for about ten nanoseconds using standard conditions, such as $T = 300$ K and $P = 1$ atm. The configuration snapshots were saved every picoseconds. In total, 10000 frames were produced and analysed.

Figure 13.2 presents the results of principal component analysis. In particular, the eigenvalues (in nanometer square, nm^2), cumulative eigenvalues, and cumulative eigenvalues (in %) of the PCA for the BPTI protein are shown. It can be seen that about the first 20 eigenvalues characterise more than 80% of the fluctuations, and thus they determine the essential subspace of the internal motions in the system.

In Fig. 13.3 is shown a dialog form for principal component analysis.

Note that before running the principal component analysis, the overall translational and rotational motions must be removed. In general, during standard MD simulations, it is expected that the protein diffuses around the simulation box in the solvent. Figure 13.4 presents a dialog form for removing the overall translational and rotational motions of a protein.

Next, we compare the stability of the PCA by calculating the inner product matrix **P**. For that, a ten ns long trajectory is split into two equal length trajectories, each of length five ns; that is, $N_w = 2$. Then, the PCA was performed for each time window to determine the subspaces of each shorter trajectory.

The results presented in Fig. 13.5(Top) show that the projections of the essential modes are nearly diagonal, indicating that all modes of the first half have high

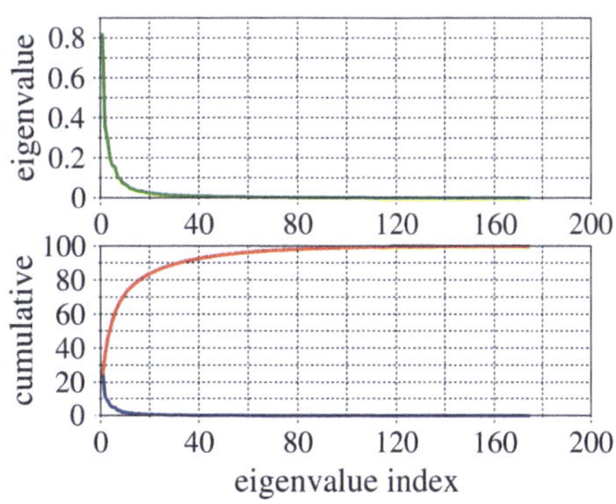

Fig. 13.2 The eigenvalues (in nm^2), cumulative eigenvalues, and cumulative eigenvalues (in %) of the PCA for the BPTI protein

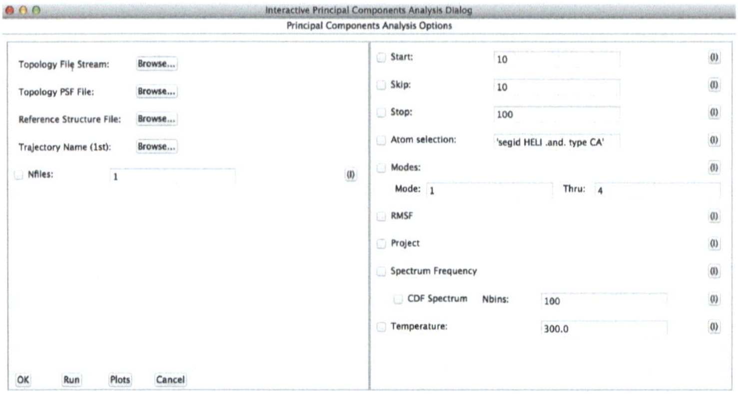

Fig. 13.3 A dialog form for principal component analysis

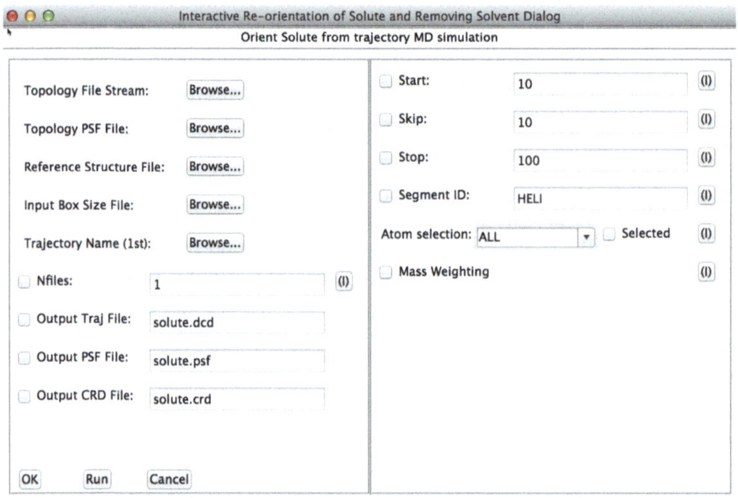

Fig. 13.4 A dialog form for removing the overall translational and rotational motion of protein

projections on the modes of second one. Thus, the directions in the configurational space of both halves of the entire trajectory are ordered similarly with respect to the total amount of fluctuations.

Furthermore, Fig. 13.5(Bottom) indicates that the first 20 modes (that is $M = 20$), which determine the essential degrees of freedom, define the overall overlap between the two subspaces given by the two time windows.

Figure 13.6 presents $\lambda_{2,i}^{\text{proj}}$ and $\rho_{2,i}$ for the first 30 modes associated with the largest eigenvalues obtained from PCA of the BPTI protein of the two halves of the trajectory. Our results indicates that essentially the two windows characterising

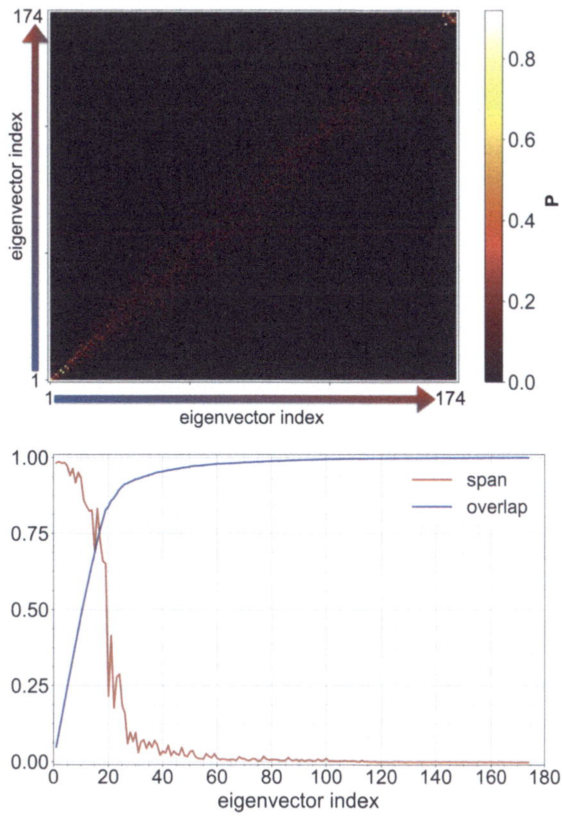

Fig. 13.5 (Top) Projection of the eigenvectors calculated from the second half of the trajectory onto those calculated from the first half. (Bottom) Comparison of the fluctuations along the first twenty eigenvectors associated with the largest eigenvalues in the two halves of the MD trajectory for the BPTI protein

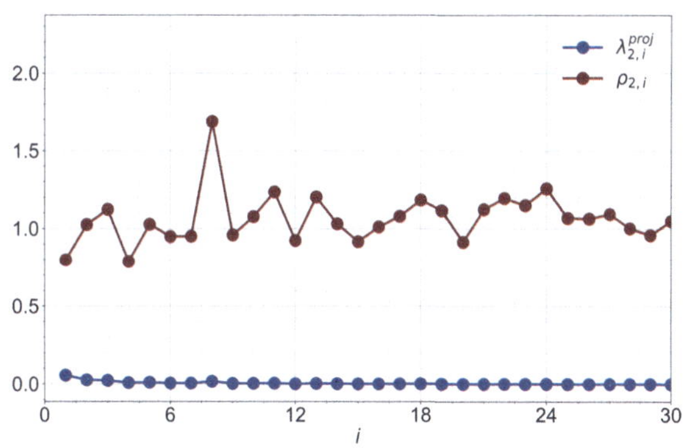

Fig. 13.6 Comparisons of the amplitudes of the fluctuations of the modes obtained from PCA of the BPTI protein of the two halves of the trajectory

two halves of the entire trajectory represent essentially the same fluctuations; in particular, $\lambda_{2,i}^{\text{proj}} \approx 0$ and $\rho_{2,i}$ fluctuates around one.

13.5 Mini Projects

Exercise 13.1 (Activity case study 1) Consider again the BPTI system in explicit solvent. The input files are under the folder *casestudies/lab5/inputs*. Run standard MD simulations for up to 20 ns at the temperature $T = 300$ K, $P = 1$ atm.

1. Analyse the PCA of the C-α and backbone atoms. What can you conclude about the significance of different modes on the overall fluctuations of the system?
2. Divide the trajectory into 4 shorter trajectories using a window of length 5 ns. What can you conclude by comparing the eigenvectors of the subspaces for each shorter trajectory?
3. Analyse using the VMD program the fluctuations for each shorter trajectory projected on the subspace extended by the first three highest modes. What do you conclude?

Exercise 13.2 (Activity case study 2) Consider now the mutated BPTI system in explicit solvent. For that, you can use the X-ray structure with PDB ID: 1YLD (Getun et al. 2008). Run standard MD simulations for about 20 ns at the temperature $T = 300$ K, $P = 1$ atm.

1. Analyse the PCA of the C-α for both systems. What can you conclude about the significance of different modes on the overall fluctuations of the system?
2. Compare the PCA modes of two systems. What can you conclude by comparing the eigenvectors of the subspaces for each system?
3. Analyse using the VMD program the fluctuations for each trajectory projected on the subspace extended by the first three highest modes. What do you conclude?
4. Run normal mode analysis for the X-ray structures of the wild- and mutated-types. Compare the normal modes of these two systems. What do you conclude from this analysis?

13.6 Summary

In this chapter, we described the principal component analysis of the biological molecules used to characterise essential degrees of freedom the system. The focus was on introducing the algorithm for performing PCA from an MD simulation trajectory. Besides, the diffusive motion in a protein was discussed in details, where the short and long diffusive motions concepts were introduced.

Furthermore, the convergence of the PCA calculations was discussed by introducing different quantities to describe the degree of the similarity between the time windows of the same trajectory. These results showed that an MD simulation length of 5 ns is sufficient to achieve the convergence of the PCA.

References

D. Aalten, A. Amadei, A. Linssen, V. Eijsink, G. Vriend, H. Berendsen, The essential dynamics of thermolysin-conformation of the hinge-bending motion and comparison of simulations in vacuum and water. Proteins **22**, 45–54 (1993)

J. Albers, J.M. Deutch, I. Oppenheim, Generalized Langevin equations. J. Chem. Phys. **54**(8), 3541–3546 (1971)

A. Amadei, A.B.M. Linssen, H.J.C. Berendsen, Essential dynamics of proteins. Proteins **17**, 412 (1993)

A. Amadei, B. de Groot, M. Ceruso, M. Paci, A. Di Nola, H. Berendsen, A kinetic model for the internal motions of proteins: diffusion between multiple harmonic wells. Proteins **35**(3), 283–292 (1999)

U. Balucani, M. Zoppi, *Dynamics of the Liquid State* (Clarendon, Oxford, 1994)

B.B. Brooks, D. Janezic, M. Karplus, Hamonic-analysis of large systems. 1. Methodology. J. Comput. Chem. **16**, 1522–1542 (1995)

A.E. Garcia, Large-amplitude nonlinear motions in proteins. Phys. Rev. Lett. **68**, 2696 (1992)

I.V. Getun, C.K. Brown, J. Tulla-Puche, D. Ohlendorf, C. Woodward, G. Barany, Partially folded Bovine pancreatic trypsin inhibitor analogues attain fully native structures when co-crystallized with S195A rat trypsin. J. Mol. Biol. **375**, 812 (2008)

N. Go, A theorem on amplitudes of thermal atomic fluctuations in large molecules assuming specific conformations calculated by normal mode analysis. Biophys. Chem. **35**, 105–112 (1990)

H. Grubmüller, Predicting slow structural transitions in macromolecular systems: conformational flooding. Phys. Rev. E **52**, 2893 (1995)

T. Ichiye, M. Karplus, Collective motions in proteins: a covariance analysis of atomic fluctuations in molecular dynamics and normal modes simulations. Proteins **11**(3), 205–217 (1991)

D. Janezic, B.B. Brooks, Harmonic analysis of large systems: II. Comparison of different protein models. J. Comput. Chem. **16**, 1543–1553 (1995)

H. Kamberaj, A theoretical model for the collective motion of proteins by means of principal component analysis. Cent. Eur. J. Phys. **9**(1), 96–109 (2011)

H. Kamberaj, A. van der Vaart, Correlated motions and interactions at the onset of the dna-induced partial unfolding of ets-1. Biophys. J . **96**, 1307–1317 (2009)

K. Karhunen, Uber Lineare Methoden in der Wahrscheinlichkeitsrechnung. Ann. Acad. Sci. Fenn. Ser. **A1**(37), 1–79 (1947)

M. Karplus, J.N. Jushick, Method for estimating the configurational entropy of macromolecules. Macromolecules **14**, 325–332 (1981)

A. Kitao, F. Hirata, N. Go, The effect of solvent on the conformation and the collective motions of protein: normal mode analysis and molecular dynamics simulations of melittin in water and in vacuum. Chem. Phys. **158**, 447–472 (1991)

O.F. Lange, H. Grubmüller, Can principal components yield a dimension reduced description of protein dynamics on long time scales? J. Phys. Chem. B **110**, 22842–22852 (2006)

O.F. Lange, L.V. Schäfer, H. Grubmüller, Flooding in gromacs: accelerated barrier crossing in molecular dynamics. J. Comput. Chem. **27**(14), 1693–1702 (2006)

W.H. Press, S.A. Teukolsky, W.T. Vetterling, B.P. Flannery, *Numerical Recipes in Fortran 90. The Art of Parallel Scientific Computing*, vol. 2, 2nd edn. (Cambridge University Press, New York, 1996)

M. Stepanova, Dynamics of essential collective motions in proteins: theory. Phys. Rev. E **76**(5), 051918 (2007)

A. Tournier, J. Smith, Principal components of the protein dynamical transition. Phys. Rev. Lett. **91**, 208106 (2003)

Index

H. Kamberaj, *Computer Simulations in Molecular Biology*, Scientific Computation,
https://doi.org/10.1007/978-3-031-34839-6

SPRINGER NATURE

GPSR Compliance

The European Union's (EU) General Product Safety Regulation (GPSR) is a set of rules that requires consumer products to be safe and our obligations to ensure this.

If you have any concerns about our products, you can contact us on ProductSafety@springernature.com

In case Publisher is established outside the EU, the EU authorized representative is:

Springer Nature Customer Service Center GmbH
Europaplatz 3
69115 Heidelberg, Germany

The manufacturer's authorised representative in the EU is Springer Nature Customer Service Centre GmbH, Europaplatz 3, 69115 Heidelberg, Germany. If you have any concerns regarding our products, please contact ProductSafety@springernature.com

Printed and bound by CPI Group (UK) Ltd, Croydon, CR0 4YY
29/07/2026
02180513-0001